HIGH ENERGY PHYSICS

Volume V

This is Volume 25-V in
PURE AND APPLIED PHYSICS
A Series of Monographs and Textbooks
Consulting Editors: H. S. W. Massey and Keith A. Brueckner
A complete list of titles in this series appears at the end of this volume.

HIGH ENERGY PHYSICS

Edited by
E. H. S. BURHOP

PHYSICS DEPARTMENT
UNIVERSITY COLLEGE
LONDON, ENGLAND

Volume V

ACADEMIC PRESS New York · London · 1972

ACADEMIC PRESS, INC.
111 Fifth Avenue, New York, New York 10003

United Kingdom Edition published by
ACADEMIC PRESS, INC. (LONDON) LTD.
24/28 Oval Road, London NW1

LIBRARY OF CONGRESS CATALOG CARD NUMBER: 66-26271

PRINTED IN THE UNITED STATES OF AMERICA

CONTRIBUTORS

A. Donnachie, Daresbury Nuclear Physics Laboratory, Daresbury, Warrington, Lancashire, England

R. Gatto, Istituto di Fisica dell'Universita, Padova, Italy

R. J. N. Phillips, Rutherford High Energy Laboratory, Chilton, Berkshire, England

G. A. Ringland, Rutherford High Energy Laboratory, Chilton, Berkshire, England

PREFACE

It was stated in the Preface to Volume IV that a further volume was planned "to include an account of some topics that were in their infancy, or in some cases barely thought of when the original work was planned." The subjects treated in this volume—photo- and electroproduction processes, and Regge phenomenology—are very topical and represent fields in which progress in recent years has been very rapid and in which there has been a need for a comprehensive account.

This volume includes also an updating of Professor Gatto's review of the present state of quantum electrodynamics which appeared in Volume II. It is truly remarkable how, despite all the sophistication and increased experimental accuracy that has since developed, no substantial evidence has emerged for the breakdown of QED.

Once again the publishers have shown an understanding attitude toward rather extensive changes in proof which are necessitated by the continuing explosive rate of growth of knowledge in these fields.

E. H. S. BURHOP

CONTENTS

CONTENTS OF OTHER VOLUMES

PHOTO- AND ELECTROPRODUCTION PROCESSES

A. Donnachie

I. General Formalism

The first section is concerned primarily with the details of pion photoproduction and electroproduction: the kinematics, the choice of invariant amplitudes and their properties, the multipole amplitude and helicity amplitude expansions of the transition matrix, the Born terms, and cross section and polarization formulas. The emphasis on the pion production reactions is a natural one, since this is where the bulk of experimental information is found and where theoretical understanding and interpretation are most developed. The differences which occur in the photoproduction of other pseudoscalar mesons (in this context η and K mesons) are listed briefly at the end.

A. KINEMATICS FOR PION PRODUCTION

We consider the reactions of photopion production

$$\gamma(K) + N(P_1) \to N(P_2) + \pi(Q) \tag{I-1}$$

and electropion production

$$e(K_1) + N(P_1) \to e'(K_2) + N(P_2) + \pi(Q). \tag{I-2}$$

The four-momentum of each particle is indicated in parenthesis, and

$$P_1{}^2 = P_2{}^2 = -M^2, \qquad Q^2 = -\mu^2, \qquad K_1{}^2 = K_2{}^2 = -m^2, \tag{I-3}$$

where M is the nucleon mass, μ is the pion mass, and m is the electron mass.

For photopion production, the S matrix is given in terms of the T matrix by

$$S_{fi} = \delta_{fi} - i(2\pi)^{-2} \, \delta^4(P_2 + Q - P_1 - K)(M^2/4K_0 Q_0 P_{10} P_{20})^{1/2} T_{fi}, \tag{I-4}$$

and the transition matrix element in turn is given by

$$T = \varepsilon_\mu J_\mu, \tag{I-5}$$

where J_μ is the nucleon electromagnetic current matrix element

$$J_\mu = {}_{\text{out}}\langle P_2 Q | j_\mu | P_1 \rangle_{\text{in}} \tag{I-6}$$

and ε_μ is the photon polarization vector.

For electropion production, the S matrix is given in terms of the T matrix by

$$S_{fi} = \delta_{fi} - i(2\pi)^{-7/2} \, \delta^4(P_2 + Q + K_2 - P_1 - K_1)$$
$$\times (M^2 m^2/2Q_0 P_{10} P_{20} K_{10} K_{20})^{1/2} T_{fi}. \tag{I-7}$$

In the one-photon exchange approximation of Fig. 1 (see page 29), the transition matrix element may again be written as in (I-5), with J_μ the nucleon electromagnetic current matrix element of (I-6), but ε_μ is now the lepton electromagnetic current and photon propagator

$$\varepsilon_\mu = \bar{u}(K_2)\gamma_\mu u(K_1)e/K^2, \tag{I-8}$$

where K is the four-momentum transfer between the electrons (that is, the four-momentum of the virtual photon)

$$K = K_1 - K_2. \tag{I-9}$$

As a consequence of this explicit separation of the pion electroproduction matrix element into a lepton part and a hadron part, it is possible to treat pion electroproduction as pion photoproduction with virtual photons whose properties (energy, mass, polarization) are completely specified by the kinematics of the electron. The lepton part of the matrix element is explicitly calculable, and the interesting object of study is the nucleon electromagnetic current matrix element of (I-6), to which discussion can be restricted for most purposes, and, at least as far as the basic formalism is concerned, photoproduction can be treated as a special case of electroproduction.

Thus, both for photoproduction and electroproduction we are considering the process of (I-1), remembering that the photon can have nonzero "mass," the squared "mass" of the virtual photon $-K^2$ being negative.

For this process, the usual Mandelstam kinematical variables are given by

$$
\begin{aligned}
s &= -(K + P_1)^2 = -(Q + P_2)^2, \\
t &= -(K - Q)^2 = -(P_2 - P_1)^2, \\
u &= -(K - P_2)^2 = -(Q - P_1)^2,
\end{aligned}
\tag{I-10}
$$

which satisfy

$$
s + t + u = 2M^2 + \mu^2 - K^2.
\tag{I-11}
$$

Two other useful invariants are

$$
v = -(P \cdot K)/m, \qquad P = \tfrac{1}{2}(P_1 + P_2); \qquad v_B = (Q \cdot K)/2M.
\tag{I-12}
$$

The terms v, v_B and s, t are related by

$$
v - v_B = (s - M^2)/2M, \qquad v_B = (t - \mu^2 + K^2)/4M.
\tag{I-13}
$$

Noninvariant quantities such as energies and angles will be labeled by a superscript L when they are laboratory frame quantities and will not be labeled when they are quantities in the pion-nucleon center-of-mass system. In the latter system, let W be the total energy and define

$$
K = (\mathbf{k}, k_0), \qquad P_1 = (-\mathbf{k}, E_1), \qquad Q = (\mathbf{q}, q_0), \qquad P_2 = (-\mathbf{q}, E_2).
\tag{I-14}
$$

Then

$$
\begin{aligned}
s &= W^2, \\
t &= 2\mathbf{k} \cdot \mathbf{q} - 2k_0 q_0 + \mu^2 - K^2, \\
u &= -2\mathbf{k} \cdot \mathbf{q} - 2k_0 E_2 + M^2 - K^2,
\end{aligned}
\tag{I-15}
$$

and the scattering angle θ is given by

$$
\cos \theta = (\mathbf{k} \cdot \mathbf{q})/(|\mathbf{k}| |\mathbf{q}|) = x.
\tag{I-16}
$$

In the following, we will use the notation

$$
k = |\mathbf{k}|, \qquad \hat{\mathbf{k}} = \mathbf{k}/k; \quad q = |\mathbf{q}|, \qquad \hat{\mathbf{q}} = \mathbf{q}/q.
\tag{I-17}
$$

The various energies and momenta in this frame are given in terms of W and K^2 by

$$K_0 = (W^2 - K^2 - M^2)/2W, \qquad E_1 = (W^2 + M^2 + K^2)/2W,$$
$$E_2 = (W^2 + M^2 - \mu^2)/2W, \qquad q_0 = (W^2 - M^2 + \mu^2)/2W,$$
$$E_1 \pm M = [(W \pm M)^2 + K^2]/2W, \qquad E_2 \pm M = [(W \pm M)^2 - \mu^2]/2W,$$
$$q = (1/2W)[(W^2 - (M + \mu)^2 W^2 - (M - \mu)^2]\}^{1/2} \qquad (\text{I-18})$$
$$= (1/2W)\{W^4 - 2(M^2 + \mu^2)W^2 + (M^2 - \mu^2)^2\}^{1/2},$$
$$K = (1/2W)\{W^4 - 2(M^2 - K^2)W^2 + (M^2 + K^2)^2\}^{1/2}.$$

B. Invariant Amplitudes for Pion Production

The most general Lorentz covariant pseudo-four-vector for the nucleon electromagnetic current can be written in terms of the eight pseudovectors

$$N_\mu{}^1 = i\gamma_5 \gamma_\mu \gamma \cdot K, \qquad N_\mu{}^2 = 2i\gamma_5 P_\mu,$$
$$N_\mu{}^3 = 2i\gamma_5 Q_\mu, \qquad N_\mu{}^4 = 2i\gamma_5 K_\mu,$$
$$N_\mu{}^5 = \gamma_5 \gamma_\mu, \qquad N_\mu{}^6 = \gamma_5 \gamma \cdot K P_\mu, \qquad (\text{I-19})$$
$$N_\mu{}^7 = \gamma_5 \gamma \cdot K K_\mu, \qquad N_\mu{}^8 = \gamma_5 \gamma \cdot K Q_\mu,$$

with $P = \frac{1}{2}(P_1 + P_2)$. That is

$$J_\mu = \sum_{i=1}^{8} B_i(s, t, u, K^2)N_\mu{}^i, \qquad (\text{I-20})$$

where the B_i are scalar amplitudes. Any other pseudovector is reducible to a linear combination of those of (I-19), none of these reductions introducing any kinematical singularities.

Since the current j_μ is conserved, it is necessary that

$$K_\mu j_\mu = 0, \qquad (\text{I-21})$$

which imposes two conditions on the amplitudes B_i, namely,

$$\tfrac{1}{2}K^2 B_1 + P \cdot K B_2 + Q \cdot K B_3 + K^2 B_4 = 0,$$
$$B_5 + P \cdot K B_6 + K^2 B_7 + Q \cdot K B_8 = 0. \qquad (\text{I-22})$$

We choose to eliminate B_3 and B_5 using these conditions, this being done in such a way as to introduce the smallest number of kinematical singularities while still keeping the parallel between electroproduction and photoproduction. The reduced current has the form

$$j_\mu = [N_\mu{}^1 - (K^2/2K \cdot Q)N_\mu{}^3]B_1 + [N_\mu{}^2 - (P \cdot K/K \cdot Q)N_\mu{}^3]B_2$$
$$+ [N^4 - (K^2/K \cdot Q)N^3]B_4 + (N_\mu{}^6 - P \cdot K N_\mu{}^5)B_6$$
$$+ (N_\mu{}^7 - K^2 N_\mu{}^5)B_7 + (N_\mu{}^8 - Q \cdot K N_\mu{}^5)B_8. \qquad (\text{I-23})$$

In this form the current is manifestly conserved, since it is easily seen from (I-23) that upon forming the four-vector product $K \cdot j$, the coefficient of each B_i is identically zero.

The condition

$$K \cdot \varepsilon = 0 \tag{I-24}$$

holds for both photoproduction and electroproduction, obviously so in the former case, since ε_μ is the photon polarization vector and in the latter case it follows immediately from (I-8) on using the Dirac equation for the electron spinors and recalling that $K = K_1 - K_2$.

Making use of this, and defining

$$F_{\mu\nu} = \varepsilon_\mu K_\nu - \varepsilon_\nu K_\mu, \tag{I-25}$$

then

$$\begin{aligned}
j_\mu \varepsilon_\mu = [\tfrac{1}{2} i\gamma_5 \gamma_\mu \gamma_\nu F_{\mu\nu} &+ i\gamma_5 (K_\mu Q_\nu/K \cdot Q)F_{\mu\nu}]B_1 \\
&+ 2i\gamma_5 (P_\mu Q_\nu/K \cdot Q)F_{\mu\nu} B_2 + 2i\gamma_5 (K_\mu Q_\nu/K \cdot Q)F_{\mu\nu} B_4 \\
&- \gamma_5 \gamma_\mu P_\nu F_{\mu\nu} B_6 + \gamma_5 K_\mu \gamma_\nu F_{\mu\nu} B_7 - \gamma_5 \gamma_\mu Q_\nu F_{\mu\nu} B_8 .
\end{aligned} \tag{I-26}$$

The set of matrices M_i defined by

$$\begin{aligned}
M_1 &= \tfrac{1}{2} i\gamma_5 \gamma_\mu \gamma_\nu F_{\mu\nu}, & M_2 &= 2i\gamma_5 P_\mu (Q - \tfrac{1}{2}K)_\nu F_{\mu\nu}, \\
M_3 &= \gamma_5 \gamma_\mu Q_\nu F_{\mu\nu}, & M_4 &= 2\gamma_5 \gamma_\mu P_\nu F_{\mu\nu} - M i\gamma_5 \gamma_\mu \gamma_\nu F_{\mu\nu}, \\
M_5 &= i\gamma_5 K_\mu Q_\nu F_{\mu\nu}, & M_6 &= \gamma_5 K_\mu \gamma_\nu F_{\mu\nu}
\end{aligned} \tag{I-27}$$

are explicitly gauge invariant, and we may write

$$j_\mu \varepsilon_\mu = \sum_{i=1}^{6} A_i(s, t, u, K^2) M_i, \tag{I-28}$$

where

$$\begin{aligned}
A_1 &= B_1 - MB_6, & A_2 &= 2B_2/(t - \mu^2), \\
A_3 &= -B_8, & A_4 &= -\tfrac{1}{2}B_6, \\
A_5 &= 1/(K \cdot Q)B_1 + 2B_4 - (2P \cdot K)/(t - \mu^2)B_2, & A_6 &= B_7 .
\end{aligned} \tag{I-29}$$

It should be noted that M_5, M_6, and the K_ν term in M_2 are identically zero for photoproduction.

It is clear that kinematical singularities have been introduced into A_2 and A_5 by the current conservation condition. However, these kinematical singularities do not influence any physically observable quantities, since when taken together, the singular part of the combination $A_2 M_2 + A_5 M_5$ is of the form

$$\{[2B_2(P \cdot K) + K^2(B_1 + 2B_4)]/(Q \cdot K)\} i\gamma_5 Q \cdot \varepsilon \tag{I-30}$$

[using (I-29) and (I-28)], the numerator of which is explicitly zero by current conservation, (I-22).

The amplitudes $A_i(s, t, u, K^2)$ are related to the matrix elements of the physical s-, t-, and u-channel processes as follows:

s channel:

$$\langle N(P_2), \pi(Q) | T | N(P_1), \gamma(K) \rangle = \bar{u}(\mathbf{P}_2) \left\{ \sum_{i=1}^{6} A_i(s, t, u, K^2) M_i \right\} u(\mathbf{P}_1).$$

(I-31)

t channel:

$$\langle N(P_2), \bar{N}(-P_1) | T | \pi(-Q), \gamma(K) \rangle = \bar{u}(\mathbf{P}_2) \left\{ \sum_{i=1}^{6} A_i(s, t, u, K^2) M_i \right\} v(-\mathbf{P}_1).$$

(I-32)

u channel:

$$\langle \bar{N}(-P_1), \pi(Q) | T | \bar{N}(-P_2), \gamma(K) \rangle = -\bar{v}(-\mathbf{P}_2) \left\{ \sum_{i=1}^{6} A_i(s, t, u, K^2) M_i \right\} v(-\mathbf{P}_1).$$

(I-33)

C. Isospin Decomposition and Crossing for Pion Production

In isotopic spin space, the electromagnetic current behaves like an isoscalar plus the third component of an isovector

$$j_\mu = j_\mu^{v3} + j_\mu^{s}.$$

(I-34)

From this, it follows that the decomposition of the invariant amplitudes in isotopic spin space has the form

$$A_i = A_i^{(+)} \delta_{\alpha 3} + A_i^{(-)} \tfrac{1}{2}[\tau_\alpha, \tau_3]_- + A_i^{(0)} \tau_\alpha \qquad (i = 1, \ldots, 6),$$

(I-35)

where τ_α and τ_3 are the Pauli spin matrices and α is the isospin index of the pion.

The isovector transition amplitudes $A_i^{(+, -)}$ may be expressed in terms of the amplitudes $A_i^{(1, 3)}$ with isospin $\tfrac{1}{2}$, $\tfrac{3}{2}$ in the final state:

$$A_i^{(1)} = A_i^{(+)} + 2A_i^{(-)}, \qquad A_i^{(3)} = A_i^{(+)} - A_i^{(-)}.$$

(I-36)

The isoscalar amplitudes $A_i^{(0)}$ always lead to a final state with isospin $\tfrac{1}{2}$.

The amplitudes for particular physical processes are

$$\begin{aligned}
\langle n\pi^+ | T | \gamma p \rangle &= \sqrt{2}[A^{(0)} + A^{(-)}], & \langle p\pi^0 | T | \gamma p \rangle &= [A^{(0)} + A^{(+)}], \\
\langle p\pi^- | T | \gamma n \rangle &= \sqrt{2}[A^{(0)} - A^{(-)}], & \langle n\pi^0 | T | \gamma n \rangle &= -[A^{(0)} - A^{(+)}].
\end{aligned}$$

(I-37)

Under the exchange of s and u (crossing), the functions $A_i^{(\pm, 0)}$ are either even or odd, and this crossing symmetry can readily be derived from the assumption of C invariance for photo- and electroproduction.

Again using α to denote the isospin index of the pion, and recalling that the charge conjugation properties of the states involved are

$$C|\pi^\alpha\rangle = (-1)^{\alpha+1}|\pi^\alpha\rangle, \qquad C|\gamma\rangle = -|\gamma\rangle, \qquad C|N\rangle = |\bar{N}\rangle, \qquad \text{(I-38)}$$

the matrix elements of the s and u channels are related by the equation

$$\langle N(p_2), \pi^\alpha(Q)|T|N(P_1), \gamma(K)\rangle = (-1)^\alpha\langle \bar{N}(P_2), \pi^\alpha(Q)|T|\bar{N}(P_1), \gamma(K)\rangle.$$
$$\text{(I-39)}$$

In terms of the amplitudes A_i, (I-39) becomes [using (I-31) and (I-33)]

$$\chi^T(2)\bar{u}(\mathbf{P}_2)\sum_{i=1}^{6} A_i(s, t, u, K^2)M_i(P_1, K, P_2, Q)u(\mathbf{P}_1)\chi(1)$$

$$= -(-1)^\alpha\chi^T(1)\bar{v}(\mathbf{P}_1)\sum_{i=1}^{6} A_i(u, t, s, K^2)M_i(-P_2, K, -P_1, Q)v(\mathbf{P}_2)\chi(2).$$
$$\text{(I-40)}$$

Here χ is the nucleon isospinor which acts on the isospin dependence (suppressed) of the A_i, and the superscript T denotes the transpose of a matrix. The u and v are related by the C matrix in the usual way:

$$v = C\bar{u}^T, \qquad \bar{v} = -u^T C^{-1} \qquad (C = \gamma_2\gamma_4).$$

Making the substitutions for the M_i's in (I-27) and restoring the isospin dependence of the A_i, one finds that $A_i^{(+, 0)}$ for $i = 1, 2, 4$ and $A_i^{(-)}$ for $i = 3, 5, 6$ are symmetric under the exchange $s \leftrightarrow u$ and that the other amplitudes are odd, that is,

$$\begin{aligned}
A_i^{(+, 0)}(s, t, u) &= A_i^{(+, 0)}(u, t, s) & (i = 1, 2, 4), \\
A_i^{(-)}(s, t, u) &= A_i^{(-)}(u, t, s) & (i = 3, 5, 6), \\
A_i^{(+, 0)}(s, t, u) &= -A_i^{(+, 0)}(u, t, s) & (i = 3, 5, 6), \\
A_i^{(-)}(s, t, u) &= -A_i^{(-)}(u, t, s) & (i = 1, 2, 4).
\end{aligned} \qquad \text{(I-41)}$$

These crossing symmetry properties can be compactly written in matrix form by defining the six-vector $\tilde{A}$ with elements $A_1, \ldots, A_6$. Then

$$\tilde{A}(s, t, u) = [\tilde{\xi}]\tilde{A}(u, t, s), \qquad \text{(I-42)}$$

where $[\tilde{\xi}]$ is the diagonal matrix

$$[\tilde{\xi}] = \xi \, \text{diag}\{1, \quad 1, \quad -1, \quad 1, \quad -1, \quad -1\}, \qquad \text{(I-43)}$$

and the parameter ξ is defined by

$$\xi = +1 \quad \text{for isospin index } +, 0; \qquad \xi = -1 \quad \text{for isospin index } -.$$

D. Multipole Decomposition of Amplitudes for Pion Production

As a first step, it is convenient to express the matrix element in the pion-nucleon center-of-mass system in terms of Pauli matrices and two-component spinors, the latter chosen to have the z axis as the axis of quantization. To effect this, an amplitude $\mathscr{F}$ is introduced by the relation

$$\bar{u}(\mathbf{P}_2) \sum_{i=1}^{6} A_i M_i u(\mathbf{P}_1) = \chi^{\dagger}(2)\mathscr{F}\chi(1) \qquad (\text{I-44})$$

for electroproduction, and by the relation

$$\bar{u}(\mathbf{P}_2) \sum_{i=1}^{4} A_i M_i u(\mathbf{P}_1) = (4\pi W/M)\chi^{\dagger}(2)\mathscr{F}\chi(1) \qquad (\text{I-45})$$

for photoproduction. The extra kinematical factor in (I-45) is simply to make the normalization of the photoproduction amplitude agree with what has become the normal convention. In the following discussion, photoproduction and electroproduction will be treated in parallel with the different normalization suppressed.

Starting from (I-44), the matrix $\mathscr{F}$ can be written as

$$\mathscr{F} = i\boldsymbol{\sigma} \cdot \boldsymbol{\varepsilon}\mathscr{F}_1 + \boldsymbol{\sigma} \cdot \hat{\mathbf{q}}\boldsymbol{\sigma} \cdot (\hat{\mathbf{k}} \times \boldsymbol{\varepsilon})\mathscr{F}_2 + i\boldsymbol{\sigma} \cdot \hat{\mathbf{k}}\hat{\mathbf{q}} \cdot \boldsymbol{\varepsilon}\mathscr{F}_3 + i\boldsymbol{\sigma} \cdot \hat{\mathbf{q}}\hat{\mathbf{q}} \cdot \boldsymbol{\varepsilon}\mathscr{F}_4$$
$$+ i\boldsymbol{\sigma} \cdot \hat{\mathbf{k}}\hat{\mathbf{k}} \cdot \boldsymbol{\varepsilon}\mathscr{F}_5 + i\boldsymbol{\sigma} \cdot \hat{\mathbf{q}}\hat{\mathbf{k}} \cdot \boldsymbol{\varepsilon}\mathscr{F}_6 - i\boldsymbol{\sigma} \cdot \hat{\mathbf{q}}\varepsilon_0\mathscr{F}_7 - i\boldsymbol{\sigma} \cdot \hat{\mathbf{k}}\varepsilon_0 \mathscr{F}_8 . \qquad (\text{I-46})$$

As in the case of the invariant amplitudes, current conservation imposes two conditions on the $\mathscr{F}_i$, namely

$$\mathscr{F}_1 + \hat{\mathbf{k}} \cdot \hat{\mathbf{q}}\mathscr{F}_3 + \mathscr{F}_5 - k_0/k\mathscr{F}_8 = 0, \qquad \hat{\mathbf{k}} \cdot \hat{\mathbf{q}}\mathscr{F}_4 + \mathscr{F}_6 - k_0/k\mathscr{F}_7 = 0. \qquad (\text{I-47})$$

Eliminating $\mathscr{F}_7$ and $\mathscr{F}_8$ yields

$$\mathscr{F} = i\boldsymbol{\sigma} \cdot \mathbf{a}\mathscr{F}_1 + \boldsymbol{\sigma} \cdot \hat{\mathbf{q}}\boldsymbol{\sigma} \cdot (\hat{\mathbf{k}} \times \mathbf{a})\mathscr{F}_2 + i\boldsymbol{\sigma} \cdot \hat{\mathbf{k}}\hat{\mathbf{q}} \cdot \mathbf{a}\mathscr{F}_3$$
$$+ i\boldsymbol{\sigma} \cdot \hat{\mathbf{q}}\hat{\mathbf{q}} \cdot \mathbf{a}\mathscr{F}_4 + i\boldsymbol{\sigma} \cdot \hat{\mathbf{k}}\hat{\mathbf{k}} \cdot \mathbf{a}\mathscr{F}_5 + i\boldsymbol{\sigma} \cdot \hat{\mathbf{q}}\hat{\mathbf{k}} \cdot \mathbf{a}\mathscr{F}_6 , \qquad (\text{I-48})$$

with

$$a_\mu = \varepsilon_\mu - (\varepsilon_0/k_0)K_\mu . \qquad (\text{I-49})$$

It is clear from (I-49) that $a_0 = 0$, that is, that a has no scalar component and hence this choice of amplitudes $\mathscr{F}_i$ corresponds to a virtual photon with transverse and longitudinal components.

Instead of eliminating $\mathscr{F}_7$ and $\mathscr{F}_8$, it is possible to eliminate $\mathscr{F}_5$ and $\mathscr{F}_6$ from (I-46). This gives

$$\mathscr{F} = i\boldsymbol{\sigma} \cdot \mathbf{b}\mathscr{F}_1 + \boldsymbol{\sigma} \cdot \hat{\mathbf{q}}\boldsymbol{\sigma} \cdot (\hat{\mathbf{k}} \times \mathbf{b})\mathscr{F}_2 + i\boldsymbol{\sigma} \cdot \hat{\mathbf{k}}\hat{\mathbf{q}} \cdot \mathbf{b}\mathscr{F}_3$$
$$+ i\boldsymbol{\sigma} \cdot \hat{\mathbf{q}}\hat{\mathbf{q}} \cdot \mathbf{b}\mathscr{F}_4 - i\boldsymbol{\sigma} \cdot \hat{\mathbf{q}}b_0\mathscr{F}_7 - i\boldsymbol{\sigma} \cdot \hat{\mathbf{k}}b_0\mathscr{F}_8 , \qquad (\text{I-50})$$

with

$$b_\mu = \varepsilon_\mu - (\varepsilon \cdot \hat{\mathbf{k}}/k)K_\mu. \tag{I-51}$$

In this case, $\mathbf{b} \cdot \mathbf{k} = 0$, that is b_μ has no longitudinal component but $b_0 \neq 0$, so that this choice of amplitudes $\mathscr{F}_i$ corresponds to a virtual photon with transverse and scalar components. It is this latter choice of amplitudes (that is, $\mathscr{F}_1, \ldots, \mathscr{F}_4, \mathscr{F}_7, \mathscr{F}_8$) which we shall adopt, first because $\mathscr{F}_7$ and $\mathscr{F}_8$ have a simpler angular momentum decomposition than $\mathscr{F}_5$ and $\mathscr{F}_6$ and second because the latter may introduce spurious singularities in numerical calculations.

It should be noted that choosing a particular set of amplitudes in this way is in effect a choice of gauge, since, because of current conservation, ε_μ can be replaced by $\varepsilon_\mu - \lambda K_\mu$ for any λ. Further, the choice of gauge does not affect the completely transverse character of the real photon in photoproduction, for one still has $a \cdot k = b \cdot k = 0$ since $K^2 = 0$. In photoproduction only the first four terms, with coefficients $\mathscr{F}_1, \mathscr{F}_2, \mathscr{F}_3, \mathscr{F}_4$ of (I-46), (I-48), or (I-50) are present in $\mathscr{F}$.

Decomposing the left-hand side of (I-44) into Pauli matrices and spinors, one can readily obtain the connection between the amplitudes $\mathscr{F}_1, \ldots, \mathscr{F}_4, \mathscr{F}_7, \mathscr{F}_8$ and the amplitudes $A_1, \ldots, A_6$. Defining the six-vector $\tilde{\mathscr{F}}$ with elements $\mathscr{F}_1, \mathscr{F}_2, \mathscr{F}_3, \mathscr{F}_4, \mathscr{F}_7, \mathscr{F}_8$, the result is

$$\tilde{\mathscr{F}}(s, t, K^2) = \{C^{-1}(s, K^2)\}\{B(s, t, K^2)\}\, \tilde{A}(s, t, K^2), \tag{I-52}$$

$$\tilde{A}(s, t, K^2) = \{B^{-1}(s, t, K^2)\}\{C(s, K^2)\}\tilde{\mathscr{F}}(s, t, K^2), \tag{I-53}$$

where the diagonal matrix $\{C(s, K^2)\}$ and the matrices $\{B(s, t, K^2)\}$ and $\{B^{-1}(s, t, K^2)\}$ are given by (I-54), (I-55), and (I-56), respectively.

$$\{C\} = \frac{2M}{(W-M)\{(E_1+M)(E_2+M)\}^{1/2}} \begin{bmatrix} 1 \\ \dfrac{(E_1+M)(E_2+M)(W-M)}{qk(W+M)} \\ \dfrac{(E_1+M)(W-M)}{qk(W+M)} \\ \dfrac{(E_2+M)}{q^2} \\ \dfrac{(E_2+M)(W-M)}{q} \\ \dfrac{(E_1+M)(W-M)}{k} \end{bmatrix},$$

$$\tag{I-54}$$

$$
\{B\} =
\begin{bmatrix}
1 & 0 & -\dfrac{K\cdot Q}{(W-M)} & (W-M)+\dfrac{K\cdot Q}{(W-M)} & 0 & \dfrac{K^2}{W-M} \\[2ex]
-1 & 0 & -\dfrac{K\cdot Q}{W+M} & (W+M)+\dfrac{K\cdot Q}{(W+M)} & 0 & \dfrac{K^2}{W+M} \\[2ex]
0 & \dfrac{(W^2-M^2+\frac{1}{2}K^2)}{(W+M)} & 1 & -1 & \dfrac{-K^2}{(W+M)} & 0 \\[2ex]
0 & -\dfrac{(W^2-M^2+\frac{1}{2}K^2)}{(W-M)} & 1 & -1 & \dfrac{K}{W-M} & 0 \\[2ex]
-(E_1-M) & \dfrac{1}{2k_0}\{k^2(3K\cdot Q+2k_0 W) - \mathbf{k}\cdot\mathbf{q}(2(W^2-M^2)+K^2)\} & K\cdot Q+q_0(W-M) & \begin{array}{l}-K\cdot Q-q_0(W+M)\\+(E_1-M)(W+M)\end{array} & q_0K^2-k_0K\cdot Q & -(E_1-M)(W+M) \\[3ex]
(E_1+M) & \dfrac{-1}{2k_0}\{k^2(3K\cdot Q+2k_0 W) - \mathbf{k}\cdot\mathbf{q}(2(W^2-M^2)+K^2)\} & K\cdot Q+q_0(W+M) & \begin{array}{l}-K\cdot Q-q_0(W+M)\\+(E_1+M)(W-M)\end{array} & -q_0K^2+k_0K\cdot Q & -(E_1+M)(W-M)
\end{bmatrix},
$$

$$(1\text{-}55)$$

$$\{B^{-1}\} =$$

$$
\begin{array}{l|cccccc}
\dfrac{1}{4W^2k_0} &
(W-M)\left\{\!\begin{array}{l}(W+M)^2-K^2\\ -\dfrac{2MK^2}{k^2}(E_1+M)\end{array}\!\right\} &
\dfrac{-k_0(W+M)}{(E_1+M)}\\ \times\{W^2-M^2+K^2\} &
2M(W+M)\\ \times\left\{K\cdot Q-\dfrac{K^2}{k^2}\,\mathbf{q}\cdot\mathbf{k}\right\} &
2M(W-M)\\ \times\left\{K\cdot Q-\dfrac{K^2}{k^2}\,\mathbf{q}\cdot\mathbf{k}\right\} &
2MK^2\dfrac{k_0}{k^2} &
2MK^2\dfrac{k_0}{k^2}\\[4ex]

\dfrac{K^2}{2k_0W^2(t-\mu^2)} &
(W-M)\\ \times\left\{1-\dfrac{(E_1+M)(W-M)}{k^2}\right\} &
\dfrac{k_0(W+M)}{(E_1+M)} &
(W^2-M^2)\\ \times\left[\dfrac{K\cdot Q}{k^2}+\dfrac{\mathbf{q}\cdot\mathbf{k}}{k^2}\right] &
(W^2-M^2)\\ \times\left[\dfrac{K\cdot Q}{k^2}+\dfrac{\mathbf{q}\cdot\mathbf{k}}{k^2}\right] &
-(W+M)\dfrac{k_0}{k^2} &
(W-M)\dfrac{k_0}{k^2}\\[4ex]

=\ \dfrac{1}{4W^2k_0} &
(W-M)\\ \times\left\{W+M-\dfrac{K^2}{k^2}(E_1+M)\right\} &
\dfrac{k_0(W+M)}{(E_1+M)} &
(W+M)\\ \times\left\{K\cdot Q+2k_0W-\dfrac{K^2}{k^2}\,\mathbf{q}\cdot\mathbf{k}\right\} &
(W-M)\\ \times\left\{K\cdot Q+2k_0W-\dfrac{K^2}{k^2}\,\mathbf{q}\cdot\mathbf{k}\right\} &
\dfrac{K^2k_0}{k^2} &
\dfrac{K^2k_0}{k^2}\\[4ex]

\dfrac{1}{4W^2k_0} &
(W-M)\\ \times\left\{W+M-\dfrac{K^2}{k^2}(E_1+M)\right\} &
\dfrac{k_0(W+M)}{(E_1+M)} &
(W+M)\\ \times\left\{K\cdot Q-\dfrac{K^2}{k^2}\,\mathbf{q}\cdot\mathbf{k}\right\} &
(W-M)\\ \times\left\{K\cdot Q-\dfrac{K^2}{k^2}\,\mathbf{q}\cdot\mathbf{k}\right\} &
\dfrac{K^2k_0}{k^2} &
\dfrac{K^2k_0}{k^2}\\[4ex]

\dfrac{\beta}{2k_0W^2(t-\mu^2)} &
(W-M)\\ \times\left\{1-\dfrac{(E_1+M)(W-M)}{k^2}\right\} &
\dfrac{k_0(W+M)}{(E_1+M)} &
(W^2-M^2)\\ \times\left[\dfrac{\delta}{\beta}-\dfrac{\mathbf{q}\cdot\mathbf{k}}{k^2}\right] &
(W^2-M^2)\\ \times\left[-\dfrac{\delta}{\beta}+\dfrac{\mathbf{q}\cdot\mathbf{k}}{k^2}\right] &
-\dfrac{(W+M)K_0}{k^2} &
\dfrac{(W-M)k_0}{k^2}\\[4ex]

\dfrac{W^2-M^2}{4W^2k_0} &
-1+\dfrac{(E_1+M)(W-M)}{k^2} &
\dfrac{k_0}{(E_1+M)} &
-\dfrac{K\cdot Q}{(W-M)}+\dfrac{(W+M)}{k^2}\mathbf{q}\cdot\mathbf{k} &
\dfrac{K\cdot Q}{(W+M)}-\dfrac{(W-M)\mathbf{q}\cdot\mathbf{k}}{k^2} &
-\dfrac{k_0}{k^2} &
-\dfrac{k_0}{k^2}
\end{array}
$$

(I-56)

where $\delta=\tfrac{1}{2}K\cdot Q+K_0W$ and $\beta=W^2-M^2+\tfrac{1}{2}K^2.$

11

Recalling the different normalization between the electroproduction and photoproduction amplitudes [(I-44) and (I-45)], the matrix $\{C\}$ must be multiplied by $4\pi W/M$ for photoproduction and K^2 set to zero. In the photoproduction case, of course, only the first four rows and columns contribute.

It is now straightforward, but somewhat tediuos, to connect $\tilde{\mathscr{F}}$ with the eigenamplitudes of parity and angular momentum, therefore we will state only the results.

For electroproduction, there are six types of transitions possible to a pion nucleon final state with angular momentum l and definite parity, which are classified according to the character of the photon, transverse or scalar (or alternatively, longitudinal) and the total angular momentum $J = l \pm \frac{1}{2}$ of the final state. In addition, the transverse photon states can either be electric, with parity $(-1)^L$, or magnetic, with parity $(-1)^{L+1}$, where L is the total orbital angular momentum of the photon. This classification scheme is shown in Table I, and the longitudinal multipoles are related to the scalar multipoles by

$$L_{l\pm} = (k_0/k)S_{l\pm}. \tag{I-57}$$

In photoproduction only the transverse states $(E_{l\pm}, M_{l\pm})$ contribute.

TABLE I

MULTIPOLE STATES FOR PION PHOTOPRODUCTION AND ELECTROPRODUCTION[a]

J	L	Parity $(-1)^l$	Multipole transition	Notation	Lowest value of l permitted
$l + \frac{1}{2}$	$L = J + \frac{1}{2} = l + 1$	$(-1)^L$	Electric 2^{l+1}	E_{l+}	0
$l - \frac{1}{2}$	$L = J - \frac{1}{2} = l - 1$	$(-1)^L$	Electric 2^{l-1}	E_{l-}	2
$l + \frac{1}{2}$	$L = J - \frac{1}{2} = l$	$-(-1)^L$	Magnetic 2^l	M_{l+}	1
$l - \frac{1}{2}$	$L = J + \frac{1}{2} = l$	$-(-1)^L$	Magnetic 2^l	M_{l-}	1
$l + \frac{1}{2}$	$L = J + \frac{1}{2} = l + 1$	$(-1)^L$	Scalar 2^{l+1}	S_{l+}	0
$l - \frac{1}{2}$	$L = J - \frac{1}{2} = l - 1$	$(-1)^L$	Scalar 2^{l-1}	S_{l-}	1

[a] The scalar multipoles are related to the longitudinal multipoles by $L_{l\pm} = (k_0/k)S_{l\pm}$.

Defining the six-vector $\tilde{M}_l$ with elements $E_{l+}, E_{l-}, M_{l+}, M_{l-}, S_{l+}, S_{l-}$, the relation between the multipole amplitudes and the amplitudes $\mathscr{F}_i$ is

$$\tilde{\mathscr{F}}(s, t, K^2) = \sum_{l=0}^{\infty} \begin{bmatrix} G_l(x) & 0 \\ 0 & H_l(x) \end{bmatrix} \tilde{\mathscr{M}}_l(s, K^2), \tag{I-58}$$

where G_l and H_l are respectively 4×4 and 2×2 matrices with elements

$$G_l = \begin{bmatrix} P'_{l+1} & P'_{l-1} & P'_{l+1} & (l+1)P'_{l-1} \\ 0 & 0 & (l+1)P'_l & lP'_l \\ P''_{l+1} & P''_{l-1} & -P''_{l+1} & P''_{l-1} \\ -P''_l & -P''_l & P''_l & -P''_l \end{bmatrix}, \tag{I-59}$$

$$H_l = \begin{bmatrix} -(l+1)P'_l & lP'_l \\ (l+1)P'_{l+1} & -lP'_{l-1} \end{bmatrix}. \tag{I-60}$$

Here the P_l are Legendre polynomials of the first kind, and a prime superscript denotes differentiation with respect to x ($x = \cos\theta$). Only the matrix G_l occurs in photoproduction.

The inverse of (I-58) is

$$\tilde{M}_l(s, K^2) = \int_{-1}^{1} dx \begin{bmatrix} D_l(x) & 0 \\ 0 & E_l(x) \end{bmatrix} \mathscr{F}(s, t, K^2), \tag{I-61}$$

where

$$D_l = \begin{bmatrix} \dfrac{1}{2(l+1)} \left\{ P_l, -P_{l+1}, \dfrac{1}{2l+1}(P_{l-1} - P_{l+1}), \dfrac{l+1}{2l+3}(P_l - P_{l+2}) \right\} \\[2ex] \dfrac{1}{2l} \left\{ P_l, -P_{l-1}, \dfrac{l+1}{2l+1}(P_{l+1} - P_{l-1}), \dfrac{l}{2l-1}(P_l - P_{l-2}) \right\} \\[2ex] \dfrac{1}{2(l+1)} \left\{ P_l, -P_{l+1}, \dfrac{1}{2l+1}(P_{l+1} - P_{l-1}), 0 \right\} \\[2ex] \dfrac{1}{2l} \left\{ P_l, -P_{l-1}, \dfrac{1}{2l+1}(P_{l-1} - P_{l+1}), 0 \right\} \end{bmatrix}, \tag{I-62}$$

$$E_l = \begin{bmatrix} \dfrac{1}{2(l+1)} \{ P_{l+1}, P_l \} \\[2ex] \dfrac{1}{2l} \{ P_{l-1}, P_l \} \end{bmatrix}. \tag{I-63}$$

E. The Helicity Formalism for Pion Production

Denoting the photon helicity by λ, the initial nucleon helicity by μ, and the final nucleon helicity by μ', then, following Jacob and Wick (1959), the angular momentum decomposition of the helicity amplitudes $f_{\mu', \lambda\mu}(\theta, \phi)$ is given by

$$f_{\mu', \lambda\mu} = \sum_J (J + \tfrac{1}{2}) \langle \mu' | T^J | \lambda\mu \rangle \exp[-i(\lambda - \mu + \mu')\phi] \, d^J_{\mu - \lambda, \mu'}. \tag{I-64}$$

Although it would appear that there are twelve such helicity amplitudes, parity reduces the number of independent amplitudes to six by the relation

$$f_{\lambda, \lambda'}(\theta, \phi) = -f_{-\lambda, -\lambda'}(\theta, \pi - \phi). \tag{I-65}$$

The normalization of the helicity amplitudes is determined by the relation

$$f_{\mu', \lambda\mu} = \langle \mu' | \mathscr{F} | \lambda\mu \rangle, \tag{I-66}$$

where $\mathscr{F}$ is the conventional center-of-mass amplitude, specified in (I-44) and (I-45).

As a first step towards obtaining the connection between the helicity formalism and the multipole formalism, the expansion of $\mathscr{F}$ in terms of spin functions, (I-50) is inserted into (I-66) and on evaluating the spin matrix elements, one finds

$$\sqrt{2}if_{\frac{1}{2}, 1\frac{1}{2}} = 2 \sin \tfrac{1}{2}\theta(\mathscr{F}_1 + \mathscr{F}_2) + \cos \tfrac{1}{2}\theta \sin \theta(\mathscr{F}_3 + \mathscr{F}_4),$$

$$\sqrt{2}if_{\frac{1}{2}, -1-\frac{1}{2}} = -2 \cos \tfrac{1}{2}\theta(\mathscr{F}_1 - \mathscr{F}_2) + \sin \tfrac{1}{2}\theta \sin \theta (\mathscr{F}_3 - \mathscr{F}_4),$$

$$\sqrt{2}if_{\frac{1}{2}, 1-\frac{1}{2}} = -\sin \tfrac{1}{2}\theta \sin \theta(\mathscr{F}_3 - \mathscr{F}_4),$$

$$\sqrt{2}if_{\frac{1}{2}, -1\frac{1}{2}} = -\cos \tfrac{1}{2}\theta \sin \theta(\mathscr{F}_3 + \mathscr{F}_4), \tag{I-67}$$

$$\sqrt{2}if_{\frac{1}{2}, 0\frac{1}{2}} = \sqrt{2} \cos \tfrac{1}{2}\theta(\mathscr{F}_7 + \mathscr{F}_8)k_0/k,$$

$$\sqrt{2}if_{\frac{1}{2}, 0-\frac{1}{2}} = \sqrt{2} \sin \tfrac{1}{2}\theta(\mathscr{F}_7 - \mathscr{F}_8)k_0/k.$$

The relation between the multipole amplitudes and the angular momentum amplitudes $\langle \mu' | T^J | \lambda\mu \rangle$ now follows from the equation for projection of the multipoles from the $\mathscr{F}_i$: (I-61), (I-62), (I-63) and the inverse of (I-67). Introducing the notation

$$\begin{aligned}
T_J(1) &= \langle \tfrac{1}{2} | T^J | -1 \tfrac{1}{2} \rangle, &\qquad T_J(2) &= \langle \tfrac{1}{2} | T^J | 1 - \tfrac{1}{2} \rangle, \\
T_J(3) &= \langle \tfrac{1}{2} | T^J | -1 - \tfrac{1}{2} \rangle, &\qquad T_J(4) &= \langle \tfrac{1}{2} | T^J | 1 \tfrac{1}{2} \rangle, \\
T_J(7) &= \langle \tfrac{1}{2} | T^J | 0 \tfrac{1}{2} \rangle, &\qquad T_J(8) &= \langle \tfrac{1}{2} | T^J | 0 - \tfrac{1}{2} \rangle,
\end{aligned} \tag{I-68}$$

then the result of this projection is

$$E_{l+} = \frac{-\sqrt{2}i}{4(l+1)} \{[l/(l+2)]^{1/2}[-T_{l+\frac{1}{2}}(1) + T_{l+\frac{1}{2}}(2)] \\
+ [T_{l+\frac{1}{2}}(3) - T_{l+\frac{1}{2}}(4)]\},$$

$$E_{(l+1)-} = \frac{-\sqrt{2}i}{4(l+1)} \{[(l+2)/l]^{1/2}[T_{l+\frac{1}{2}}(1) + T_{l+\frac{1}{2}}(2)] \\
+ [T_{l+\frac{1}{2}}(3) + T_{l+\frac{1}{2}}(4)]\}, \tag{I-69}$$

$$M_{l+} = \frac{-\sqrt{2}i}{4(l+1)} \{[(l+2)/l]^{1/2}[T_{l+\frac{1}{2}}(1) - T_{l+\frac{1}{2}}(2)] \\
+ [T_{l+\frac{1}{2}}(3) - T_{l+\frac{1}{2}}(4)]\},$$

$$M_{(l+1)} = \frac{-\sqrt{2}i}{4(l+1)} \{[l/(l+2)]^{1/2}[T_{l+\frac{1}{2}}(1) + T_{l+\frac{1}{2}}(2)]$$

$$-[T_{l+\frac{1}{2}}(3) + T_{l+\frac{1}{2}}(4)]\},$$

$$S_{l+} = [i/2(l+1)]\{T_{l+\frac{1}{2}}(5) + T_{l+\frac{1}{2}}(6)\}(k/k_0),$$

$$S_{(l+1)-} = [i/2(l+1)]\{T_{l+\frac{1}{2}}(5) - T_{l+\frac{1}{2}}(6)\}(k/k_0).$$

(I-69,
contd.)

These equations have the inverse

$$\sqrt{2}iT_{l+\frac{1}{2}}(1) = [l(l+2)]^{1/2}\{E_{l+} - M_{l+} - E_{(l+1)-} - M_{(l+1)-}\},$$

$$\sqrt{2}iT_{l+\frac{1}{2}}(2) = [l(l+2)]^{1/2}\{M_{l+} - E_{l+} - E_{(l+1)-} - M_{(l+1)-}\},$$

$$\sqrt{2}iT_{l+\frac{1}{2}}(3) = (l+2)\{M_{(l+1)-} - E_{l+} - lM_{l+} + E_{(l+1)-}\},$$

$$\sqrt{2}iT_{l+\frac{1}{2}}(4) = l\{M_{l+} - E_{(l+1)-} + (l+2)M_{(l+1)-} + E_{l+}\},$$

$$iT_{l+\frac{1}{2}}(5) = (l+1)\{S_{l+} + S_{(l+1)-}\}(k_0/k),$$

$$iT_{l+\frac{1}{2}}(6) = (l+1)\{S_{l+} - S_{(l+1)-}\}(k_0/k).$$

(I-70)

F. CONDITIONS FROM UNITARITY AND T INVARIANCE FOR PION PRODUCTION

The unitarity of the S matrix, together with time-reversal invariance when the final state differs from the initial state, leads to the condition for the R matrix.

$$\text{Im}\langle f|R|i\rangle = \tfrac{1}{2}\sum_n \langle n|R|i\rangle^\dagger\langle n|R|i\rangle, \tag{I-71}$$

where $|n\rangle$ is a complete set of states. For photoproduction and electro-production of pions, where intermediate states containing photons may be neglected, (I-71) imposes a phase condition on the individual multipole amplitudes in the region of partial-wave elasticity. This condition, namely that the phase of a multipole transition to a final pion-nucleon state is equal to the scattering phase shift (modulo π) of that pion nucleon state, was first derived by Watson (1954) and is generally known as Watson's theorem. An elegant derivation in terms of the helicity formalism has been given by Shaw (1966) and we outline this method here.

First, it is necessary to introduce the helicity amplitudes for pion-nucleon scattering from an initial state of helicity μ', angles (θ, ϕ), to a final state of helicity μ'', angles (θ', ϕ'), by

$$\langle\mu''|R|\mu'\rangle = F(\theta', \phi', \theta, \phi)$$

$$= (1/2\pi)\sum_J \sum_m (J+\tfrac{1}{2})\langle|T^J|\rangle D_m^J(\phi', \theta', -\phi')D_m^J(\phi, \theta, -\phi). \tag{I-72}$$

Then, retaining only one pion and one nucleon intermediate states in (I-71),

$$\text{Im} f_{\mu', \lambda\mu}(\theta, \phi) = \tfrac{1}{2} \sum_{\mu''} \int d\Omega' \, F^{*}_{\mu''\mu'}(\theta', \phi', \theta, \phi) f_{\mu'', \lambda\mu}(\theta', \phi'). \qquad \text{(I-73)}$$

Substituting (I-64) and (I-72) in (I-73) and using the orthonormality relation

$$\int d\Omega' D^{J*}_{m, n}(\phi', \theta', -\phi') D^{J'}_{m, n'}(\phi', \theta', -\phi') = [4\pi/(2J + 1)] \, \delta_{JJ'} \, \delta_{mm'}, \qquad \text{(I-74)}$$

then one obtains immediately the relation for the helicity amplitudes

$$\text{Im} \langle \mu' | T^{J} | \lambda\mu \rangle = \tfrac{1}{2} \sum_{\mu''} \langle \mu'' | T^{J} | \mu' \rangle^{\dagger} \langle \mu'' | T^{J} | \lambda\mu \rangle. \qquad \text{(I-75)}$$

Using (I-69) and the parity relation of the scattering amplitudes,

$$\langle \tfrac{1}{2} | T^{J} | \pm \tfrac{1}{2} \rangle = -\langle -\tfrac{1}{2} | T^{J} | \mp \tfrac{1}{2} \rangle \qquad \text{(I-76)}$$

gives immediately

$$\text{Im} \begin{pmatrix} E_{l\pm} \\ M_{l\pm} \\ S_{l\pm} \end{pmatrix} = \begin{pmatrix} E_{l\pm} \\ M_{l\pm} \\ S_{l\pm} \end{pmatrix} \tfrac{1}{2} \{ \langle \tfrac{1}{2} | T^{J} | \tfrac{1}{2} \rangle \pm \langle \tfrac{1}{2} | T^{J} | -\tfrac{1}{2} \rangle \}^{*}. \qquad \text{(I-77)}$$

But since the partial-wave scattering amplitude is given by

$$f_{l\pm} = (\text{kinematical factor}) \{ \langle \tfrac{1}{2} | T^{J} | \tfrac{1}{2} \rangle \pm \langle \tfrac{1}{2} | T^{J} | -\tfrac{1}{2} \rangle \} \qquad \text{(I-78)}$$

with

$$f_{l\pm} = [\exp(i\delta_{l\pm}) \sin \delta_{l\pm}]/q, \qquad \text{(I-79)}$$

it follows immediately that

$$\begin{pmatrix} E_{l\pm} \\ M_{l\pm} \\ S_{l\pm} \end{pmatrix} = \begin{pmatrix} |E_{l\pm}| \\ |M_{l\pm}| \\ |S_{l\pm}| \end{pmatrix} \exp(i\delta_{l\pm} + in\pi), \qquad \text{(I-80)}$$

where n is an integer.

In principle, this relation holds only up to the threshold for two-pion production, that is to a photon laboratory energy of 322 MeV (171-MeV pion laboratory energy in pion nucleon scattering). However, in practice it turns out that the effective inelastic threshold in pion-nucleon scattering is very much higher than this for most partial waves, and, consequently, the phase condition can be applied to a much higher energy. The energies at which inelasticity is first observable in pion-nucleon scattering (E_1) and at which the inelasticity parameter η first goes below 0.9 (E_2) are shown in Table II.

TABLE II

Amplitude	E_1 (MeV)[a]	E_2 (MeV)[b]	Amplitude	E_1 (MeV)[a]	E_2 (MeV)[b]
S_{11}	310	450	S_{31}	410	600
P_{11}	250	370	P_{31}	700	870
P_{13}	530	700	P_{33}	550	870
D_{13}	370	530	D_{33}	370	600
D_{15}	580	700	D_{35}	700	870
F_{15}	580	750	F_{35}	750	950
F_{17}	800	1230	F_{37}	750	1230

[a] E_1 is the lowest pion laboratory energy at which inelasticity is observable.
[b] E_2 is the lowest pion laboratory energy at which the inelasticity parameter $\eta < 0.9$.

G. THE BORN TERMS FOR PION PRODUCTION

The Born pole terms arise from the diagrams of Fig. 1 (see page 29), and their residues are derived from the general renormalized Born approximation for one-nucleon and one-pion exchange. According to the Feynman rules, for the s channel, this is

$$g\bar{u}(\mathbf{P}_2)\chi^\dagger(2)\gamma_5[i\gamma \cdot (K + P_1) - M]/[(K + P_1)^2 + M^2]\{[\tfrac{1}{2}F_1^{(s)}(K^2)\tau_2$$
$$+ \tfrac{1}{2}F_1^{(v)}(K^2)(\tfrac{1}{2}[\tau_\alpha, \tau_3] + \delta_{\alpha 3})]\gamma_\mu + [\tfrac{1}{2}F_2^{(s)}\tau_\alpha$$
$$+ \tfrac{1}{2}F_2^{(v)}(K^2)(\tfrac{1}{2}[\tau_\alpha, \tau_3] + \delta_{\alpha 3})]\sigma_{\mu\nu}K_\nu\}\varepsilon_\mu \chi(1)u(\mathbf{P}_1); \tag{I-81}$$

for the u channel it is

$$g\bar{u}(\mathbf{P}_2)\chi^\dagger(2)\{[\tfrac{1}{2}F_1^{(s)}(K^2)\tau_\alpha + \tfrac{1}{2}F_1^{(v)}(K^2)(-\tfrac{1}{2}[\tau_\alpha, \tau_3] + \delta_{\alpha 3})]\gamma_\mu$$
$$+ [\tfrac{1}{2}F_2^{(s)}(K^2)\tau_\alpha + \tfrac{1}{2}F_2^{(v)}(K^2)(-\tfrac{1}{2}[\tau_\alpha, \tau_3] + \delta_{\alpha 3})]\sigma_{\mu\nu}K_\nu\}$$
$$\times \varepsilon_\mu[i\gamma \cdot (P_2 - K) - M]/[(P_2 - K)^2 + M^2]\,\gamma_5\chi(1)u(\mathbf{P}_1); \tag{I-82}$$

and for the t channel, it is

$$ig\bar{u}(\mathbf{P}_2)\chi^\dagger(2)\gamma_5\tfrac{1}{2}[\tau_\alpha, \tau_3][(Q - K)^2 + \mu^2]^{-1}[2\varepsilon \cdot Q - \varepsilon \cdot K]F_\pi(K^2)\chi(1)u(\mathbf{P}_1). \tag{I-83}$$

Here $F_\pi(K^2)$ is the pion electromagnetic form factor, $F_1^{(v,s)}(K^2)$ are the nucleon electric form factors for isovector and isoscalar states, and $F_2^{(v,s)}(K^2)$

are the anomalous isovector and isoscalar magnetic form factors of the nucleon. The terms $F_{1,2}^{(v,s)}$ are related to the proton and neutron form factors by

$$F_{1,2}^{(v)}(K^2) = F_{1,2}^p(K^2) - F_{1,2}^n(K^2), \qquad F_{1,2}^{(s)}(K^2) = F_{1,2}^p(K^2) + F_{1,2}^n(K^2),$$
(I-84)

and for $K^2 = 0$ (that is, photoproduction), the terms $F_{1,2}^{(v,s)}$ are the charge and anomalous magnetic moments of the nucleon:

$$F_1^{(v,s)}(0) = e, \qquad F_2^{(v)}(0) = (e/2M)(\mu_p' - \mu_n'), \qquad F_2^{(s)}(0) = (e/2M)(\mu_p' + u_n').$$
(I-85)

The normalization of the pion form factor at $K^2 = 0$ is

$$F_\pi(0) = e.$$
(I-86)

The constants have the values

$$e^2/4\pi = 1/137, \qquad g^2/4\pi = 14.4, \qquad \mu_p' = 1.79, \qquad \mu_n' = -1.91.$$
(I-87)

The Born approximations (I-81), (I-82), and (I-83) give rise to poles at $s = M^2$, $u = M^2$, and $t = \mu^2$, respectively. To find the pole terms in the amplitudes A_i, the most straightforward way (although not the way usually found in the literature) is to expand the Born terms in the quantities M_i of (I-27) at the pole values $s = M^2$, $u = M^2$, and $t = \mu^2$, respectively. That the residues of (I-81), (I-82), and (I-83) are each on its own expandable in terms of M_i is due to the fact that each of the Born approximations is conserved when the exchanged particle is on the mass shell. In general, the total Born approximation is not conserved, and (I-81), (I-82), and (I-83), when taken together, cannot be expanded in terms of the quantities M_i without a trick. Replacement of ε by K in the total Born term gives rise to a term

$$-i\gamma_5 g\{F_\pi - F_1^{v}\}\tfrac{1}{2}\{\tau_\alpha, \tau_3\}.$$

It is then necessary to add the term

$$i\gamma_5(g/K^2)\{F_\pi - F_1^{v}\}(\varepsilon \cdot K)\tfrac{1}{2}\{\tau_\alpha, \tau_3\},$$

using the argument that $\varepsilon \cdot K = 0$. This has the effect of ensuring conservation of the general Born approximation consisting of the three contributions and the added term, which can now be expanded in the M_i. In either case, the result in the usual matrix notation is

$$\tilde{A}(s, t, K^2) = [(s - M^2)^{-1} + [\tilde\xi](u - M^2)^{-1}]\tilde\Gamma(t, K^2) + \tfrac{1}{2}(1 - \xi)\tilde\Gamma_t/(t - \mu^2),$$
(I-88)

where $[\tilde\xi]$ is defined by (I-43), the vector $\tilde\Gamma(t, K^2)$ has the elements

$$\Gamma_1^{(\pm, 0)} = \tfrac{1}{2}gF_1^{(v,s)}(K^2), \qquad \Gamma_2^{(\pm, 0)} = -gF_1^{(v,s)}(K^2)/(t - \mu^2),$$
$$\Gamma_3^{(\pm, 0)} = \Gamma_4^{(\pm, 0)} = -\tfrac{1}{2}gF_2^{(v,s)}(K^2), \qquad \Gamma_5^{(\pm, 0)} = \tfrac{1}{2}\Gamma_2^{(\pm, 0)}, \qquad \Gamma_6^{(\pm, 0)} = 0,$$
(I-89)

and only the fifth element of $\tilde{\Gamma}_t$ is nonzero, with the value

$$\Gamma_t = (2g/K^2)[F_\pi(K^2) - F_1^{(v)}(K^2)]. \tag{I-90}$$

H. Cross Section and Polarization Formulas for Pion Photoproduction

The differential cross section for the transition from an initial photon-nucleon state $|i\rangle$ to a final pion-nucleon state $|f\rangle$ is given by

$$d\sigma/d\Omega = (q/k)|\langle \chi_f | \mathscr{F} | \chi_i \rangle|^2, \tag{I-91}$$

where

$$\mathscr{F} = i\boldsymbol{\sigma} \cdot \boldsymbol{\varepsilon} \mathscr{F}_1 + \boldsymbol{\sigma} \cdot \hat{\mathbf{q}}\boldsymbol{\sigma} \cdot (\hat{\mathbf{k}} \times \boldsymbol{\varepsilon})\mathscr{F}_2 + i\boldsymbol{\sigma} \cdot \hat{\mathbf{k}}\hat{\mathbf{q}} \cdot \boldsymbol{\varepsilon} \mathscr{F}_3 + i\boldsymbol{\sigma} \cdot \hat{\mathbf{q}}\hat{\mathbf{q}} \cdot \boldsymbol{\varepsilon} \mathscr{F}_4.$$

It is convenient to choose a coordinate frame in which the production plane is the x, z plane. Introducing the unit vectors $\hat{\mathbf{e}}_1$, $\hat{\mathbf{e}}_2$, $\hat{\mathbf{e}}_3(=\hat{\mathbf{k}})$ in the x, y, and z directions, respectively, then for right and left circularly polarized photons

$$\boldsymbol{\varepsilon} = \boldsymbol{\varepsilon}_\pm = \mp(1/\sqrt{2})(\hat{\mathbf{e}}_1 \pm i\hat{\mathbf{e}}_2), \tag{I-92}$$

and for linearly polarized photons

$$\boldsymbol{\varepsilon} = \hat{\mathbf{e}}_1 \cos \phi + \hat{\mathbf{e}}_2 \sin \phi = (1/\sqrt{2})e^{i\phi}\boldsymbol{\varepsilon}_+ - e^{-i\phi}\boldsymbol{\varepsilon}_-. \tag{I-93}$$

1. Polarization of Final Nucleon Unobserved

Summing over the final spin states yields

$$\sum_f \langle \chi_f | \mathscr{F} | \chi_i \rangle^\dagger \langle \chi_f | \mathscr{F} | \chi_i \rangle = \langle \chi_i | \mathscr{F}^\dagger \mathscr{F} | \chi_i \rangle, \tag{I-94}$$

where

$$\begin{aligned}
\mathscr{F}^\dagger \mathscr{F} = {}& |\mathscr{F}_1|^2\{\boldsymbol{\varepsilon}^* \cdot \boldsymbol{\varepsilon} + i\boldsymbol{\sigma}(\boldsymbol{\varepsilon}^* \times \boldsymbol{\varepsilon})\} + |\mathscr{F}_2|^2\{\boldsymbol{\sigma} \cdot (\hat{\mathbf{k}} \times \boldsymbol{\varepsilon}^*)\boldsymbol{\sigma} \cdot (\hat{\mathbf{k}} \times \boldsymbol{\varepsilon})\} \\
& + |\mathscr{F}_3|^2\{\hat{\mathbf{q}} \cdot \boldsymbol{\varepsilon}^*\hat{\mathbf{q}} \cdot \boldsymbol{\varepsilon}\} + |\mathscr{F}_4|^2\{\hat{\mathbf{q}} \cdot \boldsymbol{\varepsilon}^*\hat{\mathbf{q}} \cdot \boldsymbol{\varepsilon}\} \\
& + \mathscr{F}_1^*\mathscr{F}_2\{-i\boldsymbol{\sigma} \cdot (\hat{\mathbf{k}} \times \boldsymbol{\varepsilon})\hat{\mathbf{q}} \cdot \boldsymbol{\varepsilon}^* + i\boldsymbol{\sigma} \cdot \hat{\mathbf{q}}\hat{\mathbf{k}} \cdot (\boldsymbol{\varepsilon} \times \boldsymbol{\varepsilon}^*) \\
& \quad - i\boldsymbol{\sigma} \cdot \boldsymbol{\varepsilon}^*\hat{\mathbf{q}} \cdot (\hat{\mathbf{k}} \times \boldsymbol{\varepsilon}) + (\boldsymbol{\varepsilon}^* \times \hat{\mathbf{q}}) \cdot (\hat{\mathbf{k}} \times \boldsymbol{\varepsilon})\} \\
& + \mathscr{F}_1^*\mathscr{F}_3\{i\boldsymbol{\sigma} \cdot (\boldsymbol{\varepsilon}^* \times \hat{\mathbf{k}})\hat{\mathbf{q}} \cdot \boldsymbol{\varepsilon}\} \\
& + \mathscr{F}_1^*\mathscr{F}_4\{\hat{\mathbf{q}} \cdot \boldsymbol{\varepsilon}\hat{\mathbf{q}} \cdot \boldsymbol{\varepsilon}^* + i\boldsymbol{\sigma} \cdot (\boldsymbol{\varepsilon}^* \times \hat{\mathbf{q}})\hat{\mathbf{q}} \cdot \boldsymbol{\varepsilon}\} \\
& + \mathscr{F}_2^*\mathscr{F}_3\{\hat{\mathbf{q}} \cdot \boldsymbol{\varepsilon}\hat{\mathbf{q}} \cdot \boldsymbol{\varepsilon}^* + i\boldsymbol{\sigma} \cdot \hat{\mathbf{k}}\hat{\mathbf{q}} \cdot (\hat{\mathbf{k}} \times \boldsymbol{\varepsilon}^*)\hat{\mathbf{q}} \cdot \boldsymbol{\varepsilon} \\
& \quad + i\boldsymbol{\sigma} \cdot (\hat{\mathbf{k}} \times \boldsymbol{\varepsilon}^*)\hat{\mathbf{q}} \cdot \hat{\mathbf{k}}\hat{\mathbf{q}} \cdot \boldsymbol{\varepsilon}\} \\
& + \mathscr{F}_2^*\mathscr{F}_4\{\hat{\mathbf{q}} \cdot \boldsymbol{\varepsilon}i\boldsymbol{\sigma} \cdot (\hat{\mathbf{k}} \times \boldsymbol{\varepsilon}^*)\} \\
& + \mathscr{F}_3^*\mathscr{F}_4\{\hat{\mathbf{q}} \cdot \boldsymbol{\varepsilon}^*\hat{\mathbf{q}} \cdot \boldsymbol{\varepsilon}\hat{\mathbf{q}} \cdot \hat{\mathbf{k}} + i\boldsymbol{\sigma} \cdot (\hat{\mathbf{k}} \times \hat{\mathbf{q}})\} \\
& + \text{Hermitian conjugate of the off-diagonal elements.} \tag{I-95}
\end{aligned}$$

The results of the evaluation of (I-95) for different initial states of polarization is conveniently defined in terms of the following functions:

$$\alpha = |\mathscr{F}_1|^2 + |\mathscr{F}_2|^2 - 2\cos\theta\,\mathrm{Re}(\mathscr{F}_1{}^*\mathscr{F}_2)$$
$$\quad + \sin^2\theta\,\mathrm{Re}\{\mathscr{F}_1{}^*\mathscr{F}_2 + \mathscr{F}_2{}^*\mathscr{F}_3\},$$
$$\beta = \tfrac{1}{2}\sin^2\theta\{|\mathscr{F}_3|^2 + |\mathscr{F}_4|^2 + 2\cos\theta\,\mathrm{Re}(\mathscr{F}_3{}^*\mathscr{F}_4)\},$$
$$\gamma = \mathrm{Re}\{\mathscr{F}_1{}^*\mathscr{F}_3 - \mathscr{F}_2{}^*\mathscr{F}_4\} + \cos\theta\,\mathrm{Re}\{\mathscr{F}_1{}^*\mathscr{F}_4 - \mathscr{F}_2{}^*\mathscr{F}_3\},$$
$$\delta = \mathrm{Im}\{\mathscr{F}_1{}^*\mathscr{F}_3 - \mathscr{F}_2{}^*\mathscr{F}_4\} + \cos\theta\,\mathrm{Im}\{\mathscr{F}_1{}^*\mathscr{F}_4 - \mathscr{F}_2{}^*\mathscr{F}_3\} \qquad \text{(I-96)}$$
$$\quad - \sin^2\theta\,\mathrm{Im}\{\mathscr{F}_3{}^*\mathscr{F}_4\},$$
$$\alpha' = \sin^2\theta\,\mathrm{Re}\{\mathscr{F}_1{}^*\mathscr{F}_4 + \mathscr{F}_2{}^*\mathscr{F}_3\},$$
$$\beta' = \gamma + 2\,\mathrm{Re}\{\mathscr{F}_1{}^*\mathscr{F}_2\},$$
$$\gamma' = \delta + 2\,\mathrm{Im}\{\mathscr{F}_1{}^*\mathscr{F}_2\}.$$

Unpolarized nucleon, linearly polarized photon:

$$d\sigma/d\Omega = (q/k)\{\alpha + \beta + \cos 2\phi[\alpha' + \beta']\}. \qquad \text{(I-97)}$$

Unpolarized nucleon, unpolarized photon:

$$d\sigma/d\Omega = (q/k)\{\alpha + \beta\}. \qquad \text{(I-98)}$$

Total cross section, unpolarized photon and unpolarized nucleon:

$$\sigma_{\mathrm{tot}} = (2\pi q/k)\sum_{l=0}^{\infty}\{l(l+1)[|M_{l+}|^2 + |E_{(l+1)-}|^2]$$
$$\quad + l^2(l+1)[|M_{l-}|^2 + |E_{(l-1)+}|^2]\}. \qquad \text{(I-99)}$$

2. *Polarization of Final Nucleon Observed*

The polarization of the final nucleon $\mathbf{P}_f$ is defined by

$$\mathbf{P}_f\,d\sigma/d\Omega = (q/k)\langle\chi_i|\mathscr{F}^\dagger\boldsymbol{\sigma}\mathscr{F}|\chi_i\rangle. \qquad \text{(I-100)}$$

Summing over the initial nucleon spins yields

$$\sum_i \langle\chi_i|\mathscr{F}^\dagger\boldsymbol{\sigma}\mathscr{F}|\chi_i\rangle = |\mathscr{F}_2|^2\{i\hat{\mathbf{q}}\cdot(\hat{\mathbf{k}}\times\boldsymbol{\varepsilon}^*)(\hat{\mathbf{q}}\cdot\boldsymbol{\varepsilon})\hat{\mathbf{k}} - i\hat{\mathbf{q}}\cdot(\hat{\mathbf{k}}\times\boldsymbol{\varepsilon})(\hat{\mathbf{q}}\cdot\boldsymbol{\varepsilon}^*)\hat{\mathbf{k}}$$
$$\quad - i\hat{\mathbf{q}}\cdot(\hat{\mathbf{k}}\times\boldsymbol{\varepsilon}^*)(\hat{\mathbf{q}}\cdot\hat{\mathbf{k}})\boldsymbol{\varepsilon} + iq\cdot(\hat{\mathbf{k}}\times\boldsymbol{\varepsilon})(\hat{\mathbf{q}}\cdot\hat{\mathbf{k}})\boldsymbol{\varepsilon}^*$$
$$\quad + i(\hat{\mathbf{q}}\cdot\hat{\mathbf{k}})(\hat{\mathbf{q}}\cdot\boldsymbol{\varepsilon}^*)\hat{\mathbf{k}}\times\boldsymbol{\varepsilon} - i(\hat{\mathbf{q}}\cdot\hat{\mathbf{k}})(\hat{\mathbf{q}}\cdot\boldsymbol{\varepsilon})(\hat{\mathbf{k}}\times\boldsymbol{\varepsilon}^*)\}$$
$$\quad + \mathscr{F}_1{}^*\mathscr{F}_2\{-i\hat{\mathbf{q}}\cdot(\hat{\mathbf{k}}\times\boldsymbol{\varepsilon})\boldsymbol{\varepsilon}^* + i(\hat{\mathbf{q}}\cdot\boldsymbol{\varepsilon})\hat{\mathbf{k}}\times\boldsymbol{\varepsilon}^*\}$$
$$\quad + \mathscr{F}_1{}^*\mathscr{F}_3\{i(\hat{\mathbf{q}}\cdot\boldsymbol{\varepsilon})\hat{\mathbf{k}}\times\boldsymbol{\varepsilon}^*\} + \mathscr{F}_1{}^*\mathscr{F}_4\{i(\hat{\mathbf{q}}\cdot\boldsymbol{\varepsilon})\hat{\mathbf{q}}\times\boldsymbol{\varepsilon}^*\}$$
$$\quad - \mathscr{F}_2{}^*\mathscr{F}_3\{i(\hat{\mathbf{q}}\cdot\boldsymbol{\varepsilon})\hat{\mathbf{q}}\cdot(\hat{\mathbf{k}}\times\boldsymbol{\varepsilon}^*)\hat{\mathbf{k}} + i(\hat{\mathbf{q}}\cdot\boldsymbol{\varepsilon})(\hat{\mathbf{q}}\cdot\hat{\mathbf{k}})\mathbf{k}\times\boldsymbol{\varepsilon}^*\}$$
$$\quad + \mathscr{F}_2{}^*\mathscr{F}_4\{i(\hat{\mathbf{q}}\cdot\boldsymbol{\varepsilon})\hat{\mathbf{q}}\cdot(\hat{\mathbf{k}}\times\boldsymbol{\varepsilon}^*)\hat{\mathbf{q}} - i(\hat{\mathbf{q}}\cdot\boldsymbol{\varepsilon})(\hat{\mathbf{q}}\cdot\boldsymbol{\varepsilon}^*)\hat{\mathbf{q}}\times\hat{\mathbf{k}}$$
$$\quad + i(\hat{\mathbf{q}}\cdot\boldsymbol{\varepsilon})(\hat{\mathbf{q}}\cdot\hat{\mathbf{k}})\hat{\mathbf{q}}\times\boldsymbol{\varepsilon}^*\}$$
$$\quad + \mathscr{F}_3{}^*\mathscr{F}_4\{i(\hat{\mathbf{q}}\cdot\boldsymbol{\varepsilon})(\hat{\mathbf{q}}\cdot\boldsymbol{\varepsilon}^*)\hat{\mathbf{q}}\times\hat{\mathbf{k}}. \qquad \text{(I-101)}$$

Recoil nucleon polarization, unpolarized photon:

$$(k/q)\mathbf{P}_f \, d\sigma/d\Omega = -\sin\theta\hat{e}_2 \, \text{Im}\{2\mathscr{F}_1^*\mathscr{F}_2 + \mathscr{F}_1^*\mathscr{F}_3 - \mathscr{F}_2^*\mathscr{F}_4 - \mathscr{F}_3^*\mathscr{F}_4$$
$$+ \cos\theta[\mathscr{F}_1^*\mathscr{F}_4 - \mathscr{F}_2^*\mathscr{F}_3] + \cos^2\theta\mathscr{F}_3^*\mathscr{F}_4\}. \qquad \text{(I-102)}$$

It is frequently convenient, particularly for phenomenological applications, to use an alternative formulation in terms of the helicity amplitudes and parity-conserving helicity elements. In terms of the helicity amplitudes, which we will denote here by H_i, the four quantities which are currently measurable are given by

$$d\sigma/d\Omega = (q/2k) \sum_{i=1}^{4} |H_i|^2, \qquad \text{(I-103)}$$

$$\mathbf{P}_f = -2\hat{e}_2 \, \text{Im}\{H_1 H_3^* + H_2 H_4^*\}/\sum_{i=1}^{4} |H_i|^2, \qquad \text{(I-104)}$$

$$\Sigma(\theta) = (\sigma_\perp - \sigma_\parallel)/(\sigma_\perp + \sigma_\parallel)|_{\text{linearly polarized photon}}$$
$$= 2 \, \text{Re}\{H_1 H_4^* - H_2 H_3^*\}/\sum_{i=1}^{4} |H_i|^2, \qquad \text{(I-105)}$$

$$T(\theta) = (\sigma_{\text{up}} - \sigma_{\text{down}})/(\sigma +_{\text{up}} \sigma_{\text{down}})|_{\text{target nucleon}}$$
$$= 2 \, \text{Im}\{H_1 H_2^* + H_3 H_4^*\}/\sum_{i=1}^{4} |H_i|^2. \qquad \text{(I-106)}$$

The helicity amplitudes are decomposed into parity conserving helicity elements $A_{l\pm}$, $B_{l\pm}$, the $A_{l\pm}$ being the transition amplitudes from a photon-nucleon state of helicity $\frac{1}{2}$ into the usual meson-nucleon state of orbital angular momentum l, parity $(-1)^{l+1}$ and total angular momentum $j = l \pm \frac{1}{2}$, and the $B_{l\pm}$ the corresponding amplitudes for the transition from a photon-nucleon state of helicity $\frac{3}{2}$. Explicitly the relationship between H_i and $A_{l\pm}$, $B_{l\pm}$ is

$$H_1 = 1/\sqrt{2} \sin\theta \cos\tfrac{1}{2}\theta \sum_{l=1}^{\infty} \{B_{l+} - B_{(l+1)-}\}\{P_l''(\cos\theta) - P_{l+1}''(\cos\theta)\},$$

$$H_2 = \sqrt{2} \cos\tfrac{1}{2}\theta \sum_{l=0}^{\infty} \{A_{l+} - A_{(l+1)-}\}\{P_l'(\cos\theta) - P_{l+1}'(\cos\theta)\},$$
$$\qquad \text{(I-107)}$$
$$H_3 = 1/\sqrt{2} \sin\theta \cos\tfrac{1}{2}\theta \sum_{l=1}^{\infty} \{B_{l+} + B_{(l+1)-}\}\{P_l''(\cos\theta) + P_{l+1}''(\cos\theta)\},$$

$$H_4 = \sqrt{2} \sin\tfrac{1}{2}\theta \sum_{l=0}^{\infty} \{A_{l+} + A_{(l+1)-}\}\{P_l'(\cos\theta) + P_{l+1}'(\cos\theta)\}.$$

By angular momentum conservation only the $A_{l\pm}$ can contribute in the forward and backward directions, so that $H_2(\theta = 0)$ is the only amplitude contributing in the forward direction and $H_4 (\theta = \pi)$ is the only amplitude contributing in the backward direction.

The following relations hold between the multipole amplitudes and the helicity elements:

$$(l+1)E_{l+} = A_{l+} + \tfrac{1}{2}lB_{l+}, \qquad\qquad lE_{l-} = A_{l-} - \tfrac{1}{2}(l+1)B_{l-},$$
$$(l+1)M_{l+} = A_{l+} - \tfrac{1}{2}(l+2)B_{l+}, \qquad lM_{l-} = A_{l-} + \tfrac{1}{2}(l-1)B_{l-}. \tag{I-108}$$

I. CROSS-SECTION FORMULAS FOR PION ELECTROPRODUCTION

The T matrix of (I-7) is related to the differential cross section by

$$d\sigma = [(2\pi)^5]^{-1}\, \delta(K_2 + P_2 + Q - K_1 - P_1)$$
$$\times \frac{M^2 m^2}{2} \frac{|T_{fi}|^2}{[(K_1 \cdot P_1)^2 - M^2 m^2]} \frac{d\mathbf{P}_2}{E_2} \frac{d\mathbf{K}_2}{\varepsilon_2} \frac{d\mathbf{Q}}{Q_0}. \tag{I-109}$$

Because of the one-photon-exchange approximation, we have already seen that the T matrix can be separated into a purely leptonic part and a purely hadronic part (production of a pion by a virtual photon), and the cross section can be treated similarly. This separation is extremely convenient, since it enables the simultaneous evaluation of the leptonic part in the laboratory frame and the hadronic part in the pion–nucleon center-of-mass frame. For this purpose, it is necessary to choose a set of reference axes which are unchanged by the Lorentz transformation along the photon vector $\mathbf{K}$ from the laboratory frame to the pion–nucleon center-of-mass frame. The particular choice we make is defined by

$$\hat{\mathbf{e}}_3 = \mathbf{K}/|\mathbf{K}|, \qquad \hat{\mathbf{e}}_2 = (\mathbf{K}_2 \times \mathbf{K}_1)/|\mathbf{K}_2 \times \mathbf{K}_1|, \qquad \hat{\mathbf{e}}_1 = \hat{\mathbf{e}}_3 \times \hat{\mathbf{e}}_2. \tag{I-110}$$

As a simplification to the formulas, the electron mass will be ignored wherever possible. The summation over the electron spins is straightforward, and the result is

$$\tfrac{1}{2}m^2 \sum_{\substack{\text{electron}\\\text{spins}}} |J_\mu \varepsilon_\mu|^2 = \tfrac{1}{2}K^{-4}\{K_{1_\mu} K_{2_\nu} + K_{2_\mu} K_{1_\nu} + \tfrac{1}{2}K^2 g_{\mu\nu}\} J_\mu{}^* J_\nu$$
$$= K^{-4}\{(J \cdot K_1)(J^* \cdot K_1) + \tfrac{1}{4}K^2(J^* \cdot J)\}. \tag{I-111}$$

Now in the electron Breit frame, that is, the frame in which the electron scatters backwards with no loss in energy, unpolarized relativistic electrons emit transverse polarized photons because a relativistic electron preserves its helicity, and consequently the emitted photon is an equal incoherent mixture of $+1$ and -1 helicity states. However, in this frame,

$$K_{1_\mu} K_{2_\nu} + K_{2_\mu} K_{1_\nu} + \tfrac{1}{2}K^2 g_{\mu\nu} = -K^2 \rho_{\mu\nu}, \tag{I-112}$$

with $\rho_{11} = \rho_{22} = \tfrac{1}{2}$, and all other $\rho_{\mu\nu} = 0$; that is, $\rho_{\mu\nu}$ is the density matrix of the virtual photon polarization.

The four-vector products in (I-111) can be evaluated in terms of the transverse components $J_1(=\mathbf{J} \cdot \hat{\mathbf{e}}_1)$ and $J_2(=\mathbf{J} \cdot \hat{\mathbf{e}}_2)$ of the hadron current and either the longitudinal component $J_3(=\mathbf{J} \cdot \hat{\mathbf{e}}_3)$ or the scalar component J_0, the latter two being related by gauge invariance:

$$J_3 = \mathbf{J} \cdot \hat{\mathbf{e}}_3 = \mathbf{J} \cdot \mathbf{K}/|\mathbf{K}| = J_0 K_0/|\mathbf{K}|. \tag{I-113}$$

In terms of the transverse and scalar components,

$$
\begin{aligned}
(J \cdot K_1)&(J^* \cdot K_1) + \tfrac{1}{4}K^2(J^* \cdot J) \\
&= |J_0|^2\{[(\hat{\mathbf{K}}_1 \cdot \hat{\mathbf{e}}_3)K_0/|\mathbf{K}| - |\mathbf{K}_1|]^2 - \tfrac{1}{4}K^4/|\mathbf{K}|^2\} \\
&\quad + \tfrac{1}{4}K^2\{|J_1|^2 + |J_2|^2\} + (\mathbf{K}_1 \cdot \hat{\mathbf{e}}_1)^2|J_1|^2 \\
&\quad + \mathbf{K}_1 \cdot \hat{\mathbf{e}}_2\{(\hat{\mathbf{K}}_1 \cdot \hat{\mathbf{e}}_3)K_0/|\mathbf{K}| - |\mathbf{K}_1|\}2\,\mathrm{Re}(J_0^*J_1).
\end{aligned} \tag{I-114}
$$

Evaluating in turn the terms involving lepton kinematical variables in the laboratory frame,

$$
\begin{aligned}
(J \cdot K_1)&(J^* \cdot K_1) + \tfrac{1}{4}K^2(J^* \cdot J) \\
&= \varepsilon_1 \varepsilon_2 \cos^2(\tfrac{1}{2}\psi)\{|J_0|^2K^4/|\mathbf{K}|^4 + |J_1|^2K^2/|\mathbf{K}|^2 \\
&\quad + \tan^2(\tfrac{1}{2}\psi)[|J_1|^2 + |J_2|^2] \\
&\quad - 2\,\mathrm{Re}(J_0^*J_1)[\tan^2(\tfrac{1}{2}\psi) + K^2/|\mathbf{K}|^2]^{1/2}K^2/|\mathbf{K}|^2\}, \tag{I-115}
\end{aligned}
$$

where ψ is the electron scattering angle in the laboratory frame and ε_1, ε_2 are the initial and final electron energies, respectively.

At this point, it is convenient to introduce the quantity ε, defined by

$$\varepsilon = \{1 + (2|K|^2/k^2)\tan^2(\tfrac{1}{2}\psi)\}^{-1}, \tag{I-116}$$

which is a measure of the transverse linear polarization of the virtual photon (why this is so will become apparent when the density matrix of the virtual photon is exhibited explicitly again). Then

$$
\begin{aligned}
(J \cdot K_1)&(J^* \cdot K_1) + \tfrac{1}{4}K^2(J^* \cdot J) \\
&= [K^2/(1-\varepsilon)]\{\tfrac{1}{2}(1+\varepsilon)|J_1|^2 + \tfrac{1}{2}(1-\varepsilon)|J_2|^2 \\
&\quad + \varepsilon_\mathrm{S}|J_0|^2 - [\tfrac{1}{2}\varepsilon_\mathrm{S}(\varepsilon+1)]^{1/2}2\,\mathrm{Re}(J_0^*J_1)\}, \tag{I-117}
\end{aligned}
$$

with

$$\varepsilon_\mathrm{S} = (K^2/|K|^2)\varepsilon. \tag{I-118}$$

The analogous expression for the evaluation of (I-114) in terms of the transverse and longitudinal components of the hadron current is obtained from (I-117) by the substitution of J_3 for J_0 and of ε_L for ε_S, where

$$\varepsilon_\mathrm{L} = (K^2/K_0^2)\varepsilon. \tag{I-119}$$

To exhibit explicitly the density matrix of the virtual photon polarization, (I-117) can be rewritten as

$$(J \cdot K_1)(J^* \cdot K_1) + \tfrac{1}{4}K^2(J^* \cdot J) = [K^2/(1 - \varepsilon)]\rho_{ij} J_i^* J_j^*, \qquad \text{(I-120)}$$

where

$$\rho = \begin{bmatrix} \tfrac{1}{2}(1 + \varepsilon) & 0 & -\tfrac{1}{2}\varepsilon_S(1 + \varepsilon)^{1/2} \\ 0 & \tfrac{1}{2}(1 - \varepsilon) & 0 \\ -\tfrac{1}{2}\varepsilon_S(1 + \varepsilon)^{1/2} & 0 & \varepsilon_S \end{bmatrix}, \qquad \text{(I-121)}$$

with a corresponding relation in terms of the longitudinal component of the hadron current with the implicit replacement of J_0 by J_3 in (I-120) and the explicit replacement of ε_S by ε_L in (I-121).

Recalling that the density matrix for a partially linearly polarized beam of real photons of relative strength $(1 + \varepsilon')$ in the x direction to $(1 - \varepsilon')$ in the y direction is

$$\rho = \begin{bmatrix} \tfrac{1}{2}(1 + \varepsilon') & & 0 \\ 0 & \tfrac{1}{2}(1 - \varepsilon') & 0 \\ 0 & 0 & 0 \end{bmatrix}, \qquad \text{(I-122)}$$

it is immediately obvious why ε defined in (I-116) is a measure of the transverse linear polarization of the virtual photon. The scalar (longitudinal) polarization ε_S (ε_L) is also given in terms of ε, via (I-118) [(I-119)].

Summing and averaging (I-117) over the nucleon spins, inserting back in (I-109) and extracting a factor $(4\pi W/M)$ from the hadron current to normalize the multipole amplitudes to those for photoproduction, the final result for the cross section is

$$d^5\sigma/(d\varepsilon_2 \, d\Omega_e \, d\Omega_\pi) = (\alpha/2\pi^2)(\varepsilon_2/\varepsilon_1)(|\mathbf{K}|/K^2)(1 - \varepsilon)^{-1} \, d\sigma_v/d\Omega_\pi, \qquad \text{(I-123)}$$

where $d\sigma_v/d\Omega_\pi$ represents the differential cross section for the production of a pion by a virtual photon, evaluated in the pion-nucleon center-of-mass system and is given by

$$(d\sigma_v/d\Omega_\pi)(W, K^2, \varepsilon, \theta, \phi)$$
$$= (q/k)\{A(W, K^2, \theta) + \varepsilon B(W, K^2, \theta) \sin^2 \theta \cos 2\phi + \varepsilon_S C(W, K^2, \theta)$$
$$- [\varepsilon_S(1 + \varepsilon)]^{1/2} D(W, K^2, \theta) \sin \theta \cos \phi\}. \qquad \text{(I-124)}$$

The coefficients A, B, C, D are obtained from (I-117) and the explicit expression for the hadronic current in terms of the amplitudes $\mathscr{F}_i$, (I-50).

Explicitly,

$$
\begin{aligned}
A = {} & \{|\mathscr{F}_1|^2 + |\mathscr{F}_2|^2 + \tfrac{1}{2}|\mathscr{F}_3|^2 + \tfrac{1}{2}|\mathscr{F}_4|^2 + \mathrm{Re}(\mathscr{F}_1{}^*\mathscr{F}_4) + \mathrm{Re}(\mathscr{F}_2{}^*\mathscr{F}_3)\} \\
& + \{\mathrm{Re}(\mathscr{F}_3{}^*\mathscr{F}_4) - 2\,\mathrm{Re}(\mathscr{F}_1{}^*\mathscr{F}_2)\}\cos\theta - \{\tfrac{1}{2}|\mathscr{F}_3|^2 + \tfrac{1}{2}|\mathscr{F}_4|^2 \\
& + \mathrm{Re}(\mathscr{F}_1{}^*\mathscr{F}_4) + \mathrm{Re}(\mathscr{F}_2{}^*\mathscr{F}_3)\}\cos\theta^2 - \{\mathrm{Re}(\mathscr{F}_3{}^*\mathscr{F}_4)\}\cos^3\theta, \\
B = {} & \{|\mathscr{F}_3|^2 + |\mathscr{F}_4|^2 + 2\,\mathrm{Re}(\mathscr{F}_1{}^*\mathscr{F}_4) + 2\,\mathrm{Re}(\mathscr{F}_2{}^*\mathscr{F}_3)\} \\
& + \{\mathrm{Re}(\mathscr{F}_3{}^*\mathscr{F}_4)\}\cos\theta, \\
C = {} & |\mathscr{F}_7|^2 + |\mathscr{F}_8|^2 + 2\,\mathrm{Re}(\mathscr{F}_7{}^*\mathscr{F}_8), \\
D = {} & \{2\,\mathrm{Re}(\mathscr{F}_7{}^*[\mathscr{F}_1 + \mathscr{F}_4] + \mathscr{F}_8{}^*[\mathscr{F}_2 + \mathscr{F}_3])\} \\
& + \{2\,\mathrm{Re}(\mathscr{F}_7{}^*\mathscr{F}_3 + \mathscr{F}_8{}^*\mathscr{F}_4)\}\cos\theta.
\end{aligned}
\tag{I-125}
$$

In the analogous expressions for longitudinal amplitudes, ε_S is replaced by ε_L, A and B are unchanged, and in C and D the replacements

$$
\mathscr{F}_7 \to \mathscr{F}_1 + \mathscr{F}_3\cos\theta + \mathscr{F}_5 \qquad \text{and} \qquad \mathscr{F}_8 \to \mathscr{F}_4\cos\theta + \mathscr{F}_6 \tag{I-126}
$$

are made.

Comparison with the expressions for the cross section for pion production with real photons shows that A has precisely the same form as the term for photoproduction with unpolarized photons and that B has precisely the same form as the additional term when the photons are linearly polarized. The terms C and D are additional ones, C being the contribution from the scalar (longitudinal) part of the amplitude and D being the interference term between the scalar (longitudinal) and the transverse amplitudes.

If (I-124) is integrated over the pion angles θ and ϕ, we obtain

$$
\begin{aligned}
d^3\sigma/d\varepsilon_2\,d\Omega_e &= (\alpha/2\pi^2)(\varepsilon_2/\varepsilon_1)(|\mathbf{K}|/K^2)(1-\varepsilon)^{-1}\{\sigma_T(K^2, W) + \varepsilon_S\,\sigma_S(K^2, W)\} \\
&= (\alpha/2\pi^2)(\varepsilon_2/\varepsilon_1)(|\mathbf{K}|/K^2)(1-\varepsilon)^{-1}\{\sigma_T(K^2, W) + \varepsilon_L\,\sigma_L(K^2, W)\},
\end{aligned}
\tag{I-127}
$$

where $\sigma_T(K^2, W)$ is the total cross section for transverse virtual photons and $\sigma_S(\sigma_L)$ is the total cross section for scalar (longitudinal) virtual photons. They are given in terms of the multipole amplitudes by

$$
\begin{aligned}
\sigma_T &= (2\pi q/k)\sum_{l=0}^{\infty} l(l+1)[|M_{l+}|^2 + |E_{(l+1)-}|^2] \\
&\quad + l^2(l+1)[|M_{l-}|^2 + |E_{(l-1)+}|^2], \\
\sigma_S &= (4\pi q/k)\sum_{l=0}^{\infty} (l+1)^3|S_{(l+1)-}|^2 + l^3|S_{(l-1)+}|^2, \\
\sigma_L &= (4\pi q/k)\sum_{l=0}^{\infty} (l+1)^3|L_{(l+1)-}|^2 + l^3|L_{(l-1)+}|^2.
\end{aligned}
\tag{I-128}
$$

Equation (I-127) is the general form for the cross section for inelastic electron scattering, whatever the hadron final state, and in this case $\sigma_T(K^2, W)$ and $\sigma_S(K^2, W)$ [or $\sigma_L(K^2, W)$] can be interpreted as the inelastic nucleon form factors. Unfortunately, a unique notation has not evolved for this. Akerlof *et al.* (1967) include the gauge invariant factor K^2/K_0^2 in their longitudinal cross section $\sigma_L^{(A)}$, so that

$$\sigma_L^{(A)} = K^2\sigma_L/K_0^2. \tag{I-129}$$

Gilman (1968) makes a sign change for the longitudinal part, writing

$$\sigma_{\text{trans}} = \sigma_T, \qquad \sigma_{\text{long}} = -\sigma_L^{(A)}, \tag{I-130}$$

and Adler (1966) uses inelastic form factors $\alpha(K_0, K^2)$ and $\beta(k_0, K_0^2)$ $(K_0 = \varepsilon_1 - \varepsilon_2)$ which are defined by

$$d^3\sigma/(d\varepsilon_2 \, d\Omega_e) = (4\alpha^2\varepsilon^2/K^4)[2 \sin^2(\tfrac{1}{2}\psi)\alpha(K_0, K^2) + \cos^2(\tfrac{1}{2}\psi)\beta(K_0, K^2)]. \tag{I-131}$$

Thus

$$\alpha(K_0, K^2) = (|K|/4\pi^2\alpha)\sigma_T, \qquad \beta(K_0, K) = (K^2/4\pi^2\alpha|K|)(\sigma_T + \sigma_L^{(A)}). \tag{I-132}$$

Finally, Drell and Walecka (1964) use form factors W_1 and W_2 defined by

$$\alpha = W_1/M \qquad \text{and} \qquad \beta = W_2/M. \tag{I-133}$$

J. PHOTOPRODUCTION OF η AND K MESONS

1. η Mesons

Since the initial and final baryon masses are the same in this reaction, the formalism is identical to that for pion production, with two exceptions. The first of these arises because the η meson is an isoscalar particle, and consequently only transitions to $I = \tfrac{1}{2}$ final states are allowed. This means that the decomposition of the invariant amplitudes in isotopic spin space has the simple form

$$A_i = A_i^{(s)} + A_i^{(v)}\tau_3 \qquad (i = 1, \ldots, 4), \tag{I-134}$$

and the combinations of amplitudes for the particular physical processes are

$$\langle p\eta^0|T|\gamma p\rangle = A^{(s)} + A^{(v)}, \qquad \langle n\eta^0|T|\gamma n\rangle = A^{(s)} - A^{(v)}. \tag{I-135}$$

The other difference is that this reaction has a second strongly interacting channel open at threshold, and consequently there is no equivalent to Watson's theorem for pion photoproduction. This, of course, also applies to K-meson photoproduction.

2. K *Mesons*

The formal development for these reactions (that is, $\gamma + p \rightarrow K^+ + \Lambda$, $\gamma + p \rightarrow K^+ + \Sigma^0$, $K^0 + \Sigma^+$ and the associated neutron reactions) is the same as that for pion photoproduction, that is, the choice of invariant amplitudes and Pauli amplitudes is the same, as is the multipole expansion of the latter, and the cross section and polarization formulas. However, the different kinematics arising from the difference in mass between the initial and final baryons results in explicit differences in the relations linking the invariant amplitudes to the Pauli amplitudes and in the expressions for the Born terms.

When discussing the Λ or Σ states explicitly, their masses shall be denoted by M_Λ or M_Σ, respectively, but when discussing them in general we shall use Y to denote the hyperon mass. In all cases, the nucleon mass shall be denoted by M as before, and the kaon mass by μ.

The S matrix is now given in terms of the T matrix by

$$S_{fi} = \delta_{fi} - i(2\pi)^{-2}\,\delta^4(P_2 + Q - P_1 - K)(M\,Y/4K_0\,Q_0\,P_{10}\,P_{20})^{1/2}T_{fi},$$

(I-136)

the Mandelstam kinematical variables satisfy

$$s + t + u = M^2 + Y^2 + \mu^2;$$

(I-137)

and the final state momentum q in the center-of-mass frame is given by

$$q = (1/2W)[\{W^2 - (Y + \mu)^2\}\{W^2 - (Y - \mu)^2\}]^{1/2},$$

(I-138)

with other final-state kinematical quantities being similarly obtained by the substitution of Y for M.

The connection between the Pauli amplitudes $\mathscr{F}_1, \ldots, \mathscr{F}_4$ and the invariant amplitudes $A_1, \ldots, A_4$ is again of the form

$$\tilde{\mathscr{F}}(s, t) = \{C^{-1}(s)\}\{B(s, t)\}\tilde{A}(s, t),$$

(I-139)

where the matrix $\{B(s, t)\}$ is still given by (I-55), but the matrix $\{C^{-1}(s)\}$ is now

$$\{C^{-1}(s)\} = (k/4\pi)$$

$$\times \text{diag}\left\{\left(\frac{E_2 + Y}{2W}\right)^{1/2},\ \left(\frac{E_2 - Y}{2W}\right)^{1/2},\ q\left(\frac{E_2 + Y}{2W}\right)^{1/2},\ q\left(\frac{E_2 - Y}{2W}\right)^{1/2}\right\}$$

(I-140)

[recall that (I-54) has to be multiplied by $4\pi W/M$ to obtain the required photoproduction normalization].

For the reaction $\gamma + N \rightarrow K + \Lambda$ the decomposition of the amplitudes in isotopic spin space is

$$A_\Lambda = A_\Lambda^{(s)} + \tau_3 A_\Lambda^{(v)},$$

(I-141)

and the combinations of amplitudes for the particular physical processes are

$$\langle K^+\Lambda^0 | T | p\gamma \rangle = A_\Lambda^{(s)} + A_\Lambda^{(s)}, \qquad \langle K^0\Lambda_0 | T | n\gamma \rangle = A_\Lambda^{(s)} - A_\Lambda^{(v)}. \qquad (\text{I-142})$$

For the reaction $\gamma + N \rightarrow K + \Sigma$ the decomposition of the amplitudes in isotopic spin space is of the same form as in pion photoproduction

$$A_\Sigma = \delta_{\alpha 3} A_\Sigma^{(+)} + \tfrac{1}{2}\{\tau_2, \tau_3\} A_\Sigma^{(-)} + \tau_2 A_\Sigma^{(0)}, \qquad (\text{I-143})$$

and the amplitudes for the possible physical processes are

$$\langle K^0\Sigma^+ | T | \gamma p \rangle = \sqrt{2}\{A^{(0)} + A^{(-)}\}, \qquad \langle K^+\Sigma^0 | T | \gamma p \rangle = A^{(0)} + A^{(+)},$$

$$\langle K^+\Sigma^- | T | \gamma n \rangle = \sqrt{2}\{A^{(0)} - A^{(-)}\}, \qquad \langle K^0\Sigma^0 | T | \gamma n \rangle = -\{A^{(0)} - A^{(-)}\}.$$

$$(\text{I-144})$$

II. The Resonance Region

In discussing photoproduction and electroproduction processes it is convenient to separate high energy production ($\gtrsim 2.0$-GeV photon laboratory energy or its equivalent) from production at low and intermediate energies ($\lesssim 1.5$-GeV photon laboratory energy or its equivalent). There is a natural break here, both in the data and in the theoretical approach, the high energy region being primarily the province of Regge theory and the low and intermediate energy region being the resonance domain.

The resonance region itself can be subdivided into the first resonance region ($\lesssim 450$-MeV photon laboratory energy or its equivalent) and the rest. The reason for this division is that theoretical calculation and prediction is possible for the former, but for the latter it is necessary to resort to pure phenomenology.

We shall start this section by exploring the scope and success of dispersion theory in pion photoproduction and electroproduction in the region of the first resonance, and then go on to discuss the phenomenological analyses of the data on photoproduction and electroproduction of the N* resonances. Finally a discussion of the quark model will be given since this gives a good global description of the general features observed, although at its present stage of development quantitative precision is, not surprisingly, lacking.

A. Dispersion Relation Formalism for Pion Production

Apart from the kinematical singularities which are contained in the invariant amplitudes A_2 and A_5 [see (I-29)], the analytic properties of A_i are derived from a knowledge of the possible intermediate states in each channel, poles and cuts arising from one particle and two or more particle intermediate states, respectively. The pole terms are precisely the Born terms, discussed in Section I,G.

In the s channel, there is a pole at $s = M^2$ coming from the single-nucleon intermediate state (Fig. 1a), and a cut for $s \geqslant (M + \mu)^2$ coming from the (nucleon + one pion) intermediate state (Fig. 1b), the rescattering term. This analytic structure is reflected in the u channel, there being a pole at $u = M^2$, coming from the single-nucleon exchange diagram (Fig. 1c) and a cut for $u \geqslant (M + \mu)^2$ coming from the exchanged (nucleon + one pion) state (Fig. 1d), the crossed rescattering term. In the t channel, there is a pole at

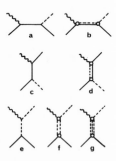

FIG. 1. The pole terms and lowest mass singularities in pion photoproduction and electroproduction.

$t = \mu^2$ coming from the single-pion exchange diagram, (Fig. 1e), a cut for $t \geqslant 4\mu^2$ in the isoscalar part of the amplitude coming from the two-pion exchange diagram (Fig. 1f) (for example, ρ-meson exchange) and a cut for $t \geqslant 9\mu^2$ in the isovector part of the amplitude coming from the three-pion exchange diagram (Fig. 1g) (for example, ω-, ϕ-meson exchange). There are, of course, other branch points in each channel at each new threshold for continuum intermediate states, but these extra branch points do not alter the analytic structure of the amplitudes in any way.

The fixed-t dispersion relations which follow from these assumed analytic properties, and from crossing symmetry (I-41) can be written in matrix notation as

$$\text{Re } \tilde{A}(s, t, K^2) = [(s - M^2)^{-1} + [\bar{\xi}](u - M^2)^{-1}]\tilde{\Gamma}(t, K^2) + \tfrac{1}{2}(1 - \xi)\tilde{\Gamma}_t/(t - \mu^2)$$

$$+ \frac{P}{\pi} \int_{(M+\mu)^2}^{\infty} ds'[(s' - s)^{-1} + [\bar{\xi}](s' - u)^{-1}] \text{ Im } \tilde{A}(s', t, K^2).$$

$$\text{(II-1)}$$

The pole terms are taken directly from (I-88), the residues being given in (I-89) and (I-90), and the symbol $P \int$ denotes the principal value integral. Apart from the pion pole term, there is no explicit representation of the t-channel exchanges in (II-1). Their contribution will be discussed in detail at a later stage.

Dispersion relations for the multipole amplitudes themselves can be derived directly from (II-1). Using (II-1), (I-61), and (I-58), one finds

$$\text{Re } \tilde{M}_l^{(T)}(s, K^2) = \int_{-1}^{1} dx \begin{bmatrix} D_l(x) & 0 \\ 0 & E_l(x) \end{bmatrix} \{C^{-1}(s, K^2)\}\{B(s, t, K^2)\}$$

$$\times [\{(s - M^2)^{-1} + [\bar{\xi}] (u - M^2)^{-1}\} (\bar{\Gamma}t, K^2)$$

$$+ \tfrac{1}{2}(1 - \xi) \tilde{\Gamma}(t, K^2)/(t - \mu^2)]$$

$$+ \frac{P}{\pi} \int_{(M+\mu)^2}^{\infty} ds' \int_{-1}^{1} dx \begin{bmatrix} D_l(x) & 0 \\ 0 & E_l(x) \end{bmatrix}$$

$$\times \{C^{-1}(s, K^2)\}\{B(s, t, K^2)\}$$

$$\times [(s' - s)^{-1} + [\bar{\xi}] (s' - u)^{-1}]\{B^{-1}(s', t, K^2)\}\{C(s', K^2)\}$$

$$\times \sum_{T'} \sum_{l'=0}^{\infty} \begin{bmatrix} G_{l'}(x') & 0 \\ 0 & H_{l'}(x') \end{bmatrix} \text{Im } \tilde{M}_{l'}^{(T')}(s', K^2), \tag{II-2}$$

where

$$x' = (kq/k'q')x + (k_0'q_0' - k_0 q_0)/k'q'. \tag{II-3}$$

In this dispersion relation, it is necessary to assume that the expansion in terms of Im $\tilde{M}_l(s', K^2)$ is convergent outside the physical region, since for $s' < s$ part of the t range is unphysical ($|x'| > 1$). This is certainly not true everywhere because of the kinematical singularities in A_2 and A_5 and because of the influence of the t cut. For photoproduction (the electroproduction case is more restrictive, and more complicated, because of the finite photon mass) it can be shown that convergence is guaranteed for any s' for photon laboratory energies of up to 450 MeV, the main restriction coming from the t cut. If the t-channel spectrum is assumed to be negligible until the vicinity of the ρ (or ω) is reached, then the convergence region is enlarged to about 550 MeV. In the complete absence of the t cut, the region of convergence goes to about 900 MeV. Since the ρ and ω contributions turn out to be very small (a point which shall be discussed in detail later), and since the usefulness of the dispersion relations as a predictive tool is limited to photon laboratory energies below about 500 MeV for quite different reasons, the validity of (II-2) need not be questioned further.

It is convenient to use the center-of-mass energy $W (W')$ as the variable in the multipole dispersion relations. Then (II-2) can be written symbolically as

$$\text{Re } \tilde{M}_l^{(T)}(W, K^2) = \tilde{M}_l^{(T)B}(W, K)$$

$$+ \frac{P}{\pi} \int_{(M+\mu)}^{\infty} dW' [\text{Im } \tilde{M}_l^{(T)}(W', K^2)/(W' - W)]$$

$$+ \frac{1}{\pi} \int_{(M+\mu)}^{\infty} dW' \sum_{T'} \sum_{l'} K_{ll'}^{(T, T')} (W, W', K^2)$$

$$\times \text{Im } \tilde{M}_{l'}^{(T')}(W', K^2), \tag{II-4}$$

where

$$\mathbf{M}_l{}^B(W, K^2) = \begin{bmatrix} E_{l+}^B(W, K^2) \\ E_{l-}^B(W, K^2) \\ M_{l+}^B(W, K^2) \\ M_{l-}^B(W, K^2) \\ S_{l+}^B(W, K^2) \\ S_{l-}^B(W, K^2) \end{bmatrix}.$$ (II-5)

The terms $E_{l\pm}^B$, $M_{l\pm}^B$, and $S_{l\pm}^B$ are the projected Born terms of (II-2). These are obtained by a series of matrix multiplications, followed by integration, the integrals giving rise to Legendre functions of the second kind,

$$Q_l(y) = \tfrac{1}{2} \int_{-1}^{1} dx[P_l(x)/(y - x)],$$

which occur with arguments $\bar{q}_0$ and $\bar{E}_2$, defined by

$$t - \mu^2 = -2kq([(2k_0 q_0 + K^2)/2kq] - x) = -2kq(\bar{q}_0 - x),$$
$$u - M^2 = -2kq([(2k_0 E_2 + K^2)/2kq] + x) = -2kq(E_2 + x).$$ (II-6)

The results are

$$E_{l+}^B(W, K^2) = \frac{\{W - M\}\{(E_1 + M)(E_2 + M)\}^{1/2}}{2M} [2(l + 1)]^{-1}$$

$$\times \left\{ \frac{2\delta_{l_0}}{(W^2 - M^2)} [\Gamma_1(K^2) + (W - M)\Gamma_3(K^2) - \xi(W + M)\Gamma_3(K^2)] \right.$$

$$- \xi[\Gamma_1(K^2) - 2M\Gamma_3(K^2)]T_{l+}(W, K^2)$$

$$- \tfrac{1}{2}(1 - \xi)[4\Gamma_1(K^2) + K^2\Gamma_t(K^2)]$$

$$\times \left[\frac{lR_l(W, K^2)}{(E_1 + M)(W - M)} - \frac{q(l + 1)R_{l+}(W, K^2)}{k(E_2 + M)(W - M)} \right]$$

$$- 2\xi[\Gamma_1(K^2) - (W + M)\Gamma_3(K^2)] \frac{lR_l{}^N(W, K^2)}{(E_1 + M)(W - M)}$$

$$+ 2\xi[\Gamma_1(K^2) + (W - M)\Gamma_3(K^2)] \frac{q(l + 1)R_{l+1}^N(W, K^2)}{k(E_2 + M)(W - M)} \right\},$$ (II-7)

$$E_{l-}^B(W, K^2) = \frac{\{W - M\}\{(E_1 + M)(E_2 + M)\}^{1/2}}{2M}(2l)^{-1}$$

$$\times \left\{ -\xi[\Gamma_1(K^2) - 2M\Gamma_3(K^2)]T_{l-}(W, K^2) \right.$$

$$+ \tfrac{1}{2}(1 - \xi)\{4\Gamma_1(K^2) + K^2\Gamma_t(K^2)\}$$

$$\times \left[\frac{(l + 1)R_l^{\pi}(W, K^2)}{(E_1 + M)(W - M)} - \frac{qlR_{l-1}^{\pi}(W, K^2)}{k(E_2 + M)(W - M)} \right]$$

$$+ 2\xi[\Gamma_1(K^2) - (W + M)\Gamma_3(K^2)]\frac{(l + 1)R_l^N(W, K^2)}{(E_1 + M)(W - M)}$$

$$\left. - 2\xi[\Gamma_1(K^2) + (W - M)\Gamma_3(K^2)]\frac{qlR_{l-1}^N(W, K^2)}{k(E_2 + M)(W - M)} \right\}, \quad \text{(II-8)}$$

$$M_{l+}^B(W, K^2) = \frac{\{W - M\}\{(E_1 + M)(E_2 + M)\}^{1/2}}{2M}[2(l + 1)]^{-1}$$

$$\times \left\{ -\xi[\Gamma_1(K^2) - 2M\Gamma_3(K^2)]T_{l+}(W, K^2) \right.$$

$$+ \tfrac{1}{2}(1 - \xi)[4\Gamma_1(K^2) + K^2\Gamma_t(K^2)]\frac{R_l^{\pi}(W, K^2)}{(E_1 + M)(W - M)}$$

$$\left. + 2\xi[\Gamma_1(K^2) - (W + M)\Gamma_3(K^2)]\frac{R_l^N(W, K^2)}{(E_1 + M)(W - M)} \right\}, \quad \text{(II-9)}$$

$$M_{l-}^B(W, K^2) = \frac{\{W - M\}\{(E_1 + M)(E_2 + M)\}^{1/2}}{2M}(2l)^{-1}$$

$$\times \left\{ \frac{2qk\delta_{l_1}}{(E_1 + M)(E_2 + M)(W - M)^2} \right.$$

$$\times [-\Gamma_1(K^2) + (W + M)\Gamma_3(K^2) - \xi(W - M)\Gamma_3(K^2)]$$

$$+ \xi[\Gamma_1(K^2) - 2M\Gamma_3(K^2)]T_{l-}(W, K^2)$$

$$- \tfrac{1}{2}(1 - \xi)[4\Gamma_1(K^2) + K^2\Gamma_t(K^2)]\frac{R_l^{\pi}(W, K^2)}{(E_1 + M)(W - M)}$$

$$\left. - 2\xi[\Gamma_1(K^2) - (W + M)\Gamma_3(K^2)]\frac{R_l^N(W, K^2)}{(E_1 + M)(W - M)} \right\}, \quad \text{(II-10)}$$

$$S_{l+}^B(W, K^2) = \frac{\{W - M\}\{(E_1 + M)(E_2 + M)\}^{1/2}}{2M}[2(l + 1)]^{-1}$$

$$\times \left\{ \frac{2k\delta_{l_0}}{(E_1 + M)(W^2 - M^2)}[-\Gamma_1(K^2) + (E_1 + M)\Gamma_3(K^2) \right.$$

$$+ \xi(W + M)\Gamma_3(K^2) + \tfrac{1}{4}(1 - \xi)k_0(W + M)\Gamma_t(K^2)]$$

$$+ \tfrac{1}{2}(1 - \xi)\frac{(2q_0 - k_0)}{(W - M)}[\tfrac{1}{2}K^2\Gamma_t(K^2) + 2\Gamma_1(K^2)]$$

$$\times \left[\frac{Q_l(\bar{q}_0)}{q(E_1 + M)} - \frac{Q_{l+1}(\bar{q}_0)}{k(E_2 + M)}\right]$$

$$+ \frac{(-1)^{l+1}\xi}{(W - M)}[-M(2q_0 - k_0)\Gamma_3(K^2) + (2q_0 - W)\Gamma_1(K^2)]$$

$$\times \left[\frac{Q_l(E_2)}{q(E_1 + M)} + \frac{Q_{l+1}(E_2)}{k(E_2 + M)}\right]$$

$$+ \frac{(-1)^{l+1}}{(W - M)}[M\Gamma_1(K^2) + (k_0 W + K^2 - \mu^2)\Gamma_3(K^2)]$$

$$\times \left[\frac{Q_l(E_2)}{q(E_1 + M)} - \frac{Q_{l+1}(E_2)}{k(E_2 + M)}\right]\Bigg\}, \tag{II-11}$$

and

$$S_{l-}^B(W, K^2) = \frac{\{W - M\}\{(E_1 + M)(E_2 + M)\}^{1/2}}{2M}(2l)^{-1}$$

$$\times \Bigg\{\frac{2q\delta_{l1}}{(E_2 + M)(W - M)}[\Gamma_1(K^2) + (E_1 - M)\Gamma_3(K^2)$$

$$+ \xi(W - M)\Gamma_3(K^2) - \tfrac{1}{4}(1 - \xi)k_0(W - M)\Gamma_t(K^2)]$$

$$+ \frac{(1 - \xi)(2q_0 - k_0)}{2(W - M)}\{\tfrac{1}{2}K^2\Gamma_t(K^2) + 2\Gamma_1(K^2)\}$$

$$\times \left[\frac{Q_l(\bar{q}_0)}{q(E_1 + M)} - \frac{Q_{l-1}(\bar{q}_0)}{k(E_2 + M)}\right]$$

$$+ \frac{(-1)^{l+1}\xi}{(W - M)}\{-M(2q_0 - k_0)\Gamma_3(K^2) + (2q_0 - W)\Gamma_1(K^2)\}$$

$$\times \left[\frac{Q_l(\bar{E}_2)}{q(E_1 + M)} + \frac{Q_{l-1}(\bar{E}_2)}{k(E_2 + M)}\right]$$

$$+ \frac{(-1)^{l+1}\xi}{(W - M)}\{M\Gamma_1(K^2) + (k_0 W + K^2 - \mu^2)\Gamma_3(K^2)\}$$

$$\times \left[\frac{Q_l(E_2)}{q(E_1 + M)} - \frac{Q_{l-1}(E_2)}{k(E_2 + M)}\right]\Bigg\}. \tag{II-12}$$

In these equations, the functions $R_l{}^N(W, K^2)$, $R_l{}^\pi(W, K^2)$, and $T_{l\pm}(W, K^2)$ are defined by

$$R_l{}^N(W, K^2) = \frac{(-1)^l}{(2l+1)} [Q_{l+1}(\bar{E}_2) - Q_{l-1}(\bar{E}_2)],$$

$$R_l{}^\pi(W, K^2) = (2l+1)^{-1}[Q_{l+1}(\bar{q}_0) - Q_{l-1}(\bar{q}_0)], \tag{II-13}$$

$$T_{l\pm}(W, K^2) = (-1)^l\left[\frac{Q_l(\bar{E}_2)}{kq} - \frac{W+M}{(E_1+M)(E_2+M)(W-M)} Q_{l\pm1}(\bar{E}_2)\right].$$

It should be recalled that $\Gamma_1(K^2)$ and $\Gamma_3(K^2)$ depend on $(\pm, 0)$ indices, which have been suppressed here for convenience.

The threshold behavior of the multipole amplitudes can be inferred directly from (II-7)–(II-12), on recalling that

$$Q_l(z) \xrightarrow[z\to\infty]{} z^{-l-1}.$$

One finds that, as $q \to 0$,

$$\begin{aligned}
E_{l+} &\sim k^l q^l, & E_{l-} &\sim k^{l-2} q^l, \\
M_{l+} &\sim k^l q^l, & M_{l-} &\sim k^l q^l, \\
S_{l+} &\sim k^l q^l, & S_{l-} &\sim k^{l-2} q^l \quad (\text{except } S_{1-} \sim kq).
\end{aligned} \tag{II-14}$$

For practical calculation, this threshold behavior must be taken into account explicitly, in particular the q dependence. It should be noted that the extraction of a kinematical factor is not unique.

Defining

$$\begin{aligned}
\mathcal{E}_{l+} &= E_{l+}/k^l q^l, & \mathcal{E}_{l-} &= E_{l-}/k^{l-2} q^l, \\
\mathcal{M}_{l+} &= M_{l+}/k^l q^l, & \mathcal{M}_{l-} &= M_{l-}/k^l q^l, \\
\mathcal{S}_{l+} &= S_{l+}/k^l q^l, & \mathcal{S}_{l-} &= S_{l-}/k^{l-2} q^l \quad (\text{except } \mathcal{S}_{1-} = S_{1-}/kq),
\end{aligned} \tag{II-15}$$

and correspondingly $\mathcal{E}_{l+}^B, \ldots, \mathcal{S}_{l-}^B$ for the Born terms, and introducing the six-vectors $\tilde{\mathcal{M}}_l$, with components $\mathcal{E}_{l+}, \ldots, \mathcal{S}_{l-}$ and $\tilde{\mathcal{M}}_l{}^B$, with components $\mathcal{E}_{l+}^B, \ldots, \mathcal{S}_{l-}^B$, the dispersion relations may be re-written as

$$\text{Re } \tilde{\mathcal{M}}_l^{(T)}(W, K^2)$$

$$= \tilde{\mathcal{M}}_l^{(T)B}(W, K^2) + \frac{P}{\pi} \int_{(M+\mu)}^\infty dW' \text{ Im } \tilde{\mathcal{M}}_l^{(T)}(W, K^2)/(W' - W)$$

$$+ \frac{1}{\pi} \int_{(M+\mu)}^\infty dW' \sum_{T'} \sum_{l'} \mathcal{K}_{ll'}(W, W', K^2) \text{Im } \tilde{\mathcal{M}}_{l'}^{(T')}(W', K^2). \tag{II-16}$$

Explicit expressions for the kernels $\mathcal{K}_{ll'}(W, W', K^2)$ have been obtained both for photoproduction and electroproduction by Berends *et al.* (1967a) and by von Gehlen (1969), but for most purposes of numerology, direct numerical evaluation of the matrix products in (II-2) is sufficient.

B. DISPERSION RELATIONS AND LOW ENERGY PHOTOPRODUCTION

1. Evaluation of the Dispersion Relations

The initial application of dispersion relations to photopion production was by Chew et al. (1956),* who evaluated the dispersion relations in what was essentially the static limit, projecting out equations for the multipole amplitudes and retaining only those which lead to s- and p-wave pion-nucleon final states. In the static limit, the dispersion relations take on a particularly simple form. The electric quadrupole transition, $E_1^{(3)}$, to the final P_{33} pion-nucleon state is consistent with zero and in the absence of this term the dispersion relations for the magnetic dipole transitions M_{1+} become identical in form to those satisfied (in the static limit) by the corresponding pion-nucleon scattering amplitudes $f_{1\pm}$. This immediately leads to the solution

$$M_{l\pm} = \frac{M_{1\pm} \text{ (static Born)}}{f_{1\pm} \text{ (static Born)}} f_{1\pm}. \tag{II-17}$$

For the $M_{1+}^{(3)}$ transition to the final P_{33} pion-nucleon state, in particular, this yields the solution

$$M_{1+}^{(3)} = (k/q^2)(e/2Mf)(\mu_p - \mu_n) \exp(i\delta_{33}) \sin \delta_{33}, \tag{II-18}$$

and, because of the smallness of the other p-wave pion-nucleon phase shifts in this energy region, it implies that the other transitions to p waves $[M_{1+}^{(1;\,0)}, M_{1-}^{(3)}, M_{1-}^{(1;\,0)}]$ are negligibly small.

Having found this solution for the resonant term, the transitions to s waves were obtained by assuming that the crossed dispersion integrals are exhausted by the $M_{1+}^{(3)}$ multipole alone. If one makes the further assumption that the direct s-wave rescattering term can be neglected,† then the solution is obtained in a particularly simple form, namely

$$E_0^{(+)}(W)/(W - M) \approx (-ef/8\pi M)(\mu_p - \mu_n)$$

$$+ (4/3\pi) \int_{(M+\mu)}^{\infty} dW' \text{ Im } M_{1+}^{(3)}(W')/k'q', \tag{II-19}$$

$$E_0^{(-)}(W) \approx (ef/4\pi)\{1 - \tfrac{1}{2}[1 + (1 - u^2)/2u \ln(1 - u)/(1 + u)]\}, \tag{II-20}$$

---

* Hereafter referred to as CGLN.

† Since the magnitude of the s-wave phase shifts is not more than 10–15° in the energy range being considered, this correction should be of the order of μ/M and consequently its omission is consistent with the static model assumptions.

where u is the velocity of the outgoing pion

$$E_{0+}^{(0)}(W)/(W - M) \approx (-ef/8\pi M)(\mu_p + \mu_n).$$ (II-21)

Although extremely simple, this solution exhibited clearly the general features of low energy photopion production. To set the scale, the value of the $M_{1+}^{(3)}$ multipole at resonance is 3.6×10^{-2}, in units $\hbar = \mu = c = 1$. From (II-19)–(II-21), it is clear that the E_{0+} multipoles will vary only slowly with energy, and to obtain a qualitative appreciation of their relative importance it is sufficient to evaluate them at threshold, where they have the values

$$E_{0+}^{(+)} = -0.65 \times 10^{-2} + \text{correction for crossed } M_{1+}^{(3)} \text{ term,}$$
$$E_{0+}^{(-)} = 2.42 + 10^{-2}, \qquad E_{0+}^{(0)} = 0.02 \times 10^{-2}.$$ (II-22)

The $E_{0+}^{(+)}$ Born term is cancelled appreciably by the crossed $M_{1+}^{(3)}$ term, which has a value at threshold of $\sim 0.4 \times 10^{-2}$. The dominance of $E_{0+}^{(-)}$ has the immediate consequence that transitions to s waves will be very large for $\gamma + p \to n + \pi^+$ and $\gamma + n \to p + \pi^-$, and in the resonance region they will contribute about 40% of the total cross section. The complete insignificance of $E_{0+}^{(0)}$ implies that the ratio $\sigma(\gamma p \to n\pi^+)/\sigma(\gamma n \to p\pi^-)$ will be close to unity, at least near threshold. The suppression of $E_{0+}^{(+)}$ means that transitions to s waves will be rather small for $\gamma + p \to p + \pi^0$ and this process should be completely dominated by the transition to the P_{33} resonance which, as we have seen, must be the $M_{1+}^{(3)}$ amplitude. If this were the only term present, then the differential cross section would be proportional to $(5 - 3\cos^2 \theta)$ [the $E_{1+}^{(3)}$ term, by itself, is proportional to $(1 + \cos^2 \theta)$].

The total cross section for $\gamma + p \to n + \pi^+$ is shown in Fig. 2, where the typical s-wave threshold rise is clearly seen, and the ratio of resonance to background is much as expected. The ratio $\sigma(\gamma n \to p\pi^-)/\sigma(\gamma p \to p\pi^+)$ is shown in Fig. 3.

For the reaction $\gamma + p \to p + \pi^0$, it is convenient to expand the differential cross section as

$$d\sigma/d\Omega = A + B\cos\theta + C\cos^2\theta.$$ (II-23)

The total cross section for this process is shown in Fig. 4, and the coefficients A, B, and C are shown in Figs. 5, 6, and 7, respectively. The absence of any appreciable s-wave term close to threshold is obvious, as is the resonance dominance of the reaction. To emphasize that this is indeed the $M_{1+}^{(3)}$ transition, the ratio C/A is shown in Fig. 8.

Although qualitatively successful, and in a restricted sense quantitatively successful too, detailed confrontation of the model with experiment indicated clearly that improvement was required. This approach was developed by Höhler and Müllensiefen (1959), Höhler et al. (1960), and Höhler and Dietz (1960), who evaluated the CGLN equations in more detail, demonstrated the importance of recoil effects, and put the theory on a sound footing. The extension to fully relativistic kinematics by Dennery (1961) and McKinley (1962) demonstrated the existence of a nonnegligible electric quadrupole transition $E_{1+}^{(3)}$, although McKinley's attempt to solve the equations via a comparison function was unsuccessful.

An alternative approach was that of Ball (1961) who worked directly in terms of the invariant amplitudes A_i, calculating Re A_i (and thereby the real part of the multipoles) assuming the imaginary part of A_i to be given entirely by Im $M_{1+}^{(3)}$, for which the CGLN result was used. This method has been discussed in detail by Höhler and Schmidt (1964), who compared it with experiment over a wide range of energies. It had the advantage over the other approaches mentioned thus far of giving a relativistic calculation, except for $M_{1+}^{(3)}$, without the complexities of fully relativistic projections. However, it still rested on the *static* $M_{1+}^{(3)}$, neglected the integrals over the imaginary parts of other multipoles, and made no attempt to estimate these imaginary parts for inclusion in cross-section formulas where, by interference with $M_{1+}^{(3)}$, they can give rise to nonnegligible effects. Still following these general lines, an improved calculation has been done by Adler (1968), the improvement consisting of a better (that is, nonstatic) treatment of $M_{1+}^{(3)}$. The *ansatz* for $M_{1+}^{(3)}$ is a natural generalization of (II-17), namely

$$M_{1+}^{(3)} = \frac{M_{1+}^{(3)}(\text{relativistic Born})}{f_{1+}^{(3)}(\text{relativistic Born})} f_{1+}^{(3)}. \qquad \text{(II-24)}$$

Apart from this, the calculation encompasses the deficiencies mentioned above.

To overcome these deficiencies, it is necessary to tackle directly the problem of solving the coupled singular integral equations which the dispersion relations constitute. Fortunately, on detailed investigation of the dispersion relations, considerable simplifications appear and the problem is not quite as formidable as it would seem at first sight. Evaluation of the kernels and study of the expected absorptive parts of the multipoles reveals, first, that the dispersion equation for the $M_{1+}^{(3)}$ is almost completely decoupled from the set, contributions from multipoles in the u-channel integral, other than $M_{1+}^{(3)}$ itself, amounting to only a few percent of the Born term, and consequently this multipole can be treated in isolation. Second, the dominant contributor to the u-channel integral for the $E_{1+}^{(3)}$ multipole is $M_{1+}^{(3)}$ and, hence, once a solution has been obtained for $M_{1+}^{(3)}$, it can be fed into $E_{1+}^{(3)}$ and this multipole then treated in isolation also. Third, the only two multipoles that are strongly coupled, and which consequently should be treated simultaneously, are $E_{0+}^{(1)}$ and $E_{0+}^{(3)}$.

Let us consider now the question of the solution of a singular integral equation (the solution of a coupled set is a direct extension of the method).

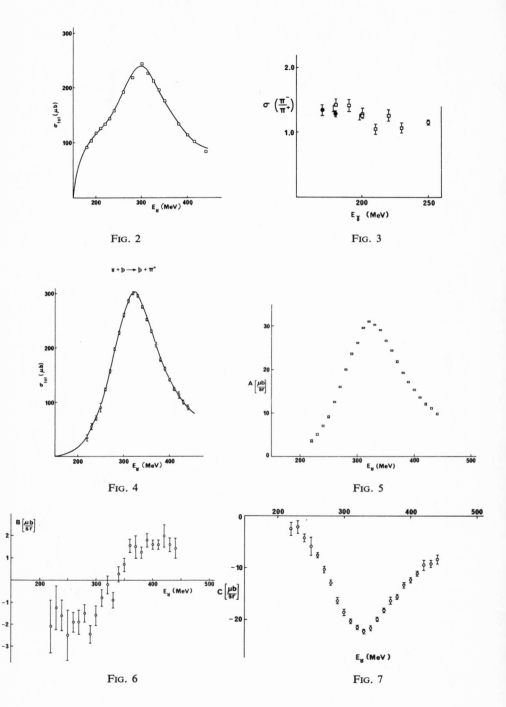

FIG. 2

FIG. 3

$\gamma + p \longrightarrow p + \pi^0$

FIG. 4

FIG. 5

FIG. 6

FIG. 7

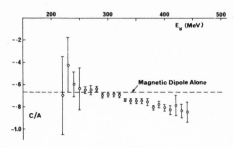

FIG. 8

One approach is to try to find solutions by iteration. In photopion production the first serious approach was that of Donnachie and Shaw (1966). In general, the iteration of a singular integral equation is a very complicated problem, but due to the fact that the static model is already a good approximation to the answer, the use of this as the initial input led to the iteration working well for the magnetic dipole $M_{1+}^{(3)}$. In carrying out the iteration procedure it was necessary to assume that the imaginary part of the amplitude goes smoothly to zero at high energies. Because of the kinematical factors which come in, the integrals converge rapidly, and when they are evaluated at low energies, consequently the contribution from the high energy part of the integral is small, and the high energy approximation causes no appreciable error in the low energy solutions. One complete loop of the iteration process was, given an approximation to the real part of the amplitude, the imaginary part was obtained in the region $(M + \mu) \leqslant W \leqslant W_I$, where W_I is the effective threshold for the onset of inelasticity, by the relation

$$\text{Im } M_{1+}^{(3)}(W) = \text{Re } M_{1+}^{(3)}(W) \tan \delta_{33}(W). \qquad (\text{II-25})$$

A high energy "tail," going to zero at infinity, was joined smoothly on to this expression, giving an approximation to the imaginary part from threshold to infinity, and allowing the integral over the imaginary part to be evaluated.

FIG. 2. The total cross section for the reaction $\gamma + p \rightarrow n + \pi^+$ in the first resonance region.

FIG. 3. The ratio $\sigma(\gamma n \rightarrow p\pi^-)/\sigma(\gamma p \rightarrow n\pi^+)$ in the first resonance region.

FIG. 4. The total cross section for the reaction $\gamma + p \rightarrow p + \pi^0$ in the first resonance region.

FIG. 5. The coefficient A in the expansion $(d\sigma/d\Omega)(\gamma p \rightarrow p\pi^0) = A + B \cos \theta + C \cos^2 \theta$.

FIG. 6. The coefficient B in the expansion $(d\sigma/d\Omega)(\gamma p \rightarrow p\pi^0) = A + B \cos \theta + C \cos^2 \theta$.

FIG. 7. The coefficient C in the expansion $(d\sigma/d\Omega)(\gamma p \rightarrow p\pi^0) = A + B \cos \theta + C \cos^2 \theta$.

FIG. 8. The ratio C/A from the expansion $(d\sigma/d\Omega)(\gamma p \rightarrow p\pi^0) = A + B \cos \theta + C \cos^2 \theta$.

The result of the integration then was added to the Born term to give a new approximation to the real part, and the procedure was repeated until a solution was found which reproduced itself sufficiently accurately. Special care had to be taken in the resonance region where the real part of the amplitude vanishes and small errors in the calculation of the real part lead to large errors in the imaginary part used in the next iteration ($\tan \delta \to \infty$ at resonance) and an unstable solution results. This problem was overcome by the imposition of the two-parametric form

$$M_{1+}^{(3)} = k(a + bW) \exp(i\delta_{33}) \sin \delta_{33}/q^2 \qquad \text{(II-26)}$$

for a short region about the resonance. The solution obtained was stable to better than 1% over the range from threshold to $E_\gamma = 700$ MeV.

The same procedure was applied to the coupled equations for $E_{0+}^{(1)}$ and $E_{0+}^{(3)}$, this time using the Born terms as the starting point and again acceptable solutions were obtained, stable to about one percent. However no reliable result was obtained for the electric quadrupole $E_{1+}^{(3)}$, where no good starting point was known. A similar calculation has been completed recently by von Gehlen (1970), although the photoproduction results were incidental to a calculation of electroproduction.

The basic philosophy behind this approach to the solution of the singular integral equations is that it is reasonable to make statements about the high energy behavior of the magnitude of the multipoles, namely, that they must decrease with energy at least as fast as some specific inverse power of the energy. A more sophisticated application of this philosophy was made by Berends et al. (1967b), who evaluated the dispersion relations using a conformal mapping technique. A discussion of these techniques in the context of dispersion relations has been given by Lovelace (1969) and by Donnachie (1967). In essence, it can be shown that there exist two sets of functions $g_k(W)$ and $h_k(W)$ ($k = 1, 2, \ldots, \infty$) with the properties

(i) $\qquad g_k(W) = \dfrac{P}{\pi} \int_{(M+\mu)}^{\infty} dW' h_k(W')/(W' - W) \qquad (k = 1, 2, \ldots, \infty);$

$$\text{(II-27)}$$

(ii) $\qquad\qquad\qquad \operatorname{Im} M_l(W) = \sum_{k=0} (a_l)_k h_k(W) \qquad \text{(II-28)}$

for any multipole M_l, and this series converges under very general conditions.

(iii)   The functions $g_k(W)$ and $h_k(W)$ have the correct threshold behavior and vanish asymptotically sufficiently rapidly to satisfy the requirements there.

Defining $\mathscr{F}_l(W)$ to be the total inhomogeneous term (that is, the Born term plus whatever part of the u channel is thought to be necessary), then the dispersion relation may be written as

$$\text{Re } M_l(W) = \mathscr{F}_l(W) + \sum_{k=1} (a_l)_k \{ g_k(W) + \tilde{g}_k(W) \}, \tag{II-29}$$

where

$$\tilde{g}_k(W) = (1/\pi) \int_{(M+\mu)}^{\infty} dW' K_{ll}(W, W') h_k(W'). \tag{II-30}$$

Truncating the series in (II-28) and (II-29) at some value $k = K$ and applying the usual relation (II-25) between the imaginary and real parts at the points $W = W_i$ $(i = 1, 2, \ldots, N)$, where the scattering phase shift is known, yields the set of linear equations

$$\sum_{k=1}^{K} (a_l)_k \{ h_k(W_i) - [g_k(W_i) + \tilde{g}_k(W_i)] \tan \delta_l(W_i) \}$$

$$= \mathscr{F}_l(W_i) \tan \delta_l(W_i) \qquad (i = 1, 2, \ldots N), \tag{II-31}$$

and the coefficients $(a_l)_k$ can be found by standard linear fitting techniques (in practice $K \ll N$). Since the equations are linear, given the errors on the scattering phase shifts θ_l, the evaluation of the eorrrs on $(a_l)_k$ is straightforward. In carrying out this procedure, it was found necessary to impose strong constraints on the high energy behavior of the amplitudes, very similar in nature to those imposed by Donnachie and Shaw (1966). One of the more interesting results of this calculation was the possibility of making an estimate of the "theoretical error" on the calculated physical quantities. Those turned out to be quite large, being typically of the order of $\pm 7\%$, on the average, on cross sections.

An alternative approach is to apply a general method of solution of singular linear integral equations, developed by Mushkelishvili (1953) and by Omnes (1958). The necessary information is the knowledge of the phase of the amplitude everywhere on the right-hand cut. The integral equation whose solution is required is of the form

$$M_l(W) = M_l^{\text{inh}}(W) + (1/\pi) \int_{(M+\mu)}^{\infty} dW' \text{ Im } M_l(W')/(W' - W), \tag{II-32}$$

where $M_l^{\text{inh}}(W)$, the inhomogeneous part of the equation, is a function whose only singularities are to the left of $W = (M + \mu)$ in the W plane, and which may be expressed as an integral over its singularities by

$$M_l^{\text{inh}}(W) = (1/\pi) \int_L dW' \ F_l^{\text{inh}}(W')/(W' - W). \tag{II-33}$$

Suppose that the phase $\delta_l(W)$ of $M_l(W)$ is known for $(M + \mu) \leqslant W < \infty$. Then

$$\text{Im } M_l(W) = \exp\{ -i\delta_l(W) \} \sin \delta_l(W) M_l(W), (M + \mu) \leqslant W < \infty. \tag{II-34}$$

As a first step, we construct the auxiliary function

$$D_l(W) = \exp\left\{(W/\pi) \int_{(M+\mu)}^{\infty} dW' \, \delta_l(W')/[W'(W' - W)]\right\} \quad \text{(II-35)}$$

(assuming that the asymptotic behavior of $\delta_l(W)$ is such that the integral converges).

For $W = W_0 \pm i\varepsilon$, with W_0 on the physical cut from $(M + \mu)$ to ∞,

$$D_l(W_0) = \exp\{\phi_l(W_0) \pm i\delta_l(W_0)\} \quad \text{(II-36)}$$

with

$$\phi_l(W_0) = (W_0/\pi)P \int_{(M+\mu)}^{\infty} dW' \, \delta_l(W')/[W'(W' - W_0)]. \quad \text{(II-37)}$$

It is convenient to normalize $D(W)$ by choosing

$$D_l(0) = 1 \quad \text{(II-38)}$$

so that asymptotically

$$D_l(W) \xrightarrow[w \to \infty]{} -(M + \mu)/[W - (M + \mu)] \exp\left[\int_{(M+\mu)}^{\infty} (dW'/W')(1 - \delta_l(W')/\pi)\right]. \quad \text{(II-39)}$$

In terms of $D_l(W)$, the solution to the integral equation is

$$M_l(W) = D_l(W)\left[(1/\pi) \int_L dW' F^{\text{inh}}(W')/[D_l(W')(W' - W)] + P_l(W)\right], \quad \text{(II-40)}$$

where $P_l(W)$ is an arbitrary polynomial. Equation (II-40) clearly has the phase required by (II-34) and the analytic properties of (II-32). If we now make further assumptions that $F_l^{\text{inh}}(W) \xrightarrow[w \to \infty]{} 0$ and $\int^{\infty} dW' \, \text{Im} \, M_l(W')$ exists, then it is necessary that $M_l(W)$ vanish as $W \to \infty$, restricting the polynomial $P_l(W)$ in (II-40) to be at most a constant.

Another way of looking at this solution is to apply Cauchy's theorem to $M_l^{\text{inh}}(W)/D_l(W)$, yielding

$$\frac{M_l^{\text{inh}}(W)}{D_l(W)} = \frac{1}{\pi} \int_L dW' \, \frac{F_l^{\text{inh}}(W')}{D_l(W')(W' - W)}$$

$$- \frac{1}{\pi} \int_{(M+\mu)}^{\infty} dW' \, \frac{\exp[-\phi_l(W')]\sin \delta_l(W')F_l^{\text{inh}}(W')}{(W' - W)}$$

$$+ \lim_{W \to \infty} \left[\frac{M_l^{\text{inh}}(W')}{D_l(W')}\right], \quad \text{(II-41)}$$

the three terms on the right-hand side of (II-41) coming, respectively, from the singularities of $M^{inh}(W)$, the physical cut, and the circle at infinity. Comparing (II-41) with (II-40), we find

$$
M_l(W) = M_l^{inh}(W) + D_l(W) \left\{ \frac{1}{\pi} \int_{(M+\mu)}^{\infty} dW' \frac{\sin \delta_l(W') \exp[-\phi_l(W')] M_l^{inh}(W')}{(W' - W)} \right.
$$

$$
\left. - \lim_{W' \to \infty} \left[\frac{F_l^{inh}(W')}{D_l(W')} \right] + P_l(W) \right\}, \tag{II-42}
$$

the $P_l(W)$, of course, being the same polynomial as that occurring in (II-40).

To go further and obtain a predictive solution, further assumptions are required. We have already mentioned the assumptions necessary to reduce $P_l(W)$ to a constant, and if we now make the further assumption that we will take that solution which falls off most rapidly as $W \to \infty$, then a unique solution is obtained. This is the position adopted by Adler (1968), who gave it some justification by applying precisely the same assumptions to the fixed-t dispersion relations for pion-nucleon scattering and showing that it gives a good result there for the resonant P_{33} amplitude. Also, by a comparison of the exact solution for the resonant multipoles of photoproduction with the corresponding solution for the resonant P_{33} amplitude, with the arbitrary constant zero in both cases, Adler (1968) was led directly to make the *ansatz* (II-24). The solution thus obtained is very close to those of Donnachie and Shaw (1966) and Berends *et al.* (1967b), and involves essentially the same assumptions about the asymptotic behavior of the multipole amplitudes.

This "exact" approach has also been adopted by Schwela *et al.* (1967). However, their solution was written down without the contribution from the circle at infinity and consequently when they neglected the arbitrary constant, the solution obtained (which they call the "particular solution") was drastically wrong and the constant term from their polynomial had to be brought in to restore the situation. An estimate of this constant was readily obtained for the $M_{1+}^{(3)}$ multipole by requiring the solution to be the same as the static $M_{1+}^{(3)}$ solution at threshold, yielding a result which differed very little from that of CGLN. As an alternative, this constant and the corresponding one for $E_{1+}^{(3)}$ have been left as free parameters and obtained by fitting to the $\gamma + p \to p + \pi^0$ experimental data (see, for example, Rollnik, 1967). The significance of these constants is clear from the above discussion; they are predominantly the contribution to the solution in (II-40) coming from the circle at infinity. A more detailed discussion of this approach has been given by Schwela (1968), Schwela and Weizel (1969), and Schwela (1969).

In an attempt to dispense with the necessity of knowing the P_{33} phase to infinity, but still to retain the structure of the exact solution, some compromise calculations have been performed, which entail introducing a cutoff of some description. This can be estimated in some way or other, as was done by Finkler (1964) and by Zagury (1966), in which case the solution is close to that of Donnachie and Shaw (1966), Berends et al. (1967b), and Adler (1968), and is only weakly dependent on the cutoff. Since all of these approaches are in effect making the same assumption, this conclusion is not at all surprising. Alternatively, the cutoff can be left as a free parameter, as proposed by Engels and Schmidt (1968), in which case the situation is precisely that of Schwela et al. (1967), although the interpretation of the parameter is somewhat different.

2. High Energy Corrections

Apart from the arbitrary constants involved in the "exact" solution for the $M_{1+}^{(3)}$ and $E_{1+}^{(3)}$ multipoles, which to some extent can take into account high energy corrections, it has been a basic assumption in the solution of the dispersion relations at low energies that the high energy behavior of the multipoles is unimportant. This assumption can be reasonably justified if there is little or no structure in the high energy amplitudes, particularly in the immediate vicinity of the low energy region. However, in photopion production there is considerable structure. Both the $D_{13}(1512)$ and $F_{15}(1690)$ resonances are strongly photoproduced, and consequently one would expect a nonnegligible modification of the direct rescattering term in the associated $E_{2-}^{(0,\ 1)}$, $M_{2-}^{(0,\ 1)}$ and $M_{2+}^{(0,\ 1)}$, $M_{2+}^{(0,\ 1)}$ multipoles. Because of its comparatively low mass, and because the scattering amplitude becomes very inelastic at low energies, the $P_{11}(1450)$ resonance can also be expected to have an effect, even if it is not photoproduced strongly. The two s-wave amplitudes appear to constitute less of a problem, although inelasticity does set in at rather low energies. The resonance most likely to cause trouble is the $S_{11}(1550)$, but since its two dominant decay channels are πN and ηN, the information available on η photoproduction immediately allows an estimate of the effect of this resonance on the $E_{0+}^{(0,\ 1)}$ multipoles at low energies, and it is found to be small

To estimate the rescattering corrections from $P_{11}(1450)$, $D_{13}(1512)$, and $F_{15}(1690)$ Berends et al. (1967b) assumed that each resonance completely dominated its associated multipoles in the resonance region, making the ansatz

$$\mathcal{M}_{l\pm}^{(T)}(W) = \lambda \mathcal{F}_{l\pm}^{(T)}(W) f_{l\pm}^{(T)}(W), \tag{II-43}$$

where $\mathcal{F}_{l\pm}^{(T)}(W)$ is the inhomogeneous part of the multipole dispersion relation, $f_{l\pm}^{(T)}$ is the corresponding scattering amplitude and λ is a constant, determined by a consistency condition at the resonance position. The strength of the $D_{13}(1512)$ and the $F_{15}(1690)$ resonance obtained in this way was not inconsistent with that estimated from total photopion production cross sections. No check could be carried out on $P_{11}(1450)$. The contribution of the rescattering terms at 500-MeV photon laboratory energy was of the order of 100% of the inhomogeneous term in the $M_1^{(0,\,1)}$ cases, about 40% of the inhomogeneous term in the $E_2^{(0,\,1)}$, $M_2^{(0,\,1)}$ cases, and only a few percent in the $E_{2+}^{(0,\,1)}$, $M_{2+}^{(0,\,1)}$ cases.

An *ansatz* somewhat similar to (II-43) has been made by Schwela and Weizel (1969) for the $E_{0+}^{(0,\,1,\,3)}$ and $M_{1-}^{(0,\,1)}$ multipoles, namely

$$M_{l\pm}^{(T)}(W) = g_{l\pm}^{(T)}(W) f_{l\pm}^{(T)}(W), \tag{II-44}$$

where $g_{l\pm}^{(T)}(W)$ is some unspecified but real function, and $f_{l\pm}^{(T)}(W)$ is again the corresponding scattering amplitude. In their approach, $g_{l\pm}(W)$ is never specified, (II-44) being used only to specify the phase $\phi_{l\pm}(W)$ of the multipoles, which is

$$\phi_{l\pm}^{(T)}(W) = \tan^{-1}\{[1 - \eta_l^{(T)}(W) \cos 2\delta_l^{(T)}(W)]/[\eta_l^{(T)}(W) \sin 2\delta_l^{(T)}(W)]\}, \tag{II-45}$$

which, of course, reduces to the scattering phase shift itself when the scattering amplitude is elastic, that is, $\eta_{l\pm}^{(T)} = 1$. This phase was then put into the Omnès-Mushkelishvili solution (II-40) for these multipoles with the assumption of the following asymptotic behavior:

$$\phi_{s_{11}}(\infty) = \pi, \tag{II-46}$$

$$\phi_{s_{31}}(\infty) = 0, \tag{II-47}$$

$$\phi_{p_{11}}(\infty) = \pi/2. \tag{II-48}$$

The assumption of (II-47) ensures that the arbitrary constant in (II-40) must be zero for the $E_{0+}^{(3)}$ multipole. However, the assumptions (II-46) and (II-48) do not imply this, and an arbitrary constant is left in each of $E_{0+}^{(0)}$, $E_{0+}^{(1)}$, $M_{1-}^{(0)}$, and $M_{1-}^{(1)}$.

Schwela and Weizel (1969) used a narrow resonance approximation to estimate the corrections to the $E_2^{(0,\,1)}$, $M_2^{(0,\,1)}$ coming from the $D_{13}(1512)$ resonance, obtaining corrections somewhat smaller than those of Berends *et al.* (1967b). The arbitrary constants which occur in the calculation of Schwela and Weizel (1969) have been evaluated by fitting to the experimental data in the region of the first resonance. The fit to the processes $\gamma + p \rightarrow p + \pi^0$ and $\gamma + p \rightarrow \pi^+ + n$ is given in Schwela and Weizel (1969) and the process $\gamma + n \rightarrow \pi^- + p$ is discussed by Schwela (1968, 1969).

3. Comparison with Experiment

Because of the onset of inelasticity in some amplitudes at a photon laboratory energy of about 400 MeV, and the consequent difficulty in solving the dispersion relations and interpreting their results, it has become conventional to restrict the comparison of dispersion relation calculations with experiment to photon laboratory energies below 450 MeV. However, since these effects occur predominantly in the amplitudes leading to the $P_{11}(1450)$, $S_{11}(1550)$, and $D_{13}(1512)$ resonances, and we have seen that the effects of these resonances can be estimated reasonably accurately, with the possible exception of the $M_{1-}^{(0, 1)}$ transitions to the $P_{11}(1450)$, the cutoff will be applied at 600 MeV, when there are appropriate data.

For this comparison, the results of only two of the many available calculations will be shown, that of Berends et al. (1967b), which epitomizes the heuristic no-parameter approach, and that of Schwela (1968) and Schwela and Weizel (1969), which presents the alternative approach and which does contain free parameters. The results of the latter are published only up to 440 MeV, so above that energy only the former are shown. The comparison is made with only a subset of the data, but it is sufficient to show clearly all the general features of the cross-section and polarization measurements, and the quality of the theoretical calculations.

a. *The Reaction* $\gamma + p \rightarrow n + \pi^+$. Differential cross sections are given in Fig. 9 for photon laboratory energies from 180 to 580 MeV, and it is clear that there is little choice between the calculations. At most energies where both calculations are available, the differences between them are less than the uncertainties among different experiments, and they give a very acceptable account of the data. The calculation of Berends et al. (1967b) is still reasonably accurate at 500 MeV, but clearly has lost its accuracy by 580 MeV.

The predictions for the asymmetry ratio, that is, the ratio $(\sigma_\parallel - \sigma_\perp)/(\sigma_\parallel + \sigma_\perp)$ for cross sections using plane-polarized photons (where $\sigma_\parallel$ is the cross section in which the plane of photon polarization and the plane of pion emission are parallel, and $\sigma_\perp$ is the cross section in which the plane of photon polarization and the plane of pion emission are perpendicular) are equally satisfactory. The excitation curves up to photon laboratory energies of 500 MeV, and at center-of-mass angles from 30 to 150° are shown in Fig. 10.

Only two data points exist for the recoil neutron polarization, and again the theoretical calculations are in agreement with each other and with the data. The general features of the recoil neutron polarization between 260 and 440 MeV are shown in Fig. 11. The most notable feature is the predicted high polarization.

b. *The Reaction* $\gamma + p \rightarrow p + \pi^0$. It is here that the first of the differences between the calculations of Berends et al. (1967b) and Schwela and Weizel (1969) appears. It should be recalled that this process is completely dominated by the $M_{1+}^{(3)}$ transition to the $P_{33}(1238)$ resonance, and that this is one of the amplitudes in which Schwela and Weizel (1969) have a free parameter, that is, they are effectively making a one-parameter fit to the data. The results for the differential cross sections are shown in Fig. 12 for photon laboratory energies from 180 to 580 MeV. It is clear that below resonance, the $M_{1+}^{(3)}$ amplitude of Berends *et al.* (1967b) is too small by about 12–15%, although above resonance the results are much more acceptable and give a reasonable description of the reaction up to 580 MeV. Neither calculation is particularly successful at very low energies, the predicted cross sections being much too flat to encompass the strong angular dependence found close to threshold. Berends *et al.* (1967c) have given a possible explanation of this in terms of explicit vector-meson exchange contributions, a point which will be discussed in the next section.

It should be noted that there is no inconsistency between the close agreement of the two calculations in $\gamma + p \rightarrow n + \pi^+$ and their disagreement in $\gamma + p \rightarrow p + \pi^0$. In the former reaction, the $M_{1+}^{(3)}$ contribution to the amplitude is suppressed by a factor of $\sqrt{2}$ compared to the latter, and it is further masked by the very large transition to s waves. The excitation curves for the asymmetry ratio at 60, 90, and 120° are shown in Fig. 13. What little data there are agree well with the theoretical calculations. Recoil nucleon polarization data are also in fair agreement with the theoretical calculations, and the results between 240 and 420 MeV are shown in Fig. 14.

c. *The Reaction* $\gamma + n \rightarrow p + \pi^-$. Schwela (1969) has obtained a good description of this reaction by fitting the free parameters in the $E_{0+}^{(0; 1)}$ and $M_{1-}^{(0; 1)}$ amplitudes to the total cross section alone. Only two parameters were involved, since the combinations $(E_{0+}^{(0)} + \frac{1}{3}E_{0+}^{(1)})$ and $(M_{1-}^{(0)} + \frac{1}{3}M_{1-}^{(1)})$ of the isoscalar and isovector amplitudes were constrained to the value obtained in the fit to the data on photoproduction from protons.

On the other hand, the calculation of Berends *et al.* (1967b) becomes extremely inaccurate above resonance. Berends and Donnachie (1969) have argued that this is due entirely to the impossibility of making a reasonable calculation of the $M_{1-}^{(0; 1)}$ amplitudes, and they found that by making their contribution sufficiently small (in effect, zero) a good description of the data could be obtained. This adjustment, which is a very drastic one, does not alter the predictions for photoproduction off protons in any way. It so happens that the $M_{1-}^{(0)}$ and $M_{1-}^{(1)}$ amplitudes, which are individually quite large in the calculation of Berends *et al.* (1967b), cancel almost completely in the combination $(M_{1-}^{(0)} + \frac{1}{3}M_{1-}^{(1)})$ (for protons), but produce a strong enhancement in the combination $(M_{1-}^{(0)} - \frac{1}{3}M_{1-}^{(1)})$ (for neutrons). Consequently, reducing both $M_{1-}^{(0)}$ and $M_{1-}^{(1)}$ makes little difference to the predictions for the proton data. Berends and Donnachie (1969) justify laying the blame entirely on $M_{1-}^{(0; 1)}$

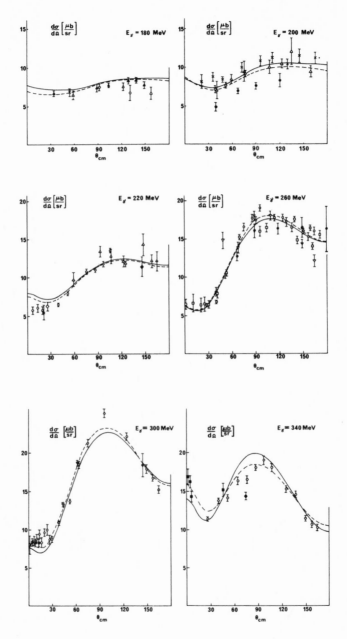

FIG. 9a–f

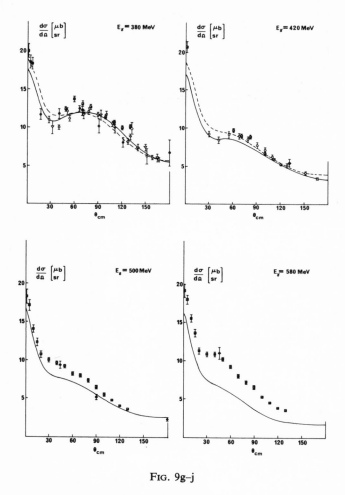

FIG. 9g–j

FIG. 9. Differential cross section for the reaction $\gamma + p \rightarrow n + \pi^+$ in the first resonance region. The solid line is the result of a dispersion relation calculation and the broken line the result of a modified dispersion relation fit.

because of the comparatively good agreement obtained with all data up to 600 MeV for photoproduction of protons, arguing that this is a reliable test of all the multipoles except, of course, $M_1^{(0, 1)}$.

Typical results for the differential cross sections are shown in Fig. 15, and, for the asymmetry ratio, in Fig. 16. It is very clear from these that the calculation of Berends *et al.* (1967b) is wrong above resonance, but there is little to choose between the modification proposed by Berends and

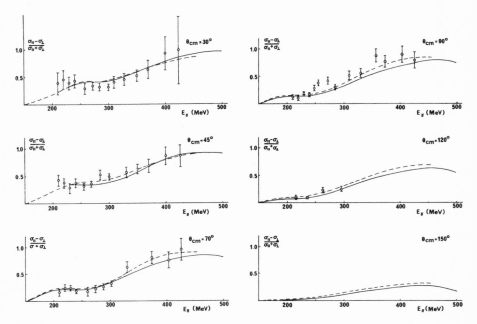

FIG. 10. Polarized photon asymmetry ratio for the reaction $\gamma + p \to n + \pi^+$ in the first resonance region. The solid line is the result of a dispersion relation calculation and the broken line the result of a modified dispersion relation fit.

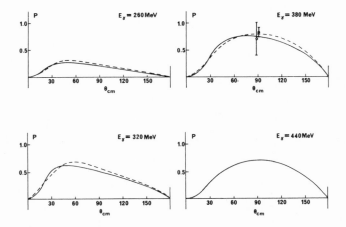

FIG. 11. Recoil neutron polarization for the reaction $\gamma + p \to n + \pi^+$ in the first resonance region. The solid line is the result of a dispersion relation calculation and the broken line the result of a modified dispersion relation fit.

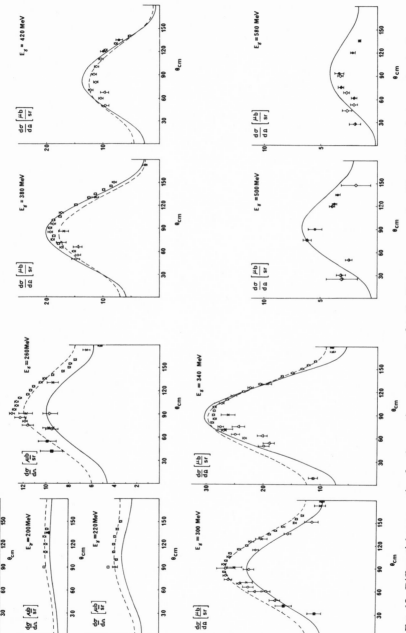

FIG. 12. Differential cross section for the reaction $\gamma + p \rightarrow p + \pi^0$ in the first resonance region. The solid line is the result of a dispersion relation calculation and the broken line the result of a modified dispersion relation fit.

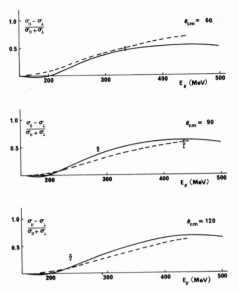

FIG. 13. Polarized photon asymmetry ratio for the reaction $\gamma + p \rightarrow p + \pi^0$ in the first resonance region. The solid line is the result of a dispersion relation calculation and the broken line the result of a modified dispersion relation fit.

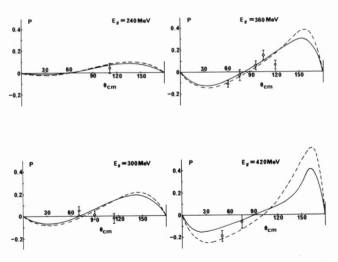

FIG. 14. Recoil proton polarization for the reaction $\gamma + p \rightarrow p + \pi^0$ in the first resonance region. The solid line is the result of a dispersion relation calculation and the broken line the result of a modified dispersion relation fit.

Donnachie (1969) and the calculation by Schwela (1969). The difference in their approach manifests itself most strongly in the asymmetry ratio, particularly at the upper end of the energy range where the calculations of Berends and Donnachie (1969) appears to give a better description than that of Schwela (1969), but the data at the moment are too sparse to justify drawing any definitive conclusion.

4. Vector Meson Exchange Corrections

In all the work which we have discussed so far, it has been tacitly assumed that the contribution to low energy photopion production from vector meson exchange (that is, ω and ρ exchange) is sufficiently small to be neglected. Comparison of the theoretical calculations with experiment certainly indicates that any such contribution must be small, with the consequence that they can be treated simply as t-channel pole terms in the fixed-t dispersion relations.

The $\gamma\pi V$ vertex (where V stands for ω or ρ) is given by

$$i\lambda_V \varepsilon_{\mu\nu\rho\sigma} \varepsilon_\mu K_\nu Q_\rho; \tag{II-49}$$

the NNV vertex is given by

$$i\bar{u}(\mathbf{P}_2)\{g_{V_1}\gamma_\mu + g_{V_2}\sigma_{\mu\nu}(P_2 - P_1)_\nu\}u(\mathbf{P}_1); \tag{II-50}$$

and the V propagator by

$$\frac{\delta_{\sigma\tau} - (Q - K)_\sigma(Q - K)_\tau/m_V^2}{(m_V^2 - t)}, \tag{II-51}$$

where m_V is the mass of V.

Only the first term of the propagator can contribute to the amplitudes because of the asymmetry of $\varepsilon_{\mu\nu\rho\sigma}$, so that the complete V exchange contribution is given by

$$(\lambda_V/t - m_V^2)\varepsilon_{\mu\nu\rho\sigma} \varepsilon_\mu K_\nu Q_\rho \bar{u}(\mathbf{P}_2)[g_{V_1}\gamma_\sigma + g_{V_2}\sigma_{\sigma\tau}(P_2 - P_1)_\tau]u(\mathbf{P}_1). \tag{II-52}$$

Expanding this expression in terms of the invariants of (I-28), the contribution of the V exchange pole to the dispersion relations for the invariant amplitudes A_1, A_2, A_3, A_4 is

$$\begin{array}{ll} A_1: \ \lambda_V g_{V_2} m_V^2/(t - m_V^2), & A_2: \ -\lambda_V g_{V_2}/(t - m_V^2), \\ A_3: \ 0, & A_4: \ -\lambda_V g_{V_1}/(t - m_V^2). \end{array} \tag{II-53}$$

Since the vector exchange is being considered as a pole term in the t-channel, the residues in the above expressions are evaluated at $t = -m_V^2$.

The products of the coupling constants $(\lambda_V g_{V_1})$ and $(\lambda_V g_{V_2})$ can be considered as independent parameters, to be determined by fitting to the low

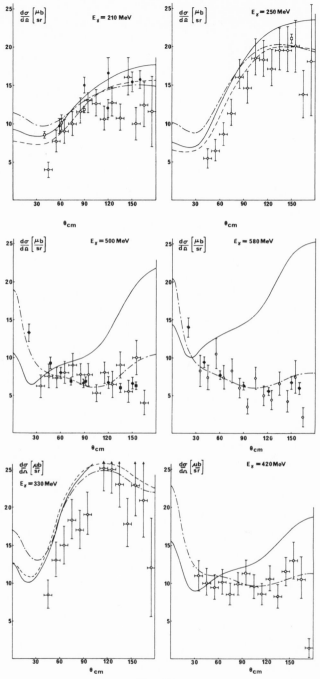

FIG. 15.

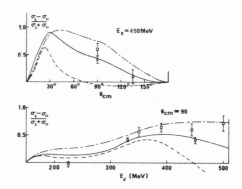

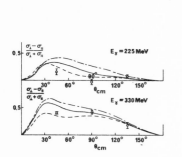

FIG. 16. Polarized photon asymmetry ratio for the reaction $\gamma + n \rightarrow p + \pi^-$ in the first resonance region. The solid line is the result of a dispersion relation calculation, the dot-dash line the result of a dispersion relation calculation with the $M_1^{(0,1)}$ multipoles set to zero and the broken line the result of a modified dispersion relation fit.

energy photoproduction data. The data must be very low energy, since the vector meson exchange affects mainly the E_{0+} and M_{1-} multipoles, and since the effect is known to be small, it is essential to look at an energy region where the conclusions will not be obscured by other uncertainties.

Turning first to charged pion photoproduction, the contribution from the ρ can be readily investigated using the π^+/π^- ratio near threshold, as was done, for example, by Donnachie and Shaw (1966). In the case of the ρ, the ρNN coupling can be obtained, for instance, from pion-nucleon scattering so that an absolute value can be obtained for the $\rho\pi\gamma$ coupling, which can most conveniently be expressed in terms of the width for the $\rho \rightarrow \pi\gamma$ decay. The value of Donnachie and Shaw (1966) corresponds to an upper limit on the $\rho \rightarrow \pi\gamma$ width of $0.2^{+0.6}_{-0.2}$ MeV, which compares well with the lowest upper limit by direct experimental measurement of about 0.6 MeV, obtained by Fidecaro *et al.* (1966). As can be gauged by these errors quoted on the $\rho \rightarrow \pi\gamma$ width, the effect of ρ exchange in charged pion production is smaller than the current experimental errors on the π^+ and π^- cross sections.

Taking the results of Donnachie and Shaw (1966), one can quickly ascertain that the ρ by itself is not sufficient to explain the discrepancy between experiment and dispersion theory in π^0 photoproduction close to threshold. This is

FIG. 15. Differential cross section for the reaction $\gamma + n \rightarrow p + \pi^-$ in the first resonance region. The solid line is the result of a dispersion relation calculation, the dot-dash line the result of a dispersion relation calculation with the $M_1^{(0,1)}$ multipoles set to zero and the broken line the result of a modified dispersion relation fit.

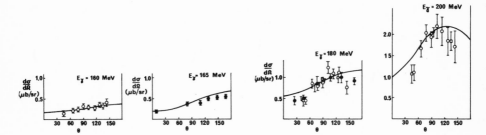

FIG. 17. Differential cross section for the reaction $\gamma + p \rightarrow p + \pi^0$ near threshold. The solid line is the result of a dispersion relation calculation modified by vector meson exchange.

a particularly favorable place in which to look for deviations, because the large s-wave multipoles cancel amost completely, leaving practically the whole cross section to be given by the p-waves, which are very small in this region. The cancellation is so extreme that at 160-MeV photon laboratory energy the π^0 cross section is less than 3 μb, while the π^+ cross section is of the order of 60 μb. Berends *et al.* (1967c) successfully explained the discrepancy in terms of ω exchange alone. In this case, λ_ω is well known from the width of the $\omega \rightarrow \pi\gamma$ decay, and $g_{\omega 1}, g_{\omega 2}$ are the unknown quantities. The values obtained by Berends *et al.* (1967c) for these quantities were in reasonable accord with those obtained elsewhere, in particular from high energy π^0 photoproduction (assuming a dominant ω exchange) and from nucleon-nucleon scattering. The resulting contribution is still sufficiently small not

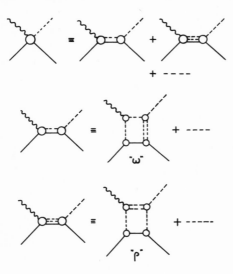

FIG. 18. The lowest mass s-channel singularities in pion photoproduction and electroproduction.

FIG. 19. Lowest order box diagram for the two-particle s-channel singularity illustrating three-pion t-channel exchange.

FIG. 20. Lowest order box diagram for the three-particle s-channel singularity illustrating two-pion t-channel exchange.

to have any noticeable effect at higher energies. The great improvement to the very low energy cross sections is shown in Fig. 17.

Although this procedure works, and the insertion of a pole in the amplitude as an approximation to the vector meson exchange contribution may appear reasonable, it has been argued by Schwela and Weizel (1969) that this is not necessarily the case and because of the method of solution of the dispersion relations via Watson's theorem there is a distinct risk of double counting, particularly for ω exchange.

The argument of Schwela and Weizel (1969) is as follows: The lowest singularities in the unitarity relation are given schematically in Fig. 18, the various terms of which can be looked at as a series of box diagrams, as illustrated in Figs. 19 and 20. The lowest singularity in this expansion of the s-channel elastic unitarity contribution is a three-pion exchange term in the t channel, and can be interpreted as a contribution of ω exchange. In other words, the ω exchange term may be considered to be a final state interaction of the meson exchange Born approximation, which is just the final state interaction evaluated when solving the dispersion relations for the multipoles using Watson's theorem, and consequently no ω-exchange term should be inserted explicitly into the dispersion relations.

The argument against this is, that while the three pion exchange term of Fig. 19 is indeed there, it is a continuum three-pion contribution and has nothing to do with the ω as such, merely acting as a background term to it, and, hence, it is justified to insert an ω pole explicitly. These considerations do not apply to the ρ-exchange contribution, since the corresponding box diagram is the lowest singularity of the inelastic s-channel unitarity graph.

Although firmly based on fundamental theoretical principles, the calculations of Schwela and Weizel (1969) and of Schwela (1969) can be considered as a semiphenomenological fit to the low energy photoproduction data. There has been a number of other such fits which have been firmly based on a foundation of dispersion theory calculations. An obvious example of this was the explicit introduction of ω exchange by Berends et al. (1967c) as an explanation of the discrepancy between the theoretical calculations and low energy π^0 photoproduction. Another example, and one more in the spirit of Schwela and Weizel (1969) was that of Donnachie and Shaw (1967) who introduced arbitrary parameters into the $E_{0+}^{(0)}$, $E_{0+}^{(1)}$, $E_{0+}^{(3)}$, $M_{1-}^{(0)}$, $M_{1-}^{(1)}$, and $M_{1-}^{(3)}$ multipoles. Although the terminology of their additive terms is different from that of Schwela and Weizel (1969) (contributions from distant left-hand singularities as opposed to the circle at infinity), the net effect is very much the same.

With the increasing availability of high precision data in the first resonance region, it has become possible to extract the multipoles from the data directly, and detailed analyses of pion photoproduction on protons have been performed by Berends and Weaver (1971a), by Noelle *et al.* (1971), and by Noelle (1971). In general the results obtained agree well with the predictions of dispersion relation calculations, although there are some discrepancies. In particular, Berends and Weaver (1971a) and Noelle *et al.* (1971) found that the $M_1^{(3)}$ multipole differed considerably from the expected value. The existence of some deviation is not surprising, since all the dispersion calculations involve some approximation, and as we have already noted, the inclusion of the effects of the higher resonances can lead to important modifications. This point was investigated by Berends and Weaver (1971a) who included all the higher resonances, taking their coupling strengths from the analysis of Walker (1969a), and showed that this was sufficient to explain qualitatively all of the discrepancies with the exception of that occurring in $M_1^{(3)}$. Berends and Weaver (1971a) also showed that the low energy π^0-polarized photon asymmetry data is almost exclusively responsible for this deviation.

A secondary cause, and also a source of some general uncertainty, lies in the π^0 differential cross sections. Any analysis is dominated by the data of Fischer *et al.* (1970), but this differs in normalization from the (much less extensive) data of Morand *et al.* (1969) (which is lower) and in normalization and shape from the (also much less extensive) data of Hilger *et al.* (1972). Although these normalization differences create a problem, the shape difference is more important. The data of Fischer *et al.* (1970) fall off rather rapidly between 90 and 60°, providing additional encouragement to the $M_1^{(3)}$ to misbehave and inducing a rather strong energy dependence into $E_{0+}^{(\pi^0)}$. Noelle (1971) prefers to let his fit follow the forward data of Hilger *et al.* (1972), in contrast to Noelle *et al.* (1971) whose fit follows the forward data of Fischer *et al.* (1970). The proton multipoles obtained by Noelle (1971) contain no serious discrepancies with the dispersion relation predictions, and no one can conclude that the proton data is understood, apart from the minor points of the normalization of the π^0 data and the low energy behavior of the π^0-polarized photon asymmetry.

The situation with the neutron data is quite different. Not only are the data less precise, but also there is much less available (for example, an almost complete lack of polarization data of any kind) and these data are subject to considerable uncertainty because of the problems of extracting free-neutron cross sections from deuterium data. Nevertheless, conventional analyses have been performed by Noelle *et al.* (1971), by Noelle (1971), by Noelle and Pfeil (1970), by Berends and Weaver (1971b), and by Schwela (1971). It is clear that the data can be fitted within the conventional formalism, but because of the data uncertainties the interpretation is not unambiguous. Further experimental information is clearly required.

Since the above are based on dispersion theory calculations, all of which are qualitatively the same, there is no significant difference in the results produced in these different approaches, although they contain small quantitative differences.

A completely different viewpoint of low energy photoproduction is contained in the purely phenomenological isobar model of Gourdin and Salin (1963) and Salin (1963), in which it is assumed that the photoproduction amplitude is given by the coherent sum of the Born terms, ρ- and ω-exchange terms and the s-channel P_{33} isobar, the coupling constants of these being taken as parameters to be fitted to experiment, except, of course, the coupling constants of the Born terms. Although reasonable fits to the data were obtained [and indeed by including the $D_{15}(1512)$ isobar, reasonable fits were obtained by Salin (1963) up to 800 MeV-photon laboratory energy], there are a number of theoretical reservations about this approach. In particular, crossed (u-channel) isobar contributions were ignored [and the dispersion relation treatment shows this to be important, a point which has been emphasized by Engels and Schmidt (1968)] although subtraction constants were introduced which compensated to some extent for this omission, and there is certainly an element of double counting, since part of the Born term, for example, is already contained within the P_{33} isobar term, a point which is exhibited explicitly by the static solution (II-18) of the dispersion relation for the $M_{1+}^{(3)}$ multipole.

An improved version of the isobar model has been developed by Pfeil (1968), who took into account the u-channel contributions of the isobars, and by inserting the isobars into the framework of fixed-t dispersion, obviated, to some extent at least, the problem of double counting. Including seven resonances, the Born terms, and ρ and ω exchange, a reasonable fit to the data was obtained to beyond 1-GeV photon laboratory energy. In the first resonance region, the results of the form of the isobar model are much closer to the dispersion relation calculations than those of Salin (1963).

C. DISPERSION RELATIONS AND LOW ENERGY PION ELECTROPRODUCTION

The techniques of solving dispersion relations for pion electroproduction are precisely the same as those already discussed for photoproduction. The static model analogue was promulgated by Fubini et $al.$ (1958), but apart from inconclusive calculations by Blankenbecler et $al.$ (1960), Dennery (1961), and Barbour (1963), there were no definitive attempts until those of Zagury (1966, 1968) and Adler (1968).

An additional complication in electroproduction is the necessity of introducing form factors. Both the proton from factors and the neutron magnetic form factor are well defined, but the neutron charge form factor is uncertain, although certainly small, and the pion form factor is unknown. The latter two affect only charged pion photoproduction to first order [they come into neutral pion photoproduction only via rescattering (that is via unitarity and crossing) and consequently reasonably direct predictions can be made about π^0 electroproduction.

Both Zagury (1966) and Adler (1968) use parametrizations of the neutron and proton Sachs form factors G_{E_p}, G_{M_p}, G_{E_n}, G_{M_n}, which are related to the Pauli form factors $F_{1,2}^v$ and $F_{1,2}^s$ (the ones appearing explicitly in our formalism) by

$$G_{M_v} = G_{M_p} - G_{M_n}, \qquad G_{M_s} = G_{M_p} + G_{M_n},$$
$$G_{E_v} = G_{E_p} - G_{E_n}, \qquad G_{E_s} = G_{E_p} + G_{E_n}; \tag{II-54}$$

$$F_1^{v,s} = [G_{E_{v,s}} + K^2 G_{M_{v,s}}/4M^2]/[1 + K^2/4M^2],$$
$$2MF_2^{v,s} = [G_{M_{v,s}} - G_{E_{v,s}}]/[1 + K^2/4M^2] \tag{II-55}$$

Following Goitein *et al.* (1967), Adler (1968) used the so-called " scaling law " fit to the Sachs form factors

$$G_{M_p}(K^2)/\mu_p = G_{M_n}(K^2)/\mu_n = G_{E_p}(K^2) = [1 + K^2/0.71 \quad (\text{GeV}/c)^2]^{-2}, \tag{II-56}$$

and assumed that

$$G_{E_n}(K^2) = 0, \qquad F_\pi(K^2) = F_1^v(K^2). \tag{II-57}$$

Zagury (1966) used the same fit to the Sachs form factors and the same assumptions about G_{E_n} and F_π for high momentum transfers [greater than $0.88 \ (\text{GeV}/c)^2$], but used a vector-meson exchange pole fit due to de Vreis *et al.* (1964) at small momentum transfers, again assuming equality between F_π and F_1^v. The sensitivity of the predictions to the neutron charge form factor and the pion form factor has been investigated by Zagury (1966, 1968). As would be expected, the predictions for π^0 electroproduction show little variation, but those for π^+ electroproduction are rather sensitive, not only in angular distributions but also in total cross sections.

Adler (1968) has argued that the calculation of Zagury (1966) is in principle in error, since in electroproduction it is necessary to make an additional subtraction to remove the kinematic singularity introduced by imposing gauge invariance. This singularity occurs at $t = \mu^2$ (at the pion pole) in the amplitude $A_5^{(-)}$, and if not removed, the residue of the pion pole will be incorrect. To remove this spurious singularity, it is necessary to modify the dispersion relation for $A_5^{(-)}$ to

$$\text{Re } A_5^{(-)}(s, t, K^2)$$
$$= A_5^{(-)\text{Born}}(s, t, K^2) + \frac{P}{\pi} \int_{(M+\mu)}^{\infty} ds' \ \text{Im } A_5^{(-)}(s', t, K^2)$$
$$\times [(s' - s)^{-1} + (s' - u)^{-1}] - (2/\pi) \int_{(M+\mu)}^{\infty} ds'/(s' - M^2 + \tfrac{1}{2}K^2)$$
$$\times \lim_{t \to \mu^2} [(t - \mu^2) \ \text{Im } A_5^{(-)}(s', t, K^2)]/(t - \mu^2). \tag{II-58}$$

In practice, however, Zagury (1968) has evaluated this correction term numerically, and found it to have little effect on his results.

The most noticeable difference between the calculations of Adler (1968) and of Zagury (1966) is in the dominant $M_{1+}^{(3)}$ multipole, that of Zagury (1966) being slightly the smaller at $K^2 = 0$, but falling off less rapidly with increasing K^2, and being appreciably the larger at high momentum transfers. This happens to be in better agreement with experiment, but since the answer depends to some extent on the cutoff chosen and also on such unknowns as the pion form factor, it is not necessarily indicative that the calculation is the more meaningful. The principal features of the calculations are the continued dominance of the magnetic dipole amplitude $M_{1+}^{(3)}$, both the electric quadrupole amplitude $E_{1+}^{(3)}$ and the corresponding scalar amplitude $S_{1+}^{(3)}$ remaining small even out to large momentum transfers, and the decreasing importance of transitions to s waves with increasing momentum transfer.

In the integral equations formed by the conventional dispersion relations, the inhomogeneous term is given by little more than the Born approximations, the corrections to it (from u-channel continuum contributions) being small in most cases, and essentially zero for the important $M_{1+}^{(3)}$. This result is based on the assumption that the u-channel contributions are dominated by the $M_{1+}^{(3)}$ multipole, this assumption itself being based on the hope that the dispersion relations will be dominated by the nearby singularities, distant singularities having little effect.

Gutbrod and Simon (1967) have argued against this, claiming that the correct inhomogeneous term should also contain at least the lowest order box diagrams. If it is assumed that one nucleon exchange is the dominant force in the πN system in the P_{33} partial wave, then the box diagrams of Fig. 21 should give an important contribution to the inhomogeneous part of the integral equation for $M_{1+}^{(3)}$.

FIG. 21. Lowest order box diagrams in pion photoproduction and electroproduction.

This model is not in accord with the concept of dominance by nearby singularities. The basic motivation for it is that in ordinary potential theory, Blankenbecler *et al.* (1960) have shown that with a Yukawa potential the left-hand cut in a partial-wave dispersion relation is calculable from the iteration of the Schrödinger equation, and Luming (1964) has demonstrated that, provided the potential is not too strong, the second iteration (the simple box diagram) gives an acceptable result.

Gutbrod and Simon (1967) find in numerical computation that the contribution of the box diagrams of Fig. 21 can have a significant effect on the results, particularly at large momentum transfers, since the K^2 dependence of the box diagrams is appreciably different from that of the Born approximation. The shape obtained for the $M_{1+}^{(3)}$ multipole is reasonable, although not quite as acceptable as in the usual dispersion relation calculation, the amplitude above resonance being distorted by the increasing importance of the box diagram contribution. On the other hand, the momentum transfer dependence is considerably improved. The $M_{1+}^{(3)}$ form factor is rather sensitive to the pion form factor (pion exchange contributes about 15% of the magnetic dipole amplitude at resonance) and with the choice of $F_\pi(K^2) = G_E(K^2)$, Gutbrod and Simon (1967) achieve a good fit to the momentum transfer dependence up to 2.5 $(\text{GeV}/c)^2$.

If one is calculating iterations of the strong interaction potential it is natural to investigate the convergence of at least the ladder series. This leads immediately to the consideration of Bethe-Salpeter theory, which has been applied to the calculation of $M_{1+}^{(3)}$ by Gutbrod (1969a, b).

One difficulty which arises in this approach is that the iteration of the one-nucleon exchange potential with the correct coupling constant does not lead to a resonance of the correct mass, and to get this right it is necessary to reduce the strength of the potential between the pion and the nucleon at short distances. This is done conventionally by modifying the propagator of the exchange nucleon with

$$\frac{i\gamma \cdot P + M}{P^2 + M^2} \to (i\gamma \cdot P + M)[(P^2 + M^2)^{-1} - (P^2 + \Lambda^2)^{-1}]. \quad \text{(II-59)}$$

The cutoff required is of the order of $\Lambda \simeq 2.0$ GeV (the precise value varies between 1.8 and 2.2 GeV, depending on the method of solution adopted).

A similar modification is to be expected in the pion exchange term, the cutoff being another free parameter. It can be specified (approximately) by choosing it to be of the order of the electromagnetic radius of the nucleon

$$(t - \mu^2)^{-1} \to (t - \mu^2)^{-1} - [t - 0.3 \quad (\text{GeV})^2]^{-1}. \quad \text{(II-60)}$$

With this choice, the resulting $M_{1+}^{(3)}$ amplitude is too large by about 10% at resonance (in the case of photoproduction). The momentum transfer dependence is again an improvement on the conventional dispersion relation approach, but this time the ρ-dominance form factor $F_\pi(K^2) = (1 + K^2/M_\rho^2)^{-1}$ is slightly preferred to the dipole fit.

The predictions of Adler (1968) for total cross sections [that is, the "single-arm" cross section $(d^2\sigma/d\Omega_e \, d\varepsilon_2)$] are compared with some of the data of Lynch *et al.* (1967) at squared momentum transfers of 0.1 ,0.2, 0.3, 0.4, 0.5, and 0.6 $(\text{GeV}/c)^2$ in Fig. 22, and, along with the prediction of Zagury (1966) compared with the data of Cone *et al.* (1967) in Fig. 23. The latter data were obtained not at fixed momentum transfer but at fixed electron scattering angle. At resonance, the values of squared momentum transfer are 1.0, 1.55, and

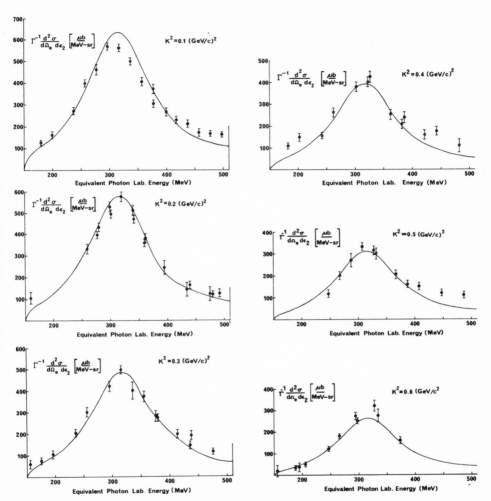

FIG. 22. Total cross sections for inelastic electron scattering in the first resonance region. The solid line is the prediction of the dispersion relation calculation of Adler (1968).

3.61 $(\text{GeV}/c)^2$, respectively. It is clear that at these higher momentum transfers, the theory is beginning to fare rather badly, particularly above resonance, where the experimental cross section is much larger relative to the cross section at the resonance peak than it is in photoproduction or in small momentum transfer electroproduction, and is a feature which is not reproduced by the calculations. The s-wave threshold is strongly marked at the lower momentum transfers, and its effect can be seen steadily to disappear as the momentum transfer increases.

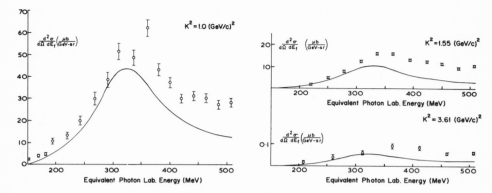

FIG. 23. Total cross sections for inelastic electron scattering in the first resonance region. The solid line is the prediction of the dispersion relation calculation of Zagury (1968).

Even at small momentum transfers, the agreement between theory and experiment is not as impressive as it appears at first sight. This becomes apparent if the transverse cross section σ_T and the longitudinal cross section σ_L are examined separately. A typical result is shown in Fig. 24, where the predictions of Adler (1968) for σ_T and σ_L at 340-MeV equivalent photon laboratory energy are compared with those extracted from the data of Lynch *et al.* (1967). The discrepancy shows up most strongly in σ_L, the experimental longitudinal cross section increasing dramatically with increasing K^2 up to $K^2 = 0.3$ $(\text{GeV}/c)^2$, then becoming consistent with zero for $K^2 \geqslant 0.4$ $(\text{GeV}/c)^2$, the theoretical cross section, on the other hand, decreasing monotonically with increasing K^2.

Electron-proton coincidence measurements of π^0 electroproduction in the first resonance region have been made by Mistretta *et al.* (1968a), by Albrecht *et al.* (1971a, b), and by Hellings *et al.* (1971) covering altogether a range of

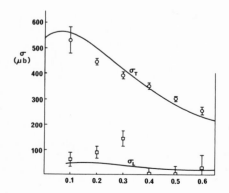

FIG. 24. The total transverse and longitudinal cross sections for inelastic electron scattering at the peak of the first resonance. The solid lines are the predictions of the dispersion relation calculation of Adler (1968).

momentum transfer from -0.05 to -1.0 $(GeV/c)^2$. The data of Albrecht
et al. (1971a, b) and of Hellings *et al.* (1971) cover the complete pion azimuthal
and polar angular ranges, but those of Mistretta *et al.* (1968a) while covering
a wide range of pion azimuthal angles are restricted to the range of pion
polar angles $100° < \theta$ (center of mass) $< 180°$.

The most convenient way in which to present the results is to plot the
coefficients of the expansion

$$(\Gamma')^{-1}(d^3\sigma/d\Omega_e \, d\varepsilon_2 \, d\Omega_\pi)$$
$$\equiv (d\sigma^V/d\Omega_\pi)$$
$$= \bar{A}_0 + \bar{A}_1 \cos \theta + \bar{A}_2 \cos^2 \theta + C_0 \sin^2 \theta \cos^2 \phi$$
$$+ [\varepsilon(\varepsilon + 1)]^{1/2}(D_0 + D_1 \cos \theta) \sin \theta \cos \phi. \qquad \text{(II-61)}$$

Figure 25a gives the coefficients at $K^2 = -0.6$ $(GeV/c)^2$ from Albrecht
et al. (1971a, b) and Hellings *et al.* (1971) as a function of W. Figure 25b
gives the coefficients at $W = 1236$ MeV from Albrecht *et al.* (1971a, b),
Hellings *et al.* (1971), and Mistretta *et al.* (1968a) as a function of K^2, along
with the photoproduction points from Fischer *et al.* (1970).

Pure magnetic dipole excitation requires that

$$\bar{A}_0 : \bar{A}_2 = -5 : 3, \qquad \bar{A}_2 = C_0, \qquad \bar{A}_1 = D_1 = D_0 = 0. \qquad \text{(II-62)}$$

The ratio $-5 : 3 : 3$ for $\bar{A}_0$, $\bar{A}_2$, C_0 is approximately fulfilled, and although
$\bar{A}_1$, D_1, and D_0 are certainly not zero, they are comparatively small, clearly
indicating the continued dominance of the magnetic dipole excitation of the
resonance. Like $\bar{A}_0$, $\bar{A}_2$, and C_0, D_1 appears to have a resonance shape,
which would be expected from $S_{1+}^{(3)}$ and $M_{1+}^{(3)}$ interference. The coefficient D_0
appears to change sign at the resonance, which can be explained by interference between the real scalar s-wave S_{0+} and the resonant magnetic dipole
term. No particular structure is evident in $\bar{A}_1$.

Without comparable data on π^+ electroproduction, it is impossible to
attempt a complete analysis on the lines of the analyses of pion photoproduction in the first resonance region. However, due to the dominance of
the magnetic dipole excitation of the resonance reasonable estimates of some
multipoles are possible with the minimum of additional assumptions, mainly
a restriction to s and p waves. The differential cross sections can be largely
explained by $|M_{1+}^{(3)}|$ and the interference terms by only those which contain
$M_{1+}^{(3)}$. These are $\text{Re}(E_{1+} M_{1+}^*)$, $\text{Re}(S_{1+} M_{1+}^*)$, $\text{Re}(S_{0+} M_{1+}^*)$, $\text{Re}(E_{0+} M_{1+}^*)$,
and $\text{Re}(M_{1-} M_{1+}^*)$. Since the angular distributions are described by six
coefficients, it thus should be possible to solve for $|M_{1+}|^2$ and these five
interference terms.

We should first recall the behavior of the corresponding interference terms
in photoproduction. There the interference term $\text{Re}(E_{0+} M_{1+}^*)$ changes sign
near resonance, as does the interference term $\text{Re}(M_{1-} M_{1+}^*)$, M_{1-} being

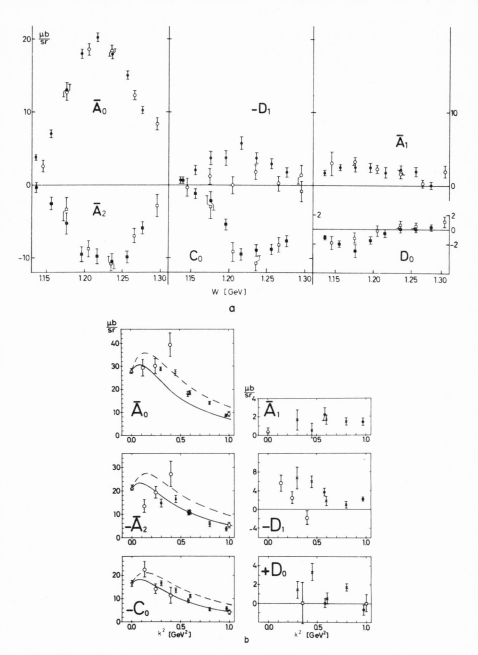

FIG. 25. (a) Angular coefficients of the differential cross section expansion (II-61) for the reaction $e^-p \to e^-p\pi^0$ at $K^2 = -0.6$ (GeV/c)^2 (data (●, ■); from Albrecht *et al.*, 1971a, b; (○, □) Hellings *et al.*, 1971). (b) Angular coefficients of the differential cross section expansion (II-61) for the reaction $e^-p \to e^-p\pi^0$ at $W = 1236$ MeV (data (●), from Albrecht *et al.*, 1971a, b; (×) Hellings *et al.*, 1971; (○) Mistretta *et al.*, 1968a; photoproduction points (△) from Fischer *et al.*, 1970).

largely real. The term $\text{Re}(E_{1+} M_{1+}^{*})$ is negative and of the order of a few percent of $|M_{1+}|^{2}$.

In electroproduction, at least for momentum transfers greater than 0.3 $(\text{GeV}/c)^{2}$, the interference term $\text{Re}(E_{0+} M_{1+}^{*})$ stays positive, but the other two photoproduction interference terms retain their general features. Considerable scalar-transverse interference $\text{Re}(S_{1+} M_{1+}^{*})$ is observed that persists across the resonance. The magnitude of this interference term indicates that $|S_{1+}^{(3)}|$ is of the order of 5–10% of $|M_{1+}^{(3)}|$. The $\text{Re}(S_{0+} M_{1+}^{*})$ interference is consistent with being zero above resonance, but contributes significantly below. All of these features are described reasonably well by the calculation of von Gehlen (1970). The agreement with the Bethe–Salpeter model of Gutbrod (1969a, b) is also quite good.

The behavior of the dominant $M_{1+}^{(3)}$ amplitude is summed up most conveniently in terms of the $\Delta(1236)$ form factor. Following Dalitz and Sutherland (1966) the magnetic moment can be defined (nonuniquely) as

$$\mu^{*} = (8\pi W_{R} q_{R} \Gamma/3M)^{1/2} |M_{1+}^{(3)}(W_{R})| k_{R}, \qquad \text{(II-63)}$$

where W_{R}, q_{R}, and k_{R} are the total energy, pion momentum, and photon momentum in the center-of-mass (cm) system at resonance and Γ is the width of the resonance. This quantity has been calculated by Bartel et al. (1968) and a comparison is made with the various theoretical calculations in Fig. 26.

Coincidence measurements in π^{+} electroproduction have also been made by Mistretta et al. (1968b) in the region of the first resonance at five momentum transfers between 0.05 and 0.6 $(\text{GeV}/c)^{2}$, again for a wide range of pion azimuthal angles but for a restricted range of pion polar angles, $0° < \theta_{\text{cm}} < 50°$. Mistretta et al. (1968b) have used the theoretical calculations of Zagury (1966) and Adler (1968) with the pion form factor as a free parameter to fit to their data. The results of this extraction of the pion form factor are shown in Fig. 27, along with those of Akerlof et al. (1967), who obtained cross sections for forward π^{+} electroproduction ($\theta_{\text{cm}} = 0°$) and used the theory of Zagury (1966) to extract the pion form factor. In each of these applications of the theory the neutron charge form factor has been set to zero, and because of the strong correlation between this and the pion form factor the true errors on the latter will be even larger than shown in Fig. 27.

The pion form factor is clearly similar to the proton form factor. Fitting the data of Fig. 27 with the single-pole expression

$$F_{\pi}(K^{2}) = (1 + K^{2}/M^{2})^{-1}$$

yields a value of 0.56 ± 0.08 GeV for M.

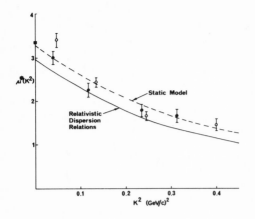

FIG. 26. The magnetic form factor of the $\Delta(1238)$ resonance compared with dispersion relation calculations.

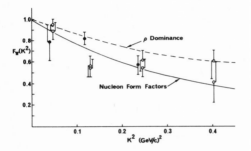

FIG. 27. The pion form factor extracted from data on the reaction $e^- + p \rightarrow e^- + n + \pi^+$, using dispersion relation calculations.

D. PHENOMENOLOGY OF PION PHOTOPRODUCTION IN THE RESONANCE REGION

As we have already seen, theoretical calculations of pion photoproduction begin to fail for a variety of reasons at photon laboratory energies above 500 MeV. Beyond that point, it is necessary to resort to phenomenology to obtain the details of the process, although the general features can be extracted from a small amount of data.

The total cross sections for π^+, π^0 and π^- photoproduction are shown in Figs. 28–30. Three resonant peaks stand out clearly. The first, of course, is $P_{33}(1236)$ and the second and the third are conventionally ascribed to $D_{13}(1512)$ and $F_{15}(1690)$. Although several other resonances exist in this energy region, detailed phenomenological analysis shows that, with the definite exception of $S_{11}(1550)$, their couplings appear to be weak and generally confirms the dominance of $P_{33}(1236)$, $D_{13}(1512)$, and $F_{15}(1690)$. Two possible exceptions are $P_{11}(1450)$ whose presence does not affect any of the peaks, and $D_{15}(1690)$ whose presence would upset the interpretation of the third resonance region. On the assumption of $P_{33}(1238)$, $D_{13}(1512)$, and $F_{15}(1690)$ dominance, the general features observed are as follows.

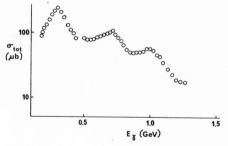

FIG. 28. Total cross section for the reaction $\gamma + p \rightarrow n + \pi^+$ below 1.5-GeV photon laboratory energy.

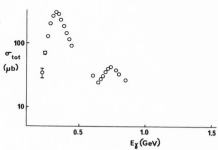

FIG. 29. Total cross section for the reaction $\gamma + p \rightarrow p + \pi^0$ below 1.5-GeV photon laboratory energy.

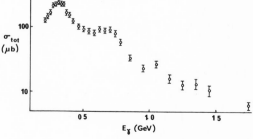

FIG. 30. Total cross section for the reaction $\gamma + n \rightarrow p + \pi^-$ below 1.5-GeV photon laboratory energy.

1. Production of the $P_{33}(1236)$ is mainly by the magnetic dipole amplitude, a point which has already been discussed in detail.

2. The $D_{13}(1512)$ is produced mainly in the helicity-$\frac{3}{2}$ amplitude. This can be inferred from the excitation curves at 0 and 180°, where the helicity-$\frac{1}{2}$ amplitude cannot contribute. The π^+ 0° and 180° cross section is shown in Fig. 31, the π^0 180° cross section in Fig. 32, and the π^- 0° cross section in Fig. 33. The lack of structure in the π^+ and π^- cross sections is very evident, and while there is some structure in the π^0 cross section, its magnitude is small and in no way reflects that observed in the total cross section. Indeed, Walker (1969) has shown that what structure there is, in particular the deep minimum at 600 MeV, can be explained in terms of the tail of the $P_{33}(1236)$ interfering with a slowly varying background which is mainly s wave, the deep

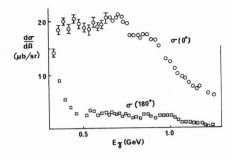

FIG. 31. The excitation curves at 0 and 180° for the reaction $\gamma + p \rightarrow n + \pi^+$.

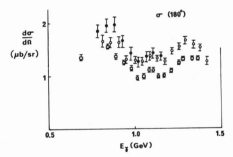

FIG. 32. The excitation curve at 180° for the reaction $\gamma + p \rightarrow p + \pi^0$.

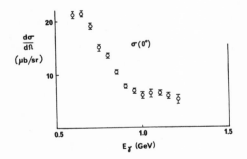

FIG. 33. The excitation curve at 0° for the reaction $\gamma + n \rightarrow p + \pi^-$.

minimum resulting from the imaginary part passing through zero at an energy where the real part is small.

3. The $D_{13}(1512)$ isovector amplitude is very much larger than the isoscalar amplitude, since the cross sections for the production of the resonance on protons and neutrons are comparable in magnitude.

4. The $F_{15}(1690)$ resonance is also produced mainly in the helicity-$\frac{3}{2}$ amplitude, the 0 and 180° cross sections on protons again showing that the helicity-$\frac{1}{2}$ amplitude is small.

5. The $F_{15}(1690)$ resonance is scarcely excited at all on neutrons, implying that the isovector and isoscalar amplitudes are approximately equal.

Detailed analyses of the photoproduction data have been undertaken by Salin (1963) (π^+, π^0 from threshold to 800-MeV photon laboratory energy), Chau *et al.* (1967) (π^+, π^0 535–850-MeV photon laboratory energy), Pfeil (1968) (π^+ from threshold to 450-MeV photon laboratory energy, π^0 from threshold to 1200-MeV photon laboratory energy), Kim (1968) (π^+ from threshold to 1200-MeV photon laboratory energy), Walker (1969a) (π^+, π^0, π^- from threshold to 1200-MeV photon laboratory energy), Moorhouse and Rankin (1970) (π^+, π^0 535–850-MeV photon laboratory energy), Rankin (1970) (π^+, π^0 900–1150-MeV photon laboratory energy), Rankin (1972) (π^+, π^0 500–1250-MeV photon laboratory energy), and Proia and Sebastiani (1970, 1971) (π^- 540–830-MeV photon laboratory energy). Most of these analyses make some pretence of model independence, although they are all much less sophisticated than corresponding pion-nucleon phase shift analyses, and to describe resonant energy dependence all use resonance models which are inserted into the analyses *a priori*. In the discussion of the results of these analyses, those features which can be considered to be reasonably model independent and well determined will be stressed, in contrast to those which are model dependent and hence poorly determined. We shall also see how it is possible to use the more reliable results of the multipole analyses together with the added constraint of fixed-t dispersion relations to extract additional (model dependent) information from the data under the assumption of resonance dominance of the imaginary parts of the photoproduction amplitudes.

Salin (1963) included the full Born term plus ρ and ω exchange as t-channel poles and $P_{33}(1236)$ and $D_{13}(1512)$ [and in an attempt to go beyond 800 MeV, $F_{15}(1690)$] in the direct channel. The parametrization of the P_{33} and D_{13} resonance terms was based on the spin-$\frac{3}{2}$ propagator

$$P_{\mu\nu} = \{(E_R - i\gamma_\lambda P_\lambda)/(E_R{}^2 + P^2)\}$$
$$\times \{\delta_{\mu\nu} - \tfrac{1}{3}\gamma_\mu\gamma_\nu + (i/3E_R)(\gamma_\mu P_\nu - \gamma_\mu P_\nu) + (2/3E_R)^2 P_\mu P_\nu\} \quad \text{(II-64)}$$

which yields

$$E_{1+} = (K_1 + K_2)/(s_R - s - i\gamma), \qquad M_{1+} = (K_2 - K_1)/(s_R - s - i\gamma),$$
$$K_1 = -[wkq/3(E_1 + M)](C_1 e/\mu), \qquad K_2 = (wkq/3)(C_2 e/\mu^2), \quad \text{(II-65)}$$

the terms C_1 and C_2 being the arbitrary parameters, and

$$E_{2-} = (2L_1 - L_2)/(s_R - s - i\gamma), \qquad M_{2-} = L_2/(s_R - s - i\gamma),$$
$$L_1 = -[wq^2/3(E_2 + M)](C_1'e/\mu), \qquad L_2 = [wk^2q/3(E_1 + M)(E_2 + M)]$$
$$\times (C_2'e/\mu^2), \quad \text{(II-66)}$$

C_1' and C_2' being the arbitrary parameters.

Despite the theoretical reservations about this particular model (omission of u-channel isobar contributions, and the existence of double counting), the model was extremely economical and gave a good fit to the data available at that time. The important role of the $D_{13}(1512)$ resonance was clearly exhibited, as was the dominance of the E_{2-} amplitude over the M_{2-} (the vanishing of the helicity-$\frac{1}{2}$ amplitude discussed above requires that $E_{2-} = 3M_{2-}$).

As we have already noted, the isobar model of Pfeil (1968) gets over some of the objections to that of Salin (1963) by encompassing it within the framework of dispersion relations. To simplify calculation, a pole approximation was used for the resonance contributions to the invariant amplitudes $A_i^{(\pm, 0)}(s, t, u)$, namely

$$A_i^{(\pm, 0)}(s, t, u) = {}^s\tilde{A}_i^{(\pm, 0)}(s, t)/(s - M_R^2 - iM_R\Gamma)$$
$$+ {}^u\tilde{A}_i^{(\pm, 0)}(u, t)/(u - M_R^2 - iM_R\Gamma), \qquad \text{(II-67)}$$

with

$$ {}^s\tilde{A}_i^{(\pm, 0)}(s, t) \simeq {}^sA_i^{(\pm, 0)}(M_R^2, t) \equiv a_i^{(\pm, 0)}(t) \qquad \text{(II-68)}$$

and

$$ {}^u\tilde{A}_i^{(\pm, 0)}(u, t) \simeq {}^u\tilde{A}_i^{(\pm, 0)}(M_R^2, t) \equiv \xi_i a_i^{(\pm, 0)}(t), \qquad \text{(II-69)}$$

where ξ_i is the appropriate crossing element [see (I-42)]. Thus

$$A_i^{(\pm, 0)}(s, t, u)$$
$$\simeq a_i^{(\pm, 0)}(t)\{(s - M_R^2 - iM_R\Gamma)^{-1} + \xi_i/(u - M_R^2 - iM_m^m\Gamma)\}. \qquad \text{(II-70)}$$

This is equivalent to making the narrow resonance approximation in the dispersion relations

$$\text{Im } A_i^{(\pm, 0)}(s, t) = \pi\delta(s - M_R^2)a_i^{(\pm, 0)}(t), \qquad \text{(II-71)}$$

and then reinserting the width into the denominator.

The elastic (pion-nucleon) partial width was taken to have the form

$$\Gamma_{El}(s) = \Gamma_0(q/q_R)^{2l+1}[(q_R^2 + x^2)/(q^2 + x^2)], \qquad \text{(II-72)}$$

where Γ_0 and x^2 are constants, and the total width to have the form

$$\Gamma_{tot} = \Gamma_0\left[\frac{1 - \exp\{-\alpha(q/q_0)^{2l+1}\}}{1 - \exp(-\alpha)}\right], \qquad \text{(II-73)}$$

with Γ_0 and α as constants.

The resonances taken into account were $P_{33}(1236)$, $P_{11}(1450)$, $D_{13}(1512)$, $S_{31}(1630)$, $D_{15}(1680)$, $F_{15}(1690)$, and $S_{11}(1710)$ in addition to the full Born terms plus ρ and ω exchange poles. The phase parameters α and the overall coupling of the resonances were left as completely free parameters, while the widths and the masses were constrained to remain close to the values obtained from the analyses of pion-nucleon scattering. A surprising omission is the $S_{11}(1550)$ resonance which must certainly be present, since it is clearly observed in η photoproduction, and its two dominant decay modes are πN and ηN.

An adequate fit to the π^0 data was obtained up to 1200-MeV photon laboratory energy. Some degree of double counting is still taking place, although to a less serious extent than that of Salin (1963). That the model is not completely accurate is indicated by a comparison of the results for the resonance couplings obtained by Pfeil (1968) with those obtained by Kim (1968), who applied the identical model to an all-over fit of π^+ photoproduction. Appreciable differences emerge.

The above approaches used a specific model for the background to the resonances. By contrast, the approach of Chau et al. (1967) was purely phenomenological, the background being arbitrary and containing free parameters to be determined in the fit to the data. Their parametrization is based on the K-matrix formalism. Following Dalitz (1963), and assuming that one eigenphase of the multichannel system resonates, then the expression they obtained for the photoproduction amplitude corresponding to a multipole or a helicity state was

$$\Gamma_{if} = C_{ri}C_{rf}\left[\frac{\frac{1}{2}\Gamma\exp(2i\delta_\infty)}{W_R - W - i(\frac{1}{2}\Gamma)} + \exp i\delta_\infty \sin\delta_\infty\right]$$
$$+ \sum_{\alpha \neq r} C_{\alpha i}C_{\alpha f}\exp(i\delta_\alpha)\sin\delta_\alpha. \qquad (\text{II-74})$$

Here the δ_α are the nonresonant (background) phases, δ_∞ is the phase of the coupling to all channels, and the partial widths are given by

$$\Gamma_i = \Gamma C_{ri}^2, \qquad \Gamma_f = \Gamma C_{rf}^2. \qquad (\text{II-75})$$

Over a restricted energy range in the vicinity of the resonance, it is reasonable to suppose that the background from the nonresonant eigenstates is constant, so that (II-74) can be rewritten as

$$\Gamma_{if} = \frac{\frac{1}{2}(\Gamma_i\Gamma_f)^{1/2}\exp(2i\delta_\infty)}{W_R - W - i(\frac{1}{2}\Gamma)} + \frac{(\Gamma_i\Gamma_f)^{1/2}}{\Gamma}\exp(i\delta_\infty)\sin\delta_\infty + A_{if} + iB_{if}. \qquad (\text{II-76})$$

Within this model, Γ_π, Γ, W_R and δ_∞ can be found from elastic scattering, since

$$\Gamma_{\pi\pi} = \frac{\frac{1}{2}\Gamma_\pi\exp(2i\delta_\infty)}{W_R - W - i(\frac{1}{2}\Gamma)} + \frac{\Gamma_\pi}{\Gamma}\exp(i\delta_\infty)\sin\delta_\infty + A_{\pi\pi} + iB_{\pi\pi}, \qquad (\text{II-77})$$

which can be fitted to the pion-nucleon scattering data. This leaves three parameters $C_{r\gamma}$, $A_{\pi\gamma}$, $B_{\pi\gamma}$ for each resonant amplitude in pion photoproduction.

The resonances considered by Chau et al. (1967), were $P_{11}(1450)$, $D_{13}(1512)$, and $S_{11}(1550)$. The multipoles leading to nonresonant states were parametrized in one of two ways. For final states which are appreciably inelastic (S_{31}, P_{13}, D_{33}) the simple parametrization

$$(qk)^{1/2}M = A + iB \qquad (\text{II-78})$$

was used, with A and B constants. For those which were elastic (P_{31}, P_{33}), a parametrization based on Watson's theorem was used:

$$(qk)^{1/2}M = Ce^{i\delta} \sin \delta - M(\pi), \tag{II-79}$$

where C is a constant, and $M(\pi)$ is the projection of the pion pole term multiplied by $(qk)^{1/2}$ and evaluated at 700 MeV, the latter approximation being made since the projection of a t-channel pole on to a given state varies slowly with energy. There was no need to worry about subtraction in other partial waves, since they all contained an arbitrary background.

The pion pole itself was treated approximately, its contribution to the amplitude $A_2^{(-)}$ being taken as

$$A_2^{(-)} = -2egG(t)/[(t - \mu^2)(s - M^2)\bar{\jmath}, \tag{II-80}$$

where $G(t)$ is a form factor

$$G(t) = 1 - \lambda(t - \mu^2)/(t - x^2) \tag{II-81}$$

with λ and x arbitrary parameters in the fit.

With 33 parameters available, a good fit was obtained and the value for coupling of the $S_{11}(1550)$ resonance was reasonably consistent with that obtained from η photoproduction. However, the results do not tie in very well with what would be expected from dispersion relation calculations [unlike those of Pfeil (1968) and Kim (1968), which do so by construction]. In particular, the coupling of the $P_{11}(1450)$ resonance was very much larger than would be expected on the basis of these calculations. Further, Moorhouse and Rankin (1970) [whose work supersedes that of Chau *et al.* (1967)] have pointed out that the fixed-phase resonant coupling is not necessarily correct, and the possibility that the photoproduction phase is different from the scattering phase should be considered, that is

$$\Gamma_{if} = \tfrac{1}{2}(\Gamma_i\Gamma_f)^{1/2}e^{i\phi}/[W_R - W - i(\tfrac{1}{2}\Gamma)] + A_{if} + iB_{if}, \tag{II-82}$$

where ϕ is now an arbitrary phase, and four parameters are now required to specify each resonance.

Moorhouse and Rankin (1970) have fitted the data [including new data not available to Chau *et al.* (1967)] in the range 535–850-MeV photon laboratory energy, allowing for the possibility of an arbitrary resonant coupling phase and incorporating explicitly the η channel in the $S_{11}(1550)$ resonance, but otherwise using the same parametrization as Chau *et al.* (1967). Of particular importance, they have investigated thoroughly the uncertainties in resonance couplings. They found that, with the help of the η-photoproduction data, the $S_{11}(1550)$ coupling was determined fairly well, although this is not without the qualification of making a preferred choice of the η-production parameters.

Both the $D_{13}(1512)$ couplings, on the other hand, were considered to be well determined, although a considerable range in the phase was permissible. There is some correlation with $S_{11}(1550)$, but virtually none with the $P_{11}(1450)$ coupling. The latter coupling was poorly determined by the data. The fixed phase model yielded four solutions for the coupling, one of which was very small (consistent with dispersion relation predictions) and one of which, not unnaturally, lay close to that of Chau *et al.* (1967). A third solution even had a different sign for the coupling. The results obtained even within this restricted model depended very much on the width chosen for the $P_{11}(1450)$. With a narrow width, for example, only a solution with small coupling could be found. When this resonance was treated in the variable-phase model, a continuum of solutions was obtained, linking the Chau *et al.* (1967) large coupling with the dispersion-type very small coupling, that is, the modulus of this resonance was undertermined by the data within significant limits.

Rankin (1970) extended this model to higher energies to extract the $D_{15}(1680)$ and $F_{15}(1690)$ couplings. Although in the various fits attempted some or all of the other resonances in this region were included, little could be said about their couplings and their principal effect was to cause variations of the $D_{15}(1680)$ and $F_{15}(1690)$ couplings within the two basic solution types which were found. In the first of these two types, the helicity-$\frac{3}{2}$ coupling of $F_{15}(1690)$ was almost zero, the helicity-$\frac{1}{2}$ $F_{15}(1690)$ amplitude and the two helicity amplitudes of $D_{15}(1680)$ dominating the process. In this case, the lack of structure already noted in the forward and backward directions comes from interference between the two resonances. In the second type, exactly the opposite situation occurred. The helicity-$\frac{3}{2}$ coupling of $F_{15}(1690)$ was dominant, the helicity-$\frac{1}{2}$ coupling and both the $D_{15}(1680)$ couplings being small. This was in accord with our previous prejudices, and can be preferred for two sounder reasons. The χ^2 obtained was somewhat better for this solution and the D_{15} dominant solution did not appear to extrapolate as well to lower energies, the interference with $D_{13}(1512)$ not being correct, although this could possibly be corrected by the inclusion of small amounts of other states. Even within the one class, there was a considerable range of coupling of the F_{15} possible, depending on the background used.

In the analysis of Walker (1969a), the background was taken to be predominantly that coming from the electric Born term. The use of the electric Born is certainly preferable to that of the total Born term, since, as we have already noted, the magnetic part of the Born term is appreciably cancelled by the crossed $P_{33}(1236)$ contribution (in a dispersion sense). Small *ad hoc* additions to the background were allowed where necessary. The analysis was of all π^+, π^0, π^- data from threshold to 1200-MeV photon laboratory energy, and the resonances taken into account were $P_{33}(1236)$, $P_{11}(1450)$, $D_{13}(1512)$,

$S_{11}(1550)$, $D_{15}(1680)$, and $F_{15}(1690)$. The parametrization was in terms of the helicity-$\frac{1}{2}$ and -$\frac{3}{2}$ elements $A_{l\pm}$, $B_{l\pm}$, in terms of which the multipoles $E_{l\pm}$, $M_{l\pm}$ are given by

$$
\begin{aligned}
(l+1)E_{l+} &= A_{l+} + \tfrac{1}{2}lB_{l+}, \\
(l+1)M_{l+} &= A_{l+} - \tfrac{1}{2}(l+2)B_{l+}, \\
(l+1)E_{(l+1)-} &= A_{(l+1)-} - \tfrac{1}{2}(l+1)B_{(l+1)-}, \\
(l+1)M_{(l+1)-} &= A_{(l+1)-} + \tfrac{1}{2}lB_{(l+1)-}.
\end{aligned}
\tag{II-83}
$$

The Breit–Wigner form used for the resonant amplitudes was

$$
A(W) = A(M_R)(k_R q_R/kq)^{1/2} W_R \Gamma^{1/2} \Gamma_\gamma^{1/2}/(s_R - s - iW_R), \tag{II-84}
$$

where

$$
\begin{aligned}
\Gamma &= \Gamma_0(q/q_R)^{2l+1}(q_R^2 + x^2)/(q^2 + x^2), \\
\Gamma_\gamma &= \Gamma_0(k/k_R)^{2j_\nu}(k_R^2 + x^2)/(k^2 + x^2)^{j_\gamma},
\end{aligned}
\tag{II-85}
$$

l being the pion angular momentum and j_γ the photon angular momentum appropriate to the multipole concerned. The constant x was fixed at 160 MeV for the $P_{33}(1236)$ resonance and at 350 MeV for all others.

The results obtained included a weakly coupled $P_{11}(1450)$, dominant helicity-$\frac{3}{2}$ $D_{13}(1512)$ and dominant helicity-$\frac{3}{2}$ $F_{15}(1690)$, the magnitude of the latter two being reasonably consistent with those of Moorhouse and Rankin (1969) and of Rankin (1970), while $D_{15}(1580)$ was weakly coupled. This solution ties in well with the results of the dispersion calculations, which one would have expected by its construction, unless the *ad hoc* background terms had turned out to be exceptionally large.

With the availability of a considerable quantity of new data, Rankin (1972) has reapplied the model of Moorhouse and Rankin (1970) to proton data between 500 and 1250 MeV, and Proia and Sebastiani (1970, 1971) have applied essentially the same model to π^- photoproduction between 540 and 830 MeV.

There is general agreement on the photoexcitation of $D_{13}(1512)$ on protons, the different analyses producing the same quantitative results. The dominance of the helicity-$\frac{3}{2}$ coupling over the helicity-$\frac{1}{2}$ coupling is well established. Both couplings are well determined and the helicity-$\frac{1}{2}$ coupling, although small, is definitely nonzero. The same is true for the excitation of $S_{11}(1540)$ on protons, all the later analyses producing precisely the same results. However, even with more data, the $P_{11}(1470)$ coupling still causes considerable difficulty, different models producing quite different answers for the coupling. If the conventional position and width of this resonance are used in the fits, then Rankin (1972) has shown that its coupling to the $p\gamma$ channel is undetermined within significant limits, a continuum of equally good solutions existing with the value of this coupling ranging from zero to a size which corresponds to a total cross section for $\gamma p \to P_{11}(1470) \to \pi^0 p$ of ~ 5 μb.

A similar, although less clear situation exists for the photoexcitation of these resonances on neutrons. The analyses of Walker (1969a) and of Proia and Sebastiani (1970, 1971) produce similar values for the D_{13}^0 (1512) coupling and show that this resonance is predominantly excited by isovector photons. Like D_{13}^+ (1512), D_{13}^0 (1512) is produced dominantly in a helicity-$\frac{3}{2}$ state, but whether the helicity-$\frac{1}{2}$ coupling is zero, or small but nonzero is still an open question. There are substantial differences in the $S_{11}^0(1512)$ coupling, but it appears that the excitation is primarily isovector. Since the $D_{13}^0(1512)$ is produced dominantly in a helicity-$\frac{3}{2}$ state, it cannot contribute to the 180° cross section and consequently structure which is observed in the data of Fujii et al. (1971) at 750 MeV is most easily interpreted as being due to the $S_{11}(1540)$ coupling. This interpretation requires a large $S_{11}^0(1540)$ coupling, favoring the solution of Proia and Sebastiani (1970, 1971). The biggest difference in the solutions lies in the coupling of $P_{11}^0(1470)$, Walker (1969a) and Proia and Sebastiani (1970, 1971) being at the two extremes of having respectively zero and a value comparable to the upper limit found on protons by Rankin (1972).

The third resonance region is less well understood than the second, partly because of the greater number of resonance that can be present and partly because of deficiencies in the experimental data. When looking at the third resonance region in isolation with no form of continuity imposed, we have seen that it is possible to obtain solutions to the proton data with either F_{15}^+ (1690) or D_{15}^+ (1680) as the dominant resonance. However, the D_{15}^+ dominant solutions are discouraged when continuity is imposed, a conclusion substantiated by resonance dominance in dispersion relations. In the F_{15}^+ dominant solution, the resonance is produced almost exclusively in the helicity-$\frac{3}{2}$ state. The total cross section for the helicity $\frac{1}{2}$ coupling in $\gamma p \to F_{15}^+(1690) \to \pi^0 p$ is at most 0.4 μb, while the total cross section for the helicity-$\frac{3}{2}$ coupling in $\gamma p \to F_{15}^+$ (1690) $\to \pi^0 p$ is at least 8.8 μb and could be appreciably larger.

As we have already noted, photoexcitation of either D_{15} (1680) or F_{15} (1690) on neutrons is weak. No analysis has been performed utilizing all the data now available, and little is known about the photonic couplings of the neutral resonances in this region. There is some indication that the D_{15} (1680) coupling is to be preferred to F_{15} (1690), but this interpretation must be treated with caution. There is no indication of which helicity state, if either, is preferred.

Further resonance information can be extracted from the data by the phenomenological use of fixed-t dispersion relations. Devenish et al. (1972) have calculated the integrals over the imaginary parts of the invariant amplitudes using the known reliable resonance imaginary parts obtained from the multipole analyses. They then used the invariant amplitudes obtained from the dispersion relations in this way to reevaluate the experimental quantities. These are compared with experimental data, and discrepancies between this calculation and experiment are removed by the addition of other resonance imaginary parts to the fixed-t dispersion relations. The additional terms involve resonances which are known to contribute to pion-nucleon scattering but about which the multipole analyses of the photoproduction data have given little information. The assumption that the imaginary parts of the invariant amplitudes can be saturated with resonances is motivated by duality, but can be checked *a posteriori*. In many of the multipole analyses it turns out that the imaginary background is much smaller than both the resonance imaginary part and the real background, except for π^0 photoproduction in the third resonance region. In terms of the model, one would expect this to be reinterpreted as a sum of resonance imaginary parts.

The first step was to go to the fourth resonance region to determine the coupling parameters of the $F_{37}(1940)$ resonance, the presence of which in forward π^+ photoproduction was qualitatively established by Engels *et al.* (1968). Without the inclusion of this resonance, the dispersion relation predictions for the near-forward π^+-photoproduction cross section are too large and too smooth, lacking a characteristic real-part resonance-background interference wiggle that is present in the data. Inclusion of the $F_{37}(1940)$ and the nearby $G_{17}(2190)$ resonances remedies these defects. The $F_{37}(1940)$ couplings are reasonably well determined and are consistent with the vanishing of the electric multipole E_{3+} for this resonance. The excitation of the $G_{17}(2190)$ resonance appears to be dominantly in the helicity $\frac{3}{2}$ amplitude, which was previously suggested by Engels *et al.* (1968).

If one now goes to lower energies, including the fourth resonance region terms in the dispersion relations along with those of the second and third resonance regions, and taking the zero coupling solution for $P_{11}(1470)$, a discrepancy is again found to exist between the dispersion relation predictions and the near-forward π^+ data. The model fails to reproduce three features of the data, namely real-part resonance-background interference wiggles at about 0.56 GeV and about 1.0 GeV and a sharp drop in the cross section at 0.7 GeV. The latter effect is caused by the cusp at the η threshold, and a calculation of the π^+ differential cross section incorporating this in a K-matrix formalism can reproduce it quite satisfactorily. It can be simulated in the resonance model by decreasing the width of the $S_{11}(1540)$ resonance from 150 to 80 MeV and changing its position to 1505 MeV. The only possible candidate for the lower interference wiggle is the $P_{11}(1470)$ resonance and the data can be accommodated by taking the upper allowed limit for this resonance coupling.

The situation in the region of the upper wiggle is complicated by the number of possible resonance candidates available. However, the situation can be simplified if one looks at the forward π^0 data, where there is little evidence of any structure. This leads to the conclusion that there are at least two resonances present of different isotopic spin. There are three possible $I = \frac{3}{2}$ resonance candidates, $S_{31}(1630)$, $D_{33}(1670)$, and $P_{33}(1690)$, and three possible $I = \frac{1}{2}$ resonance candidates, $S_{11}(1700)$, $P_{11}(1750)$, and $D_{13}(1730)$. The latter is a weak resonance candidate in pion-nucleon phase shift analysis, so fixing on $S_{11}(1700)$ and $P_{11}(1750)$ as the $I = \frac{1}{2}$ candidates, Devenish et al. (1972) found that pairing either of them with $S_{31}(1630)$ or $D_{33}(1670)$ provides a sufficient explanation. As we shall see below, there is also evidence for the photoexcitation of $P_{11}(1750)$ and possibly $S_{11}(1700)$ from the data on $\gamma p \rightarrow K^+\Lambda$ and $\gamma p \rightarrow \eta^0 p$.

Application of the model of Devenish et al. (1972) to the neutral resonances is much less certain, but it is possible to extract some information. In particular, the $D_{15}^0(1680)$ is favored as the contributor to the third resonance region and has approximately equal excitation in the helicity $\frac{1}{2}$ and helicity $\frac{3}{2}$ states.

The final results of Devenish et al. (1972) are given in Table III in terms of the Walker (1969a) quantity A_0, defined in Eq. (II-84), in units of reciprocal giga electron volts. The corresponding electromagnetic decay widths for the charged resonances are given in mega electron volts, and, to give an indication of the physical consequences of these couplings, their contribution to the total cross section for $\gamma p \rightarrow \pi^0 p$.

The general features of the data in the second and third resonance regions are shown in Figs. 34–40. The curves have been calculated from the fit of Walker (1969).

E. PION ELECTROPRODUCTION IN THE RESONANCE REGION

Apart from the first resonance region, no information on single-pion electroproduction is available, the experiments being those in which only the scattered electron is detected and yielding $d^2\sigma/(d\Omega_e \, dE_2)$ as a function of the energy E_2 of the scattered electron (or equivalently of W, the invariant mass of the hadronic system), that is, effectively measuring the total cross section for the absorption of virtual photons. Typical data sets are shown in Fig. 40. If the momentum transfer is not too high, three peaks are clearly observed and it is natural to identify them with the resonance peaks observed in pion photoproduction [$P_{33}(1236)$, $D_{13}(1512)$, and $F_{15}(1690)$ or a mixture of $F_{15}(1690)$ and $D_{15}(1680)$]. The continuum excitation falls off more slowly with increasing momentum transfer than does the excitation of the specific resonance states, and at sufficiently high momentum transfer the spectra are completely dominated by the continuum.

If each peak is assumed to be due to a single resonance, then it is possible to isolate each resonance cross section by fitting the data with resonance Breit-Wigner shapes plus a background which is typically assumed to be a

TABLE III

ELECTROMAGNETIC COUPLINGS AND DECAY WIDTHS OF THE NUCLEON RESONANCES

Resonance	$A_l\pm$ (GeV^{-1})		$B_l\pm$ (GeV^{-1})		$\Gamma_{R^+ \to p\gamma}$ (MeV)			$\sigma_T(\gamma p \to \pi^0 p)$ (μb)
	π^+	π^-	π^+	π^-	Helicity $\tfrac{1}{2}$	Helicity $\tfrac{3}{2}$	Total	
$P_{11}(1450)$	4.7×10^{-2}	?	—	—	0.39	—	0.39	5.2
$D_{13}(1512)$	1.0×10^{-2}	0	7.3×10^{-2}	-5.8×10^{-2}	0.01	0.52	0.54	19.5
$S_{11}(1540)$	3.3×10^{-2}	-4.1×10^{-2}	—	—	0.15	—	0.15	2.6
$D_{15}(1680)$	0	$-1.0 - 10^{-2}$	-0.7×10^{-2}	0.7×10^{-2}	<0.01	0.03	0.03	0.7
$F_{15}(1690)$	0	0	3.0×10^{-2}	0	<0.03	0.31	0.31	13.0
$S_{11}(1700)$	7.7×10^{-2}	?	—	—	0.35	—	0.35	2.7
$P_{11}(1750)$	-3.3×10^{-2}	0	—	—	1.53	—	1.53	2.7
$G_{17}(2190)$	1.3×10^{-3}	?	7.8×10^{-3}	?	0.01	1.63	1.64	2.2
$S_{31}(1630)$	1.7×10^{-2}	-1.7×10^{-2}	—	—	0.31	—	0.31	2.6
$D_{33}(1670)$	6.7×10^{-3}	?	—	—	0.15	—	0.15	2.6
$F_{37}(1940)$	-5.8×10^{-3}	5.8×10^{-3}	4.5×10^{-3}	-4.5×10^{-3}	0.06	0.12	0.18	4.2

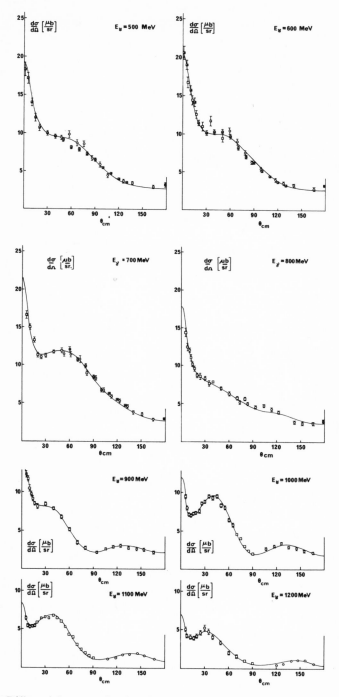

FIG. 34. Differential cross section for the reaction $\gamma p \rightarrow n + \pi^+$ between 0.5- and 1.2-GeV photon laboratory energy. The curves are calculated from the fit of Walker (1969).

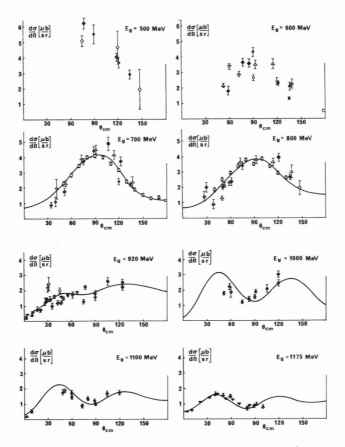

FIG. 35. Differential cross section for the reaction $\gamma + p \rightarrow p + \pi^0$ between 0.5- and 1.2-GeV photon laboratory energy. The curves are calculated from the fit of Walker (1969).

polynomial in W multiplied by a threshold factor $\{W - (M + \mu)\}^{1/2}$. The choice of this latter factor is made simply because it is known that for photo-production, and for electroproduction at small momentum transfers, the background to the $P_{33}(1236)$ resonance is dominantly s-wave. The results of such extractions appear to be insensitive to the order of the polynomial chosen for the background.

Of course, such a procedure is so beset by assumptions as to be at best imprecise and at worst very misleading. First, there is the assumption of a single resonance per peak which is certainly the only possibility for the first peak, $P_{33}(1236)$, but not for the second or the third. Photoproduction

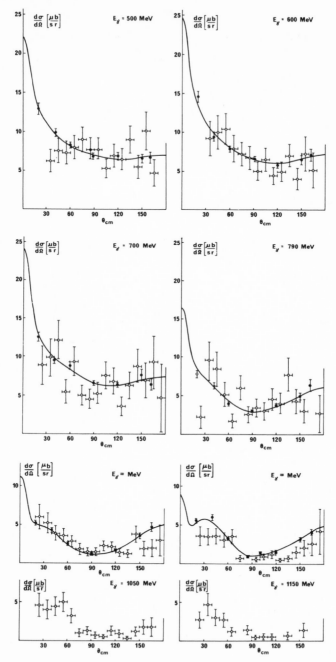

FIG. 36. Differential cross section for the reaction $\gamma + n \rightarrow p + \pi^-$ between 0.5- and 1.2-GeV photon laboratory energy. The curves are calculated from the fit of Walker (1969).

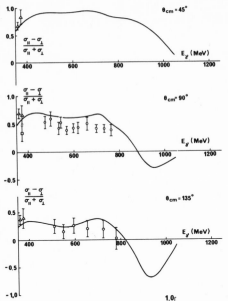

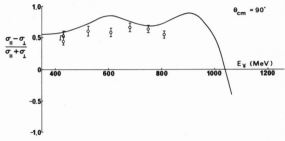

FIG. 37. Polarized photon asymmetry ratio for the reaction $\gamma + p \rightarrow n + \pi^+$ between 0.5- and 1.2-GeV photon laboratory energy. The curves are calculated from the fit of Walker (1969).

FIG. 38. Polarized photon asymmetry ratio for the reaction $\gamma + p \rightarrow p + \pi^0$ between 0.5- and 1.2-GeV photon laboratory energy. The curve is calculated from the fit of Walker (1969).

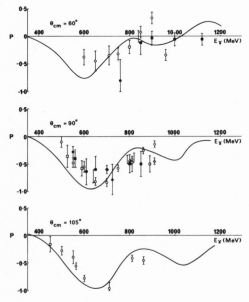

FIG. 39. Recoil proton polarization for the reaction $\gamma + p \rightarrow p + \pi^0$ between 0.5- and 1.2-GeV photon laboratory energy. The curves are calculated from the fit of Walker (1969).

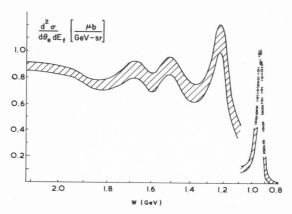

FIG. 40. Representative double differential cross section for inelastic electron scattering, clearly showing the first, second, and third resonances. The momentum transfer squared is 0.6 $(\text{GeV}/c)^2$ at the peak of the first resonance. The elastic peak is reduced by a factor of 4. The horizontal scale gives the invariant mass of the hadronic system W.

certainly can be used as a guide as to what would be expected, and this would indicate the dominance of the second peak by the $D_{13}(1512)$, although it is conceivable that the momentum transfer dependence of either (or both) $P_{11}(1450)$ and $S_{11}(1550)$ or the scalar/longitudinal excitation of either of these resonances are such as to alter this conclusion. Against this possibility, it can be argued that the position and width of 1503 ± 10 MeV and 77 MeV found for this peak by Bloom *et al.* (1968) and quoted by Panofsky (1968) make it very unlikely. For the third peak, it is not clear that it is dominated by $F_{15}(1690)$, even for photoproduction. Since both $F_{15}(1690)$ and $D_{15}(1680)$ have much the same position and width, and are believed to be appreciably narrower than other resonances in this mass region, the position and width of 1691 ± 10 MeV and 102 MeV found for the third peak by Bloom *et al.* (1968) can be taken to imply that if the third peak is not dominated by a single resonance, then, at worst, it is no more than a superposition of $D_{15}(1680)$ and $F_{15}(1690)$. Second, the assumption that the background varies smoothly with W, apart from at the pion threshold, is rather drastic, there being the possibility of other strong variations at other thresholds and, in particular, as Clegg (1969) has pointed out, there is as strong as s-wave rise at threshold in the reaction $\gamma + p \rightarrow \pi^- + \Delta^{++}$ as there is in single-pion photoproduction.

As suggested by Clegg (1969), a convenient measure of the intensity of production of a resonance is the virtual photon total cross section

$$\sum = (\Gamma_t)^{-1} \, d^2\sigma/(d\Omega \, dE_1) = \sigma_T + \varepsilon_S \sigma_S \qquad (\text{II-86})$$

due to the resonance, evaluated at the peak of the resonance. Here Γ_t is the usual lepton kinematical factor of (I-125).

The results for the first, second, and third peaks are shown in Figs. 41–43. The electroproduction points are from the fitting of the scattered-electron spectra by Bartel *et al.* (1968), the coincidence measurements of Imrie *et al.* (1968) and Ash *et al.* (1967), and measurements of the production integrated over the resonance by Cone *et al.* (1967) and Bloom *et al.* (1968). The conversion to the peak cross section using the assumed width in fitting has been taken from the evaluation of Clegg (1969). The photoproduction point for the first peak is taken from the fit of Schwela (1969) to the data of Fischer *et al.* (1970). The photoproduction points for the second and third peaks are taken from the analyses of Walker (1969), Moorhouse and Rankin (1970), and Rankin (1970), assuming complete dominance of the peaks by $D_{13}(1512)$ and $F_{15}(1690)$, and correcting for the known inelasticities of the resonances.

Even with the extreme assumptions made, the amount of information which can be extracted from these cross sections is very little. Assuming the photoproduction result of magnetic dipole dominance of the $P_{33}(1236)$ peak to hold for all momentum transfers [and we have seen already that this is indicated theoretically and is certainly true for momentum transfers up to $0.4 \ (\text{GeV}/c)^2$], then the cross section of Fig. 43 can be turned immediately into the magnetic dipole form factor, as defined in (II-63). The general features of this quantity have already been shown in Fig. 26. An interesting result is that this form factor is falling off with increasing momentum transfer appreciably faster than the proton form factors. Since any contribution from electric quadrupole or scalar/longitudinal excitation of the resonance is purely additive in Σ, the values for $\mu^*(K^2)$ above $0.5 \ (\text{GeV}/c)^2$ in Fig. 26 are an upper limit on this quantity, and the departure from the proton form factor shape may be even greater than indicated.

The situation is more complicated for the second and third peaks. Clegg (1969) has investigated a variety of simple fits to these cross sections, each corresponding to the assumption of excitation of the resonance by a single multipole (electric, magnetic, or scalar/longitudinal) or to the combination of electric and magnetic multipoles implied by the photoproduction analyses. None of these work particularly well by themselves for $D_{13}(1512)$. However, a combination of electric dipole and magnetic quadrupole in the photoproduction ratio of 3:1 and normalized to the photoproduction point, together with an (arbitrarily normalized) scalar/longitudinal excitation, does give a reasonable representation of the data, particularly for momentum transfers below $2 \ (\text{GeV}/c)^2$. This has the form

$$\Sigma = (A + Bk^4)G_{MV}^2 + CK^2 G_{MT}^2, \tag{II-87}$$

with A and B chosen to give the correct mixture of the electric dipole and magnetic quadrupole excitation at $K^2 = 0$. This fit is shown in Fig. 42.

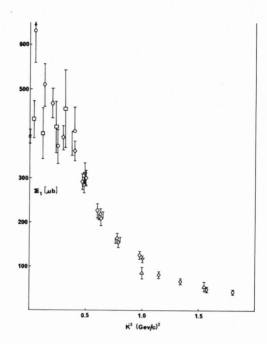

FIG. 41. The total cross section $\sigma_T + \varepsilon\sigma_S$ for production of the $\Delta(1238)$ resonance, at the resonance peak.

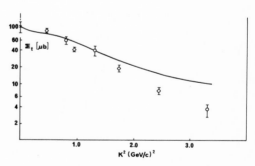

FIG. 42. The total cross section $\sigma_T + \varepsilon\sigma_S$ for production of the N(1512) resonance at the resonance peak.

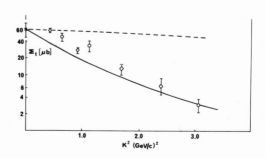

FIG. 43. The total cross section $\sigma_T + \varepsilon\sigma_S$ for production of the N(1690) resonance at the resonance peak.

For the $F_{15}(1690)$ resonance, the best individual description is for electric quadrupole excitation,

$$\sum = k^2 G_{MV}^2, \tag{II-88}$$

while the combination of electric quadrupole and magnetic octupole excitation

$$\sum = (Ak^2 + Bk^6)G_{MV}^2, \tag{II-89}$$

with A and B chosen to reproduce the ratio required by the $F_{15}(1690)$ dominant solution in photoproduction is a complete failure. Both these curves are shown in Fig. 43, normalized to the photoproduction point. Any addition of scalar/longitudinal excitation to the latter makes the situation even worse, but improves it when added to the former.

Superficially, this would appear to imply that the $D_{13}(1512)$ situation is comparatively simple but that the $F_{15}(1690)$ situation is rather complicated, and does not look at all like what one would expect from photoproduction. However, this is not necessarily true, since the procedures adopted above are very arbitrary. There is no reason why the threshold factor dependence should be retained for large momentum transfers, and if one is willing to change the size factor [encouraged in this by the behavior of the $P_{33}(1236)$ form factor], any number of reasonable fits to the data can be obtained. Obviously, much more detailed experimental information must become available before any real progress can be achieved. At the moment, one is not only unsure about which particular multipole excitation (or combination of multipole excitations) is being investigated, and it is not even clear for the third peak which resonance (or combination of resonances) is being excited.

A theoretical calculation of the momentum transfer dependence of the resonance peaks has been performed by Walecka and Zucher (1968) and by Pritchett et al. (1969). The latter, which is a natural extension of the former and supersedes it, is an attempt to go beyond the πN channel as the only important state in resonances, and takes into account the $\pi\Delta$ channel [Δ being $P_{33}(1236)$] as a component of the resonant states.

It is assumed that the linear combination

$$|N^*\rangle = \cos\varepsilon|\pi N\rangle + \sin\varepsilon|\pi\Delta\rangle \tag{II-90}$$

has a large resonant eigenphase ξ_1,

$$\tan\xi_1 = \tfrac{1}{2}\Gamma/(W - W_R), \tag{II-91}$$

while the orthogonal combination has an eigenphase ξ_2 whose sine is effectively zero,

$$\xi_2 \simeq n\pi, \tag{II-92}$$

this being the simplest possible model to make of an elastic resonance. Then the elastic amplitude is given by

$$f_{11} = [\eta \exp(2i\alpha) - 1]/2iq = \cos^2 \varepsilon \exp(i\xi_1) \sin \xi_1, \qquad \text{(II-93)}$$

and this can be used to determine both ε and ξ_1.

This description of the resonances implies that the inelastic decay proceeds entirely via the $\pi\Delta$ channel. For the resonances of main interest, namely, $D_{13}(1512)$, $D_{15}(1680)$, and $F_{15}(1690)$, this is a not unreasonable assumption. However, Walecka and Sucher (1968) also considered $S_{11}(1550)$ and $S_{11}(1710)$, for which this is not a good assumption, particularly $S_{11}(1550)$, where it is known that the ηN channel is not only the dominant inelastic channel but is probably *the* dominant channel. One of the free parameters in the theory is the ωNN coupling constant, ω exchange appearing as an important contributor in the second and third resonance regions, and Walecka and Sucher (1968) find two possible solutions with different signs for $g_{\omega NN}$, the magnitude in both cases being consistent with other determinations. In the solution with $g_{\omega NN}$ positive, $S_{11}(1550)$ and $S_{11}(1710)$ are the dominant resonances, a possibility which has been ruled out on experimental grounds. In the other solution, with $g_{\omega NN}$ negative, the s waves die out rapidly and the third peak is dominated by $F_{15}(1690)$, satisfying our previous prejudices. Consequently we shall consider only this latter type of solution, and the difficulty associated with $\pi\Delta$ decay of $S_{11}(1550)$ and $S_{11}(1710)$ will not arise.

Denoting the amplitudes for the electroproduction of a resonance in the πN and $\pi\Delta$ channels by $A_1(W,K^2)$ and $A_2(W, K^2)$ respectively, then the unitarity of the S matrix (to first order in e) yields coupled Omnès-Mushkelishvili equations, which can be written in the form

$$A_1(W, K^2)$$

$$= A_1^{\text{LHS}}(W, K^2) + \frac{1}{\pi} \int_{W_1}^{W_2} dW' \frac{\exp(-i\xi_1) \sin \xi_1 A_1(W', K^2)}{(W' - W - i\varepsilon)}$$

$$+ \frac{1}{\pi} \int_{W_2}^{\infty} dW' \frac{\exp(-i\xi_1) \sin \xi_1 \cos \varepsilon [\cos \varepsilon A_1(W, K^2) + \sin \varepsilon A_2(W, K^2)]}{(W' - W - i\varepsilon)},$$

$$\text{(II-94)}$$

$$A_2(W, K^2)$$

$$= A_2^{\text{LHS}}(W, K^2) + \frac{1}{\pi} \int_{W_1}^{W_2} dW' \frac{\exp(-i\zeta_1) \sin \xi_1 \sin \varepsilon \cos \varepsilon \, A_1(W', K^2)}{(W' - W - i\varepsilon)}$$

$$+ \frac{1}{\pi} \int_{W_2}^{\infty} dW' \frac{\exp(-i\xi_1) \sin \xi_1 \sin \varepsilon \{\cos \varepsilon A_1(W, K^2) + \sin \varepsilon A_2(W, K^2)\}}{(W' - W - i\varepsilon)}.$$

$$\text{(II-95)}$$

Here W_1 and W_2 are the thresholds for the πN and $\pi \Delta$ channels and $A_1^{\text{LHS}}(W, K^2)$ and $A_2^{\text{LHS}}(W, K^2)$ are multipole projections of gauge-invariant sets of Feynman graphs which are thought to play an important role in the excitation process.

Paralleling the line of argument given for the solution of the uncoupled Omnès-Mushkelishvili equation [see Section II,2, (II-32)–(II-40)], Pritchett *et al.* (1969) show that it is plausible to write down the solution as

$$A_{\text{res}}(W, K^2) = \{\cos \varepsilon A_1^{\text{LHS}}(W, K^2) + \sin \varepsilon A_2^{\text{LHS}}(W, K^2)\}/D(W), \quad \text{(II-96)}$$

where $D(W)$ has the usual form of the denominator function, but is expressed in terms of the real eigenphase shift ξ_1,

$$D(W) = \exp\left\{-[(W - W_0)/\pi] \int_{W_1}^{\infty} dW' \xi_1(W')/[(W' - W_0)(W' - W - i\varepsilon)]\right\},$$
$$D(W_0) = 1. \qquad\qquad\qquad\qquad\qquad\qquad\qquad\qquad\qquad\qquad\qquad \text{(II-97)}$$

The subtraction in (II-97) is necessary to ensure convergence of the integral, since the indications are that $\lim_{W \to \infty} \xi_1 = \pi/2$. The subtraction point W_0 is an arbitrary parameter, and the absolute normalization of the resonance amplitude is expressed in terms of it.

For the πN channel, the set of Feynman graphs chosen to represent $A_1^{\text{LHS}}(W, K^2)$ comprised the usual Born terms of s- and u-channel nucleon exchange and t-channel pion exchange plus t-channel ω exchange and u-channel Δ exchange. For the $\pi \Delta$ channel, the Feynman graphs comprised the corresponding nucleon and pion Born terms, u-channel Δ exchange and a contact term. These diagrams are shown in Fig. 44.

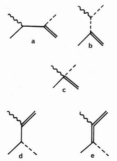

Fig. 44. The Born terms for the reaction $\gamma + N \to \pi + \Delta$.

It was assumed that all form factors were proportional to the nucleon form factors, for which the usual scaling law

$$G_{E_p} = G_{M_p}/\mu_p = G_{M_n}/\mu_n = -(4M^2/K^2)(G_{E_n}/\mu_n) \qquad \text{(II-98)}$$

was assumed, with the usual dipole empirical fit. An alternative scaling law, consistent with existing data

$$G_{E_n} = [-K^2/(4M^2 + K^2)]G_{M_n} \qquad (\text{II-99})$$

was investigated, but produced no qualitative alteration in the conclusions. It should be noted that for $P_{33}(1236)$, this model is very similar to that of Adler (1968).

Since the results are by construction proportional to the proton form factors, it is more convenient to compare the theory with the ratio of inelastic to elastic scattering rather than to compare it with the inelastic cross section itself. The comparison is shown in Figs. 45–47, the theoretical normalization in all curves being determined from the fit to the data of Bloom et al. (1968). In Fig. 45, the transverse and scalar/longitudinal excitation are shown separately, but elsewhere the combined cross section is given.

The general behavior of the momentum transfer dependence of the $P_{33}(1236)$ and $D_{13}(1512)$ peaks is reproduced quite well, except for the former at small momentum transfers. In particular, the photoproduction point is not attained. This is not at all surprising, since we have already seen that the momentum transfer dependence of this form factor is appreciably different from the proton form factor. The curve for $D_{13}(1512)$ does get within one standard deviation of the photoproduction point. By contrast, the $F_{15}(1690)$ curve is very bad, failing to reach the photoproduction point by an order of magnitude. The latter two results support the crude conclusions drawn from the naive numerological discussion of these peaks, namely, that the situation for $D_{13}(1512)$ is rather straightforward but that for $F_{15}(1690)$ is not, but again this cannot be considered to be conclusive.

Lacking any further electroproduction data, the model cannot be given a more stringent test there, and since it only is concerned with the leading resonances, too close a comparison with the photoproduction data would be unfair. However, one possible test is to compare the ratio of magnetic to electric excitation of the resonances predicted by the model for photopion production with those actually found in the photoproduction analyses. Quantitatively

	Experiment	Model	
$D_{13}(1512)M_{2-}/E_{2-}$	0.4 ± 0.1	0.11	
$F_{15}(1690)M_{3-}/E_{3-}$	0.5 ± 0.2	-0.07	(II-100)

The ratios calculated from the model are sensitive to the coupling $g_{\omega NN}$ and to the mixing parameters of the $\pi\Delta$ state, and it is quite conceivable that minor adjustments of these parameters will decrease the discrepancy. Even as things stand, the result for $D_{13}(1512)$ once again looks not too unreasonable, but that for $F_{15}(1690)$ looks otherwise.

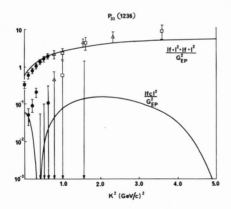

FIG. 45. Comparison of the data with the two-channel dispersion relation predictions for the excitation of the Δ(1238) resonance in inelastic electron scattering.

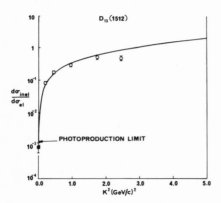

FIG. 46. Comparison of the data with the two-channel dispersion relation predictions for the excitation of the N(1512) resonance in inelastic electron scattering.

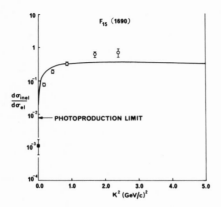

FIG. 47. Comparison of the data with the two-channel dispersion relation predictions for the excitation of the N(1690) resonance in inelastic electron scattering.

Finally, the model predicts that the $D_{13}(1512)$ peaks should be dominantly excited by scalar/longitudinal excitation beyond 4.0 $(GeV/c)^2$, and the $F_{15}(1690)$ peak beyond 1.0 $(GeV/c)^2$. Both these predictions can be tested in noncoincidence experiments.

F. Photoproduction of η Mesons in the Resonance Region

The dominant feature of η photoproduction is the rapid rise of the total cross section immediately above threshold, followed by an equally rapid fall. This structure is ascribed to the $S_{11}(1550)$ resonance, an assignation borne out by the near isotropy of the differential cross section across the peak and the observation of precisely the same effect, with the same explanation, in η production by pions. The latter reaction indicates that the dominant decay mode of the $S_{11}(1550)$, other than πN, is ηN and consequently the various analyses of pion-nucleon scattering, pion photoproduction, η photoproduction, and η production by pions can be checked for consistency of the various partial widths involved. This consistency is found, but the constraint is not a very tight one in view of the rather large uncertainties involved in the analyses.

The small anisotropy which is observed is consistent with s-wave and p-wave interference, that is

$$d\sigma/d\Omega = A + B \cos \theta, \tag{II-101}$$

but the data are not sufficiently precise to exclude higher terms. The forward-backward asymmetry has been measured explicitly by McNeely et al. (1969) and in a preliminary fit they obtain $B = 0.12 \pm 0.01$ $\mu b/sr$ at the peak of the resonance. For comparison, $A = 1.2 \pm 0.06$ $\mu b/sr$ at the peak of the resonance.

There is evidence for a second, much smaller, peak at 1.2-GeV photon laboratory energy. This is seen most clearly in the data of Booth et al. (1969). Subsequent data of Booth et al. (1971) show evidence of quite strong anisotropy at the peak of the resonance, an isotropy which is again consistent with s-wave and p-wave interference, but again the data cannot exclude higher terms.

The most straightforward explanation of this peak is in terms of the production of the $P_{11}(1750)$ resonance, since the mass of this resonance is right in the middle of the peak and by interference with the tail of the $S_{11}(1550)$ can produce the necessary anisotropy. Other $I = \frac{1}{2}$ nucleon resonances close to this energy region are those of $D_{15}(1680)$, $F_{15}(1690)$, and $S_{11}(1710)$. In the first two cases the masses, which are very well known, are much too low to account for the peak (they occur precisely at the minimum of the data). Current prejudice, as we have seen, would exclude $D_{15}(1680)$ completely from photoproduction off protons, and the analyses of η production with pions by Deans and Holladay (1967) and Carreras and Donnachie (1970) indicates that neither of these resonances has any significant coupling to the

η-nucleon channel. The mass of $S_{11}(1710)$ is not so well known and it could conceivably be sufficiently high to account for the peak, the anisotropy in this case arising from the interference between the resonance and a p-wave background. Preliminary measurements of the recoil proton polarization at 90° center of mass and between 0.8 and 1.1 GeV by Heusch *et al.* (1969) are consistent with s-wave and p-wave interference, although again this is not the only possible interpretation.

Since both the η meson and the deuteron are isoscalar particles, only the isoscalar part of the amplitude can contribute to coherent η photoproduction on deuterium, and comparison of this process with incoherent η photoproduction off deuterium and with photoproduction off protons can yield, at least in principle, a complete determination of the relative isovector and isoscalar couplings. Some data on the coherent reaction have been obtained by Anderson and Prepost (1969). Comparing this with the incoherent production data of Bacci *et al.* (1969) and with the proton data, they concluded that the production near threshold is dominated by the isoscalar amplitude and that the isovector part is consistent with zero until about 80 MeV (photon laboratory energy) above threshold, beyond which the coherent photoproduction data (and by inference the isoscalar amplitude) is, in turn, consistent with zero.

This changeover occurs below the peak of the $S_{11}(1550)$ resonance and implies that the near-threshold region is more complicated than the simple $S_{11}(1550)$ dominance model would allow. One possible explanation is that the effect arises from one of the lower resonances, and the two possibilities are the $P_{11}(1450)$ resonance (if it is photoproduced at all) and that of $D_{13}(1512)$. In a detailed analysis of the data on η production by pions, Davies and Moorhouse (1967) found that the anisotropy observed in that reaction just above threshold could be explained entirely in terms of the large $S_{11}(1550)$ contribution interfering with a very small contribution from $D_{13}(1512)$, the partial decay width to πN of the latter being less than 1 MeV, and no contribution at all necessary from $P_{11}(1450)$. This result has been confirmed by Deans (1969) in a pole-plus-resonance fit; the contribution from $P_{11}(1450)$ was not statistically significant while that from the $D_{13}(1512)$ was. All this points to $D_{13}(1512)$ resonance being the mechanism behind this effect, but we have seen from the analyses of photopion production that the coupling of $D_{13}(1512)$ is predominantly isovector, and unless the isoscalar coupling is larger than estimated there, this resonance cannot be invoked. In addition, Davies and Moorhouse (1967) found a rather small ηN width for $D_{13}(1512)$, and it is improbable that, even if the isospin structure is changed, a sufficient contribution to the cross section could be obtained.

This threshold problem apart, the data on η photoproduction can be explained readily in terms of a resonance-plus-background model. In the model of Deans and Holladay (1968) background was given by a nucleon pole and vector meson pole and included the $D_{13}(1512)$, $S_{11}(1550)$, and $F_{15}(1690)$ resonances. A good description of the data then available was obtained up to 1.1 GeV. Bajpai and Donnachie (1969) took as their background the Regge exchange of the ω and B mesons (the parameters of which were obtained from a fit to high energy η photoproduction) together with the $S_{11}(1550)$ and $P_{11}(1750)$ resonances and obtained a good fit to the data up to 1.5 GeV, beyond which the cross section is given by Regge exchange only.

A development of this calculation has been made by Donnachie (1970), using a Regge background of nondegenerate ρ and ω exchange. This is preferred to the ω- and B-exchange background used by Bajpai and Donnachie (1969) on account of the data of Booth *et al.* (1971) at 1.83 GeV, which are completely consistent with the former model but not the latter. In the fit, the data of Bloom *et al.* (1968) above 1.0 GeV was excluded since the normalization is incompatible with that of Booth *et al.* (1969, 1971). The latter two experiments were preferred because they are mutually consistent, and, as noted above, agree well at the high energy end with what is expected from Regge exchange.

The only resonance states necessary to fit the data are $S_{11}(1550)$ and $P_{11}(1750)$, although some extra *p*-wave (or higher) background near threshold cannot be excluded. The only other state which may give some contribution is $S_{11}(1710)$, and its inclusion improves the fit to the data between 1.0 and 1.1 GeV. The small forward-backward asymmetry observed by McNeely *et al.* (1969) is easily reproduced in this model. The predicted recoil proton polarization at 90° center of mass and between 0.8 and 1.1 GeV is of the order of 0.25, which is smaller than the mean value obtained by Heusch *et al.* (1969), but not incompatible with their results in view of the large errors on the latter. A typical excitation curve is shown in Fig. 48, while curves showing the differential cross section are given in Fig. 49. The curves of Figs. 48 and 49 are fitted to the data of Booth *et al.* (1969, 1971).

G. PHOTOPRODUCTION OF K MESONS IN THE RESONANCE REGION

Although the cross section for the reaction $\gamma + p \rightarrow K^+ + \Lambda^0$ rises steeply at threshold, implying *s*-wave dominance of the near-threshold region, the differential cross sections rapidly become anisotropic and have the form

$$d\sigma/d\Omega = A + B \cos \theta + C \cos^2 \theta. \qquad (II-102)$$

A is typically of the order of 0.15 μb/sr between 1.0 and 1.5 GeV; B increases steadily from about 0.05 μb/sr at 1.0 GeV to about 0.12 μb/sr at 1.2 GeV and then decreases slowly, a pattern which is followed by the coefficient C which is approximately zero at 1.0 GeV, is about 0.06 μb/sr at 1.2 GeV, and then decreases slowly. This particular pattern of anisotropy, together with the

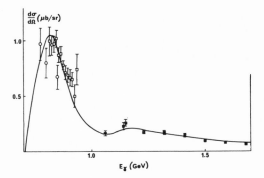

FIG. 48. The 90° excitation curve for η photoproduction. The curve is from the fit described in the text.

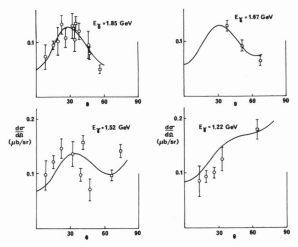

FIG. 49. Differential cross sections for η photoproduction. The curves are from the fit described in the text.

fact that it is at a maximum in the vicinity of 1.2 GeV, points very strongly to the cause being interference of the $P_{11}(1750)$ resonance, the large s-wave background implied by the threshold behavior, although at this level a d-wave and s-wave interference cannot be ruled out.

In an analysis of the reaction $\pi^- + p \to K^0 + \Lambda^0$, Wagner and Lovelace (1971) showed that the reaction proceeded dominantly through the $S_{11}(1710)$ and $P_{11}(1750)$ resonances, with a possible contribution from $D_{13}(1730)$—a resonance about which there is still considerable doubt. This result clearly favors the s-wave and p-wave interference interpretation of photoproduction data.

In an analysis in terms of a single resonance state plus Born amplitude (nucleon pole in the *s* channel; K, K* poles in the *t*-channel and Λ, Σ poles in the *u* channel) Thom (1966) was unable to distinguish at that time between a *p*-wave and *d*-wave resonance. More recently, in an analysis in which the multipole amplitudes were parametrized as resonances on an arbitrary, but slowly varying, background, Orito (1969) sided firmly with *s*-wave and *p*-wave interference. This conclusion was reached largely as a consequence of the data available on the Λ polarization, the availability of which is one of the attractions of this reaction.

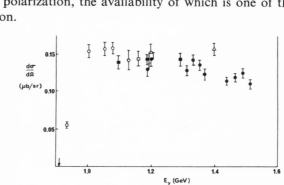

FIG. 50. The 90° excitation curve for the reaction $\gamma + p \rightarrow K^+ + \Lambda^0$.

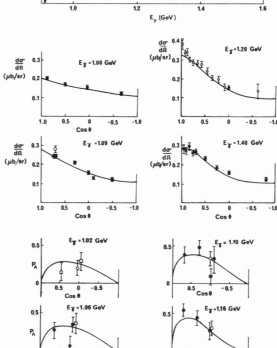

FIG. 51. Differential cross sections for the reaction $\gamma + p \rightarrow K^+ + \Lambda^0$. The curves are from the fit described in the text.

FIG. 52. Recoil λ polarization in the reaction $\gamma + p \rightarrow K^+ + \Lambda^0$. The curves are from the fit described in the text.

Orito (1969) found a considerable contribution from the $P_{11}(1750)$ resonance interfering with a strong s-wave background, but he was unable to separate out any definite structure in this background. With an increased data set, and using the same model as Orito (1969), Schorsch *et al.* (1971) showed that there was a considerable contribution from $S_{11}(1710)$ in addition to that from $P_{11}(1750)$. There was no evidence for either the $P_{13}(1860)$ or $D_{13}(1730)$ resonances.

An alternative model has been developed by Renard and Renard (1971). They used an isobar model, including both the s-channel and u-channel poles and resonances. For the $\gamma p \to K^{+}\Lambda^{0}$ transition this means the nucleon and nine $I = \frac{1}{2}$ resonances in the s channel and Λ, Σ^{0} and three resonances in the u channel. With the number of parameters available, it was not difficult to obtain a good fit to the data, but the significance of the numbers obtained is not clear. For example, in the $\gamma p \to K^{+}\Lambda$ transition there is an obvious disagreement with the $\pi^{-}p \to K^{0}\Lambda^{0}$ analysis of Wagner and Lovelace (1971), Renard and Renard (1971) requiring an appreciable contribution from F_{15} (1960). The 90° excitation curve, clearly showing the s-wave threshold rise, is shown in Fig. 50, and typical differential cross sections and Λ polarizations are shown in Figs. 51 and 52, respectively.

The 90° excitation curve for the reaction $\gamma + p \to K^{+} + \Sigma^{0}$ is shown in Fig. 53, which exhibits the general features of this reaction, namely, a steady

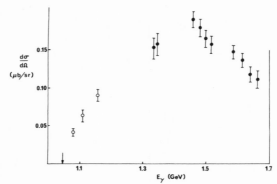

FIG. 53. The 90° excitation curve for the reaction $\gamma + p \to K^{+} + \Sigma^{0}$.

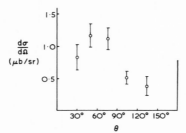

FIG. 54. Typical differential cross section for the reaction $\gamma + p \to K^{+} + \Sigma^{0}$.

increase of the cross section to a maximum around 1.45-GeV photon laboratory energy (that is, 1.90 GeV center-of-mass energy), followed by an equally
steady decrease. This is precisely what is observed in the reaction $\pi^+ + p \rightarrow$
$K^+ + \Sigma^+$, which is completely dominated by the $F_{37}(1930)$ resonance (see
Donnachie, 1968) and an interpretation of the $\gamma + p \rightarrow K^+ \Sigma^0$ data in precisely the same terms is an obvious conclusion. This interpretation is enhanced
by the differential cross section which, apart from very close to threshold,
exhibits a very high degree of anisotropy, requiring terms up to $\cos^4 \theta$ and
being consistent with almost pure f wave. A typical cross section is shown in
Fig. 54.

H. The Quark Model in the Resonance Region

One possible interpretation of the observed baryon spectrum is in terms
of the excitations of a basic three quark system. Faiman and Hendry (1968)
have investigated this interpretation in detail, assuming that the quarks obey
para-Fermi statistics of the third order, and that the three quarks interact
basically via simple harmonic forces, that is, the baryons are eigenstates of the
Hamiltonian

$$H = \sum_j (\mathbf{P}_j/2M_Q) + \tfrac{1}{2}M_Q \omega^2 \sum_{i<j} (\mathbf{r}_i - \mathbf{r}_j)^2. \qquad \text{(II-103)}$$

For the purpose of labeling the states (but not, however, for doing calculations such as decay widths) it is customary, and slightly more convenient, to
use the related shell-model Hamiltonian

$$H_{\text{SM}} = \sum_j (\mathbf{P}_j^2/2M_Q) + \tfrac{1}{2}M_Q \omega^2 \sum_j \mathbf{r}_j^2, \qquad \text{(II-104)}$$

the eigenfunctions of which are products of three one-body harmonic
oscillator wave functions, and conventional shell-model notation is used to
describe the various levels.

The complete wave functions are determined by combining the spatial
wave functions with $SU(6)$ wave functions which involve the quark spin and
isospin in such a way that the overall wave function is symmetric. This
demands that a spatial wave function of symmetric, mixed symmetric, or
antisymmetric character must be combined with the 56, 70, or 20 representation of $SU(6)$, respectively. The resultant allowed spectrum in the three lowest
levels is then as follows:

lowest level $(1s)^3$: $56 (L = 0^+)$;

second level $(1s)^2(1p)$: $70 (L = 1^-)$;

third level $(1s)^2(2s)$, $(1s)^2(1d)$, $(1s)(1p)^2$: $56 (L = 0^+)$, $70 (L = 0^+)$,

$56 (L = 2^+)$, $70 (L = 2^+)$,

$20 (L = 1^+)$. $\qquad \text{(II-105)}$

The third level yields an embarrassing profusion of states which are not observed. However, if it is postulated that only the *56* ($L =$ even$^+$) and *70* ($L =$ odd$^-$) states be populated, then a precise classification of the observed N* spectrum is obtained, and is shown in Table IV.

TABLE IV

CLASSIFICATION OF THE PION-NUCLEON RESONANCES IN THE *l*-EXCITATION
QUARK MODEL WITH PARA-FERMI STATISTICS[a]

		$(1s)^2(2s), (1s)^2(1d), (1s)(1p)^2$	
$(1s)^3$, $56 (L = 0^+)$	$(1s)^2(1p)$, $70 (L = 1^-)$	$56 (L = 0^+)$	$56 (L = 2^+)$
$8^{1/2}$ $P_{11}(939)$ $10^{3/2}$ $P_{33}(1236)$			
		$8^{1/2}$ $P_{11}(1450)$	
	$8^{1/2} \begin{cases} D_{13}(1512) \\ S_{11}(1550) \end{cases}$		
	$10^{1/2} \begin{cases} S_{31}(1630) \\ D_{33}(1670) \end{cases}$		
	$8^{3/2} \begin{cases} D_{15}(1680) \\ \\ S_{11}(1710) \\ D_{13}(1730) \end{cases}$	$10^{3/2}$ $P_{33}(1690)$	$8^{1/2} \begin{cases} F_{15}(1690) \\ \\ P_{13}(1850) \end{cases}$
			$10^{3/2} \begin{cases} F_{35}(1880) \\ P_{31}(1900) \\ P_{33} \text{ (unobserved)} \\ F_{37}(1940) \end{cases}$

[a] The $P_{11}(1750)$ must be associated with a higher level, a possibility being $l = 4$, $56, L = 0^+$.

It should be noted that the two S_{11}'s and the two D_{13}'s are mixtures of two eigenstates:

$$S_{11}(1710) = \cos \theta_s \, S_{11}(8^{3/2}) + \sin \theta_s \, S_{11}(8^{1/2}),$$
$$S_{11}(1550) = -\sin \theta_s \, S_{11}(8^{3/2}) + \cos \theta_s \, S_{11}(8^{1/2}) \tag{II-106}$$

and

$$D_{13}(1730) = \cos \theta_d \, D_{13}(8^{3/2}) + \sin \theta_d \, D_{13}(8^{1/2}),$$
$$D_{13}(1512) = -\sin \theta_d \, D_{13}(8^{3/2}) + \cos \theta_d \, D_{13}(8^{1/2}). \tag{II-107}$$

With the above postulate on the populated states, all other resonances are unmixed eigenstates.

Further justification of the model can be given by using it to calculate properties of the resonances and in particular partial decay widths. Faiman and Hendry (1968) calculated the πN and $\pi \Delta$ [$\Delta \equiv P_{33}(1236)$] decay widths of the N^* states assuming that the emission takes place via a one-quark deexcitation. The nonrelativistic form of the interaction is

$$H_{int} = \sum_j (f_Q/\mu)(\boldsymbol{\sigma}_j \cdot \mathbf{q})(\boldsymbol{\tau}_j \cdot \boldsymbol{\pi})\exp(-i\mathbf{q} \cdot \mathbf{r}_j)[(2E_\pi)^{1/2}]^{-1}, \qquad \text{(II-108)}$$

where f_Q is a measure of the strength of the quark-pion coupling, μ being any standard mass [Faiman and Hendry (1968) took it to be the pion mass], E_π is the energy of the emitted pion, and $\mathbf{q}$ is its momentum. Since the wave functions for the particles are known, the amplitude for a decay process to lowest order in f_Q is obtained by taking the matrix element of this interaction operator between the relevant N^* and the N or Δ wave functions and integrating over all space.

For the unmixed states there are only two parameters involved: $f_Q{}^2/4\pi$ and α^2 ($= M\omega$) which Faiman and Hendry (1968) determined so as to give $\Gamma(\Delta \to N\pi)$ exactly and simultaneously yield a best fit to the $N\pi$ decays of the other resonances. This gave

$$f_Q{}^2/4\pi = 0.055, \qquad \alpha^2 = 0.10 \quad (\text{GeV}/c)^2. \qquad \text{(II-109)}$$

Given these numbers, the mixing angles θ_s and θ_d then were obtained from the observed πN decay widths of $S_{11}(1550)$ and $D_{13}(1512)$. The πN widths of $S_{11}(1710)$ and $D_{13}(1730)$ can then be predicted, as can the $\pi \Delta$ widths of all four resonances. Two solutions are obtained for θ_s and θ_d, namely

$$\theta_s \simeq 35° \quad \text{or} \quad 90°, \qquad \theta_d \simeq 35° \quad \text{or} \quad 127°. \qquad \text{(II-110)}$$

The solution $\theta_s \simeq 90°$ corresponds to $S_{11}(1550)$ being entirely $\{8^{3/2}\}$ and Moorhouse (1966) has shown that N^* states originating from the $\{8^{3/2}\}$ cannot be photoexcited from a proton. We have seen that there is strong experimental evidence that the $S_{11}(1550)$ resonance is photoproduced, and consequently this solution can be ruled out. No such separation of the two values of θ_d can be made, but it is tempting to choose $\theta_d \simeq 35°$, thus enabling the S_{11} and D_{13} widths to be accounted for by a similar amount of mixing, implying a possible common mechanism. Taking this solution, the predictions for the decay widths are shown in Table V. On the whole, the results are very reasonable and in particular the small $\pi \Delta$ mode of $S_{11}(1550)$ is encouraging, since the principal inelastic mode is known to be ηN.

Armed with these pointers to the effectiveness of the model, the $N\gamma$ modes can be calculated in exactly the same way. In this case, the interaction Hamiltonian for one-quark deexcitation is

$$H_{int} = \sum_j Q_j[-ig\boldsymbol{\sigma}_j \cdot (\mathbf{k} \times \mathbf{A}) + (\mathbf{p}_j + \mathbf{p}_j') \cdot \mathbf{A}](e/2M), \qquad \text{(II-111)}$$

TABLE V

PREDICTIONS OF THE SYMMETRIC QUARK MODEL FOR THE
πN AND $\pi \Delta$ WIDTHS OF THE N* STATES

Resonance	$\Gamma_{N\pi}$ (MeV)[a]	$\Gamma_{\Delta\pi}$ (MeV)	$\Gamma_{N\pi}$ (Expt.)	Γ_{tot} (Expt.)
$P_{33}(1236)$	120*	—	120	—
$P_{11}(1450)$	45	?	140	260
$D_{13}(1512)$	60*	16	60	115
$S_{11}(1550)$	37*	6	34	100
$S_{31}(1630)$	24	121	40	160
$D_{33}(1670)$	28	55	35	230
$D_{15}(1680)$	33	112	65	145
$F_{15}(1690)$	100	?	75	125
$P_{33}(1690)$	80	?	30	280
$S_{11}(1710)$	231	132	190	280
$D_{13}(1730)$	109	197	?	?

[a] The values marked with an asterisk were fitted exactly in the determination of the parameters.

where eQ_j is the charge of the jth quark, p_j and $p_j{}'$ are, respectively, the initial and final momenta of the quark emitting a photon of momentum $\mathbf{k}$; $geQ_j/2M_Q$ represents the magnetic moment of a bound quark; and the electromagnetic field $\mathbf{A}$ is given by

$$\mathbf{A}(\mathbf{r}_j) = (4\pi)^{1/2} \sum [\varepsilon/(2k_0)^{1/2}]\{a_k{}^\dagger \exp(-i\mathbf{k}\cdot\mathbf{r}_j) + a_k \exp(i\mathbf{k}\cdot\mathbf{r}_j)\}, \qquad \text{(II-112)}$$

ε being a unit polarization vector.

Two immediate conclusions can be drawn, even without evaluation of the transition matrix elements. First, the $P_{33}(1236)$ resonance can only be excited by magnetic dipole radiation. Since both the nucleon and $P_{33}(1236)$ are in the oscillator ground state with $L = 0$, the photon orbital angular momentum must be zero and only its spin can be involved in the transition. Thus, the photon has angular momentum 1 and positive parity, that is, it is magnetic dipole radiation. Second, the decay $D_{15}(1680) \rightarrow p\gamma$ is forbidden, since it involves a transition from a three-quark state in the mixed $SU(6)$ representation 70 with a quark spin $S = \frac{3}{2}$ and orbital angular momentum $L = 1$ to a state with a symmetric $SU(6)$ representation 56 with $S = \frac{1}{2}, L = 0$. In the absence of mixing, the same would be true for $S_{11}(1710)$ and $D_{13}(1730)$. The latter results were first noted by Moorhouse (1966).

With α^2 already determined from the πN and $\pi \Delta$ modes, Faiman and Hendry (1969) specified the quark magnetic moment $\mu_Q = eg/2M_Q$ from the known radiative decay width of the $P_{33}(1236)$ resonance. Taking $\Gamma(\Delta^+ \rightarrow p\gamma)$ as 0.65 MeV yielded

$$\mu_Q = 0.18 \quad \text{GeV}^{-1}, \qquad \text{(II-113)}$$

which is somewhat larger than the value obtained if μ_Q is equated with μ_p, the proton magnetic moment, which is 0.13 GeV^{-1}.

Explicitly, with $B = \mu_Q \exp(-k^2/3\alpha^2)E/M_R$,

$$\Gamma[P_{33}(1236) \to N\gamma] = \tfrac{16}{9}k^3 B = 0.65 \quad \text{MeV},$$

$$\Gamma[P_{11}^{+}(1450) \to p\gamma] = \tfrac{1}{27}(k^7 B/\alpha^4) = 0.13 \quad \text{MeV},$$

$$\Gamma[P_{11}^{0}(1450) \to n\gamma] = \tfrac{4}{243}(k^7 B/\alpha^4) = 0.06 \quad \text{MeV},$$

$$\Gamma[D_{13}^{+}(1512) \to p\gamma] = \tfrac{2}{9}(k^5 B/\alpha^2)[1 - (2\alpha^2/k^2)g^{-1} + (4\alpha^4/k^4)(g^2)^{-1}]\cos^2 \theta_d$$
$$= 0.58[1 - 0.89/g + 0.80/g^2] \quad \text{MeV},$$

$$\Gamma[D_{13}^{0}(1512) \to n\gamma] = \tfrac{2}{27}(k^5 B/\alpha)\{[\tfrac{14}{15}\sin^2 \theta_d + \tfrac{1}{3}\cos^2 \theta_d]$$
$$+ [(\sqrt{10}/15)\sin \theta_d \cos \theta_d - (2^{*2}/k^2)\cos \theta_d\sqrt{10}\sin \theta_d$$
$$+ \cos \theta_d]g^{-1} + (12\alpha^4/4)\cos^2 \theta_d(g^2)^{-1}\}$$
$$= 0.12[1 - 3.04/g + 2.53/g^2] \quad \text{MeV},$$

$$\Gamma_{11}^{0}[S(1550) \to p\gamma] = \tfrac{2}{9}(k^5 B/\alpha^2)[1 + (2\alpha^2/k^2)g^{-1}]^2 \cos^2 \theta_s$$
$$= 0.42[1 + 0.83/g]^2 \quad \text{MeV},$$

$$\Gamma[S_{11}^{0}(1550) \to n\gamma] = \tfrac{2}{81}(k^5 B/\alpha^2)[(\sin \theta_s + \cos \theta_s) + (6\alpha^2/k^2)\cos \theta_s g^{-1}]^2$$
$$= 0.14[1 + 1.46/g]^2 \quad \text{MeV},$$

$$\Gamma[D_{15}^{+}(1680) \to p\gamma] = 0,$$

$$\Gamma[D_{15}^{0}(1680) \to n\gamma] = \tfrac{2}{45}(k^5 B/\alpha^2) = 0.20 \quad \text{MeV},$$

$$\Gamma[F_{15}^{+}(1690) \to p\gamma] = \tfrac{2}{135}(k^7 B/\alpha^4)[1 - (4\alpha^2/k^2)g^{-1} + (12\alpha^4/k^4)(g^2)^{-1}]$$
$$= 0.25[1 - 1.18/g + 1.02/g^2] \quad \text{MeV},$$

$$\Gamma[F_{15}^{0}(1690) \to n\gamma] = \tfrac{8}{1215}(k^7 B/\alpha^4) = 0.11 \quad \text{MeV},$$

$$\Gamma[S_{11}^{+}(1710) \to p\gamma] = \tfrac{2}{9}(k^5 B/\alpha^2)[1 + (2\alpha^2/k)g^{-1}]^2 \sin^2 \theta_s$$
$$= 0.36[1 + 0.56/g]^2 \quad \text{MeV},$$

$$\Gamma[S_{11}^{0}(1710) \to n\gamma] = \tfrac{2}{81}(k^5 B/\alpha^2)[(\sin \theta_s - \cos \theta_s) + (6\alpha^2/k)\sin \theta_s g^{-1}]^2$$
$$= 0.03[1 - 3.92/g]^2 \quad \text{MeV},$$

$$\Gamma[D_{13}^{+}(1730) \to p\gamma] = \tfrac{2}{9}(k^5 B/\alpha^2)[1 - (2\alpha^2/k^2)g^{-1} + (4\alpha^4/k^4)(g^2)^{-1}]\sin^2 \theta_d$$
$$= 0.36[1 - 0.59/g + 0.34/g^2] \quad \text{MeV},$$

$$\Gamma[D_{13}^{0}(1730) \to n\gamma] = \tfrac{2}{27}(k^5 B/\alpha)\{[(14/15)\cos^2 \theta_d + \tfrac{1}{3}\sin^2 \theta_d$$
$$- (\sqrt{10}/15)\sin \theta_d \cos \theta_d]$$
$$+ (2\alpha^2/k^2)\sin \theta_d(\sqrt{10}\cos \theta_d) - \sin \theta_d g^{-1}$$
$$+ (12\alpha^4/k^4)\sin^2 \theta_d(g^2)^{-1}\}$$
$$= 0.24[1 + 1.07/g + 0.52/g^2] \quad \text{MeV}.$$

Of the five radiative widths which are immediately determined, $\Gamma[D_{15}^{+}(1680)]$ is consistent with the preferred solution of the data analyses, $\Gamma[D_{15}^{0}(1680)]$

and $\Gamma[F^0_{15}(1690)]$ are not well known, although the latter is believed to be very small, and both $\Gamma[P^+_{11}(1470)]$ and $\Gamma[P^0_{11}(1470)]$ are within the present wide range of values allowed for experiment.

The problem is now to choose g. For a pure Dirac particle g would equal 1 and it is not unreasonable to suppose that $g \geqslant 1$. Certainly g cannot be very much less than 1, since all the radiative widths which depend on g could then achieve hadronic proportions. Specifying only that $g > 1$ allows a range of values to be obtained for each radiative width, a range which in most cases is not too great considering the uncertainties in the calculation. Explicitly,

$$
\begin{aligned}
0.28 \quad &\text{MeV} < \Gamma[D^+_{13}(1512)] < 0.38 \quad \text{MeV},\\
0.01 \quad &\text{MeV} < \Gamma[D^0_{13}(1512)] < 0.12 \quad \text{MeV},\\
0.42 \quad &\text{MeV} < \Gamma[S^+_{11}(1550)] < 1.41 \quad \text{MeV},\\
0.14 \quad &\text{MeV} < \Gamma[S^0_{11}(1550)] < 0.85 \quad \text{MeV},\\
0.16 \quad &\text{MeV} < \Gamma[F^+_{15}(1690)] < 0.25 \quad \text{MeV}, \qquad \text{(II-114)}\\
0.36 \quad &\text{MeV} < \Gamma[S^+_{11}(1710)] < 0.88 \quad \text{MeV},\\
0.0 \quad &\text{MeV} < \Gamma[S^0_{11}(1710)] < 0.25 \quad \text{MeV},\\
0.26 \quad &\text{MeV} < \Gamma[D^+_{13}(1730)] < 0.36 \quad \text{MeV},\\
0.24 \quad &\text{MeV} < \Gamma[D^0_{13}(1730)] < 0.65 \quad \text{MeV}.
\end{aligned}
$$

These results are quite reasonable for $\Gamma[D^+_{13}(1512)]$ and $\Gamma[F^+_{15}(1690)]$, but $\Gamma[S^+_{11}(1550)]$ is too large. Further, the radiative widths of the neutral $D_{13}(1512)$ and $S_{11}(1550)$, compared to their charged counterparts, are smaller than one would have expected, but conversely the radiative width of the neutral $F_{15}(1690)$ is somewhat larger than we would like. The large, and apparently damning values for $\Gamma[S^+_{11}(1710)]$ and $\Gamma[D^+_{13}(1730)]$ are not a problem, since both depend critically on the choice of mixing angle and consequently can be reduced easily.

Although numerically rather shaky in places, the symmetric quark model is clearly capable of giving the more obvious qualitative features of photo-production. More insight into this can be gained if the amplitudes for the πN decays of the resonances are studied, rather than the partial widths. For example, Copley *et al.* (1969) have investigated the helicity-$\frac{1}{2}$ amplitudes for $D_{13}(1512)$ and $F_{15}(1690)$, and have shown that

$$
A_{1/2}[D_{13}(1512)] \propto [(k^2/\alpha^2) - g^{-1}], \qquad A_{1/2}[F_{15}(1690)] \propto [(k^2/2\alpha^2) - g^{-1}].
$$
$$\text{(II-115)}$$

Experimentally, we have seen that both of these amplitudes are small and are consistent with zero. Both amplitudes can vanish simultaneously if

$$
k^2[D_{13}(1512)] \simeq \tfrac{1}{2} k^2[F_{15}(1690)], \qquad \text{(II-116)}
$$

which is approximately true, the former having the value 0.22 GeV^2 and the latter 0.17 GeV^2, which is sufficient to make both amplitudes small for suitable α and g. The values required are somewhat different from those we have used above, since $\alpha^2 = 0.1$ would require $g \simeq 0.5$, and $g > 1$ would require $\alpha^2 \gtrsim 0.2$.

Carrying this argument further, Walker (1969b) has shown that if the requirements of the photoproduction data are looked at in isolation, then the symmetric quark model can provide a reasonable description, with the choice of parameters $\mu_Q = \mu_p$, $g = 1.0$, and $\alpha^2 = 0.136$. The results of this

TABLE VI

COMPARISON OF EXPERIMENTAL AND SYMMETRIC QUARK MODEL
VALUES OF THE HELICITY AMPLITUDES $A_{1/2}$ AND $A_{3/2}$[a]

Resonance	Target (proton or neutron)	$A_{1/2}$ (GeV)$^{-1/2}$		$A_{3/2}$ (GeV)$^{-1/2}$	
		Model	Experiment	Model	Experiment
$P_{33}(1236)$	p	−0.10	−0.14	−0.17	−0.24
$D_{13}(1512)$	p	−0.04	−0.03 ± 0.02	0.11	0.15 ± 0.03
	n	−0.03	—	−0.11	−0.13 ± 0.05
$S_{11}(1540)$	p	0.16	0.08 ± 0.04		
	n	−0.11	−0.10 ± 0.05		
$D_{15}(1680)$	p	0	0	0	0.04 ± 0.04
	n	0.04	—	0.05	—
$F_{15}(1690)$	p	−0.01	0	0.07	0.14 ± 0.05
	n	0.04	—	0	0

[a] The results of the quark model calculation are taken from Walker (1969b), the experimental values are based on the analysis of Walker (1969a), and the errors are estimated from the conclusions of Moorhouse and Rankin (1970).

calculation are given in Table VI in terms of the amplitudes $A_{1/2}$ and $A_{3/2}$ defined by

$$A_{l\pm} = \mp [\pi^{-1}(k/q)(2J+1)^{-1}(M/M_R)(\Gamma_\pi/\Gamma)\Gamma^{-1}]^{1/2} C_{\pi N} A_{1/2},$$
$$B_{l\pm} = \pm [\pi^{-1}(k/q)(2J+L)^{-1}(M/M_R)(\Gamma_\pi/\Gamma)\Gamma^{-1}]^{1/2} \qquad \text{(II-117)}$$
$$\times [[(2J-1)(2J+3)]^{-1}]^{1/2} C_{\pi N} A_{3/2},$$

where $C_{\pi N}$ is the appropriate Clebsch-Gordan coefficient for the resonance decay into a specific πN state. The radiative width is given by

$$\Gamma_\gamma = (k^2/\pi)(M/M_R)[2/(2J+1)][|A_{3/2}|^2 + |A_{1/2}|^2]. \qquad \text{(II-118)}$$

As can be seen from the table, the price of improving the $D_{13}(1512)$, $S_{11}(1550)$, and $F_{15}(1690)$ states is a considerable worsening in the $P_{33}(1236)$ state. Apart from this, the description offered is quite remarkable.

The model, as it stands, gives immediately a prediction for the form factors of the resonances, and a discussion of this aspect of the symmetric quark model with the simple harmonic oscillator potential has been given by Thornber (1968a). From the expressions given above, it is obvious that all the form factors will be proportional to $\exp(-k^2/3\alpha^2)$. This ultimately falls off much too rapidly with increasing momentum transfer, although the multiplicative polynomial in k is sufficient to delay this rapid decrease for momentum transfer up to 0.5 or 1.0 $(\mathrm{GeV}/c)^2$, and consequently the model agrees reasonably well with the data at small momentum transfers.

This exponential dependence is a natural consequence of the harmonic oscillator potential, since the Fourier transform of the Gaussian harmonic oscillator wave function is just another Gaussian. It is interesting to ask the question of which input potential will yield a dipole fit to the form factors (or at least to the elastic proton form factors). Since the Fourier transform of $\exp(-ar)$ is proportional to $(1 + k^2/a\alpha^2)^{-2}$, a $1/r$ potential (that is, a Coulomb potential) will achieve this, since the ground state wave function is just an exponential. This potential has been studied in detail by Thornber (1968b). It should be noted that the price paid to achieve the correct form factors is a heavy one, since the mass spectrum predicted by the $1/r$ potential does not agree with experiment particularly well, and apart from $P_{33}(1236)$, the normalization of the states is completely wrong. Even for the nucleon and $P_{33}(1236)$ the momentum transfer is still not ideal, although very much improved. Taking the nucleon as an example, the reason for this is that while the form factor obtained is of the correct dipole form, it involves the three-momentum transfer rather than the four-momentum transfer and ultimately falls off too quickly. A typical comparison is made in Fig. 55, where the dipole fit to $|G_{E_p}|^2$ is compared with the predictions of the symmetric quark

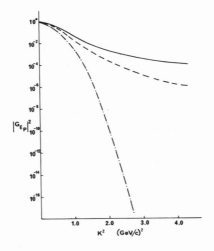

FIG. 55. Comparison of a dipole form factor (solid line) with the predictions of the symmetric quark model with the singular Coulomb potential (broken line) and with the simple harmonic oscillator potential (dot-dash line).

model with the singular Coulomb potential and the simple harmonic oscillator potential.

Each of the above calculations has had to assume para-Fermi statistics for the quarks since the ground-state wave function has a symmetric space part. A calculation of the nucleon form factor using normal Fermi statistics for the quarks with an explicitly antisymmetric three-particle ground-state wave function has been made by Meyer (1969). The *ansatz* for the wave function was

$$\psi(\mathbf{r}_1, \mathbf{r}_2, \mathbf{r}_3) = (R^6)^{-1}(\mathbf{r}_1{}^2 - \mathbf{r}_2{}^2)(\mathbf{r}_2{}^2 - \mathbf{r}_3{}^2)(\mathbf{r}_3{}^2 - \mathbf{r}_1{}^2)f(R), \quad \text{(II-119)}$$

with

$$R^2 = \mathbf{r}_1{}^2 + \mathbf{r}_2{}^2 + \mathbf{r}_3{}^2, \qquad f(R) = 48(6/\pi^3 R^3)^{1/2}(R_{3/2})^{-1} \exp(-R/R_0),$$
$$\text{(II-120)}$$

the constant coming from the normalization of the wave function. This yields a form factor

$$F(k^2) = [32/5(y + 2)^8](19y^4 + 48y^3 + 88y^2 + 80y + 40) \quad \text{(II-121)}$$

with

$$y = (1 + \tfrac{1}{6}k^2 R_0{}^2)^{1/2} - 1.$$

The arbitrary constant R_0 was obtained by matching to the proton root-mean-square radius, yielding a value of $R_0{}^2 = 0.3f^2$.

As it stands, the expression for the form factor is not very good for large momentum transfers, but, unlike previous calculations, it is not too small but too big (asymptotically being $14/k^4$ as compared with the dipole fit $0.5/k^4$). This is a situation which is easily remedied by assuming that the quarks themselves have a form factor of vector meson dominance type, an assumption which is by no means unreasonable. Taking the nucleon form factor as the product of the Schrödinger form factor, such a vector dominance quark form factor yields a result in excellent agreement with experiment, and is shown in Fig. 56.

FIG. 56. Comparison of the nucleon form-factor data with the calculation of Meyer in the quark model, assuming Fermi statistics and an explicitly antisymmetric three-particle ground-state wave function.

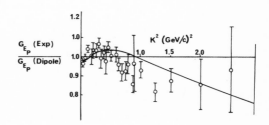

This argument has been extended by Krammer (1969) trivially to $P_{33}(1236)$ [which is the other member of the 56 ($L = 0^+$) supermultiplet] and nontrivially to $D_{13}(1512)$ [and indeed in general to the 70 ($L = 1^-$) supermultiplet]. The *ansatz* of the wave function of the latter was

$$\psi^{L=1}(\mathbf{r}_1, \mathbf{r}_2, \mathbf{r}_3) = (R^6)^{-1}(\mathbf{r}_1 - \mathbf{r}_2)(\mathbf{r}_2 - \mathbf{r}_3)(\mathbf{r}_3 - \mathbf{r}_1)48(6/\pi^3 R_0{}^5)^{-1}$$
$$\times (R^{3/2})^{-1} \exp(-R/r_0)\tfrac{1}{3}(2\mathbf{r}_3 - \mathbf{r}_1 - \mathbf{r}_2), \qquad \text{(II-122)}$$

the extra factor just representing the position vector of particle 3 in the three-particle rest frame. The resultant form factors for $P_{33}(1236)$ are

$$(M_R/M)|f_c|^2 = 0, \qquad (M_R/M)(|f_+|^2 + |f_-|^2) = \tfrac{16}{9}\mu_a{}^2k^2[F_2(kR_0)]^2.$$
$$\text{(II-123)}$$

For $D_{13}(1512)$ they are

$$(M_R/M)(|f_+|^2 + |f_-|^2) = (2\mu_Q{}^2/R_0{}^2)[\{(k^2R_0/3)F_1(kR_0)\}$$
$$- (2/3g(k)^2R_0/3)F_1(kR_0)F_2(kR_0)$$
$$+ (4/9g^2)\{F_2(kR_0)\}^2]. \qquad \text{(II-124)}$$

For $S_{11}(1550)$ they are

$$(M_R/M)|f_c|^2 = \tfrac{1}{18}k^2R_0{}^2\{F_1(kR_0)\}^2,$$
$$(M_R/M)(|f_+|^2 + |f_-|^2) = (\mu^2/R_0)[(k^2R_0/3)F_1(kR_0) + (2/3g)F_2(kR_0)]^2,$$
$$\text{(II-125)}$$

where

$$F_1(kR_0) = [4(2128y^4 + 2464y^3 + 6720y^2 + 5824y + 4480)]/35(y + 1)(y + 2)^9,$$
$$F_2(kR_0) = [32(19y^4 + 48y^3 + 88y^2 + 80y + 40)]/5(y + 2)^8,$$
$$\text{(II-126)}$$

with, as before,

$$y = (1 + \tfrac{1}{6}k^2R_0{}^2)^{1/2} - 1.$$

Taking $\mu_Q = \mu_p$ (as usual) and R_0 as obtained from the nucleon form factor leaves only g as a free parameter and gives a unique prediction for the $P_{33}(1236)$ resonance. This is shown in Fig. 57, and it is an extremely encouraging result. In particular the improved agreement with the photoproduction point should be noted.

The situation in the second resonance region is not so satisfactory. As in previous models it is possible to get reasonable values for $D_{13}(1512)$ (a photoproduction cross section of 46 μb for $g = 1$ and 37 μb for $g = 30$), but the values for $S_{11}(1550)$ are too large (30 μb for $g = 30$ and 130 μb for $g = 1$!).

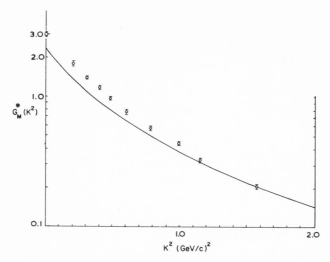

FIG. 57. Comparison of the $\Delta(1238)$ form-factor data with the calculation of Krammer in the quark model, assuming Fermi statistics and an explicitly antisymmetric three-particle ground-state wave function.

The latter, of course, can be reduced by mixing, but this has the consequence of yielding a rather large cross section for $S_{11}(1710)$. In addition, the momentum-transfer dependence is incorrect and at 5 $(GeV/c)^2$ the theoretical cross section is too small by at least a factor of 10.

III. High Energy Production

The mechanism of high energy meson photoproduction is that of Regge pole exchange, with the associated Regge cuts. This section begins with an outline of the formalism, followed by its application to forward and backward photoproduction of pseudoscalar mesons and the forward production of vector mesons. The vector dominance model is first introduced in the discussion of vector meson photoproduction on complex nuclei, and is then extended and applied to meson photoproduction generally. The section ends with a brief account of the recent developments in high energy electroproduction, a field of study which is only now beginning to develop.

A. REGGE POLE FORMALISM

The general approach to reggeization (see, for example, Collins and Squires, 1968) is to start with the t-channel reaction

$$1 + 2 \rightarrow 3 + 4$$

and to construct kinematically well-behaved amplitudes $F_{\lambda_3\lambda_4;\lambda_1\lambda_2}(s, t)$ from the usual t-channel helicity amplitudes $f_{\lambda_3\lambda_4;\lambda_1\lambda_2}(s, t)$ of Jacob and Wick (1959). The result has the general form

$$F^{\pm}_{\lambda_3\lambda_4;\,\lambda_1\lambda_2}(s, t) = K^{\pm}_{\lambda_3\lambda_4;\,\lambda_1\lambda_2}(t) \{\hat{f}_{\lambda_3\lambda_4;\,\lambda_1\lambda_2}(s, t) \pm \hat{f}_{-\lambda_3-\lambda_4;\,\lambda_1\lambda_2}(s, t)\}, \qquad \text{(III-1)}$$

where the $K^{\pm}_{\lambda_3\lambda_4;\,\lambda_1\lambda_2}(t)$ are kinematical factors depending on the masses and spins of the particles involved, and the $f_{\lambda_3\lambda_4;\,\lambda_1\lambda_2}(s, t)$ are given in terms of the helicity amplitudes by

$$\hat{f}_{\lambda_3\lambda_4;\,\lambda_1\lambda_2}(s, t) = [\cos \tfrac{1}{2}\theta_t]^{-|\lambda+\mu|}[\sin \tfrac{1}{2}\theta_t]^{-|\lambda-\mu|} \hat{f}_{\lambda_3\lambda_4;\,\lambda_1\lambda_2}(s, t), \qquad \text{(III-2)}$$

where $\lambda = \lambda_1 - \lambda_2$, $\mu = \lambda_3 - \lambda_4$, and θ_t is the center-of-mass scattering angle in the t channel.

The next step is to write the $F_{\lambda_3\lambda_4;\,\lambda_1\lambda_2}(s, t)$ in terms of the parity conserving partial-wave helicity amplitudes, $g^{J\pm}_{\lambda_3\lambda_4;\,\lambda_1\lambda_2}(s, t)$. The expansion is

$$F^{\pm}_{\lambda_3\lambda_4;\,\lambda_1\lambda_2}(s, t) = \sum_J (2J + 1)\{g^{J\pm}_{\lambda_3\lambda_4;\,\lambda_1\lambda_2}(s, t)l^{J\pm}_{\lambda,\mu}(\theta_t) + g^{J}_{\lambda_3\lambda_4;\,\lambda_1\lambda_2}(\theta_t)l^{J-}_{\lambda,\mu}(\theta_t)\}. \qquad \text{(III-3)}$$

The functions $l^{J\pm}_{\lambda,\mu}(\theta)$ are related to the rotation function $d^J_{\lambda,\mu}(\theta)$ by

$$l^{J\pm}_{\lambda,\mu}(\theta) = \tfrac{1}{2}\{[\sqrt{2}\cos \tfrac{1}{2}\theta]^{-|\lambda+\mu|}[\sqrt{2}\sin \tfrac{1}{2}\theta]^{-|\lambda-\mu|}d^J_{\lambda,\mu}(\theta)$$
$$\times (-1)^{\lambda+\lambda_M}[\sqrt{2}\sin \tfrac{1}{2}\theta]^{-|\lambda+\mu|}[\sqrt{2}\cos \tfrac{1}{2}\theta]^{-|\lambda-\mu|}d^J_{\lambda,\mu}(\theta)\}, \qquad \text{(III-4)}$$

where $\lambda_M = \max(|\lambda|, |\mu|)$, and asymptotically $(s \to \infty)$ they behave as

$$l^{J\pm}_{\lambda,\mu}(\theta_t) \sim (\cos \theta_t)^{J-\lambda_M}, \qquad l^{J-}_{\lambda,\mu}(\theta_t) \sim (\cos \theta_t)^{J-\lambda_M-1}. \qquad \text{(III-5)}$$

Thus after reggeization, $l^{J\pm}_{\lambda,\mu}(\theta)$ gives the leading contribution, and the quantum numbers of $g^{J\pm}_{\lambda_3\lambda_4;\,\lambda_1\lambda_2}(s, t)$ associated with it become the quantum number of the Regge poles contributing to $F^{J\pm}_{\lambda_3\lambda_4;\,\lambda_1\lambda_2}$. Following through the standard procedures, the amplitudes can be written at suitably high energies in the leading powers of s, as

$$F_{\lambda_3\lambda_4;\,\lambda_1\lambda_2} = \sum_\alpha G_{\lambda_3\lambda_4;\,\lambda_1\lambda_2}(t, \alpha)\xi_\alpha\, C_{\lambda_3\lambda_4;\,\lambda_1\lambda_2}(\alpha)(s/s_0)^{\alpha-\lambda_M}, \qquad \text{(III-6)}$$

the summation running over all Regge trajectories $\alpha(t)$ contributing to the particular helicity state. Here ξ_α is the usual signature factor:

$$\xi_\alpha = \{1 \pm \exp[-i\pi\alpha(t)]\}/\sin[\pi\alpha(t)]. \qquad \text{(III-7)}$$

The term $G_{\lambda_3\lambda_4;\,\lambda_1\lambda_2}(t, \alpha)$ is the residue function (which may contain factors of $(\alpha + 1)$, $(\alpha + 2), \ldots$ to cancel the zeros of $\sin(\pi\alpha)$ at negative integer values of α) and $C_{\lambda_3\lambda_4;\lambda_1\lambda_2}(\alpha)$ contains factors $(\alpha - 2)$, $(\alpha - 1)$, and α raised to suitable powers, depending on the choosing mechanism of the trajectory at the sense, nonsense points $\alpha = m$, $m - 1$, 0 in the process being considered. The precise details of these terms for different reactions will be given in the following section.

In discussing the details of the reggeization of the amplitude and the applications, the order of the particles in the s- and t-channel reactions will be taken to be

$$s \text{ channel: } \gamma + F \rightarrow B + F', \qquad t \text{ channel: } \gamma + \bar{B} \rightarrow \bar{F} + F', \quad \text{(III-8)}$$

B being the meson, mass μ (pseudoscalar or vector) photoproduced and F, F' being the initial and final fermions, masses M_F and $M_{F'}$ (not necessarily the same) in the s channel.

Many of the necessary kinematical quantities have been given already in Section I. Additional ones required here are

$$\cos \theta_s = \frac{2st + s^2 - s(M_F^2 + M_{F'}^2 + \mu^2) + M_F^2(M_F^2 - \mu^2)}{(s - M_F^2)[\{s - (M_{F'} + \mu)^2\}\{s - (M_{F'} - \mu)^2\}]^{1/2}},$$

$$\sin \theta_s = \frac{2\sqrt{s}[\phi(s, t)]^{1/2}}{(s - M_F^2)[\{s - (M_{F'} + \mu)^2\}\{s - (M_{F'} - \mu)^2\}]^{1/2}},$$

$$\cos \theta_t = \frac{2st + t^2 - t(M_F^2 + M_{F'}^2 - \mu^2) - \mu^2(M_F^2 - M_{F'}^2)}{(t - \mu^2)[\{t - (M_F + M_{F'})^2\}\{t - (M_F - M_{F'})^2\}]^{1/2}},$$

$$\sin \theta_t = \frac{2\sqrt{t}[\phi(s, t)]^{1/2}}{(t - \mu^2)[\{t - (M_F + M_{F'})^2\}\{t - (M_F - M_{F'})^2\}]^{1/2}},$$

(III-9)

with

$$\phi(s, t) = -s^2 t - t^2 s - st(M_F^2 + M_{F'}^2 + \mu^2) - s\mu^2(M_{F'}^2 - M_F^2)$$
$$- tM_F^2(M_{F'}^2 - \mu^2) + M_F^2\mu^2(M_{F'}^2 - M_F^2 - \mu^2). \quad \text{(III-10)}$$

Since the Regge pole parametrization gives only the t-channel helicity amplitudes, it is necessary in general to obtain the crossing matrix relating these to the s-channel helicity amplitudes. Following Cohen-Tannoudji *et al.* (1968) or Ader *et al.* (1968) the result for our choice of the order of particles is

$$f^{(s)}_{\lambda_3\lambda_4; \lambda_1\lambda_2}(s, t) = \exp[i\pi(\lambda_2 - \lambda_3)] \sum_{\lambda_i'} d^{s_1}_{\lambda_1\lambda_1'}(\chi_1)\, d^{s_2}_{\lambda_2\lambda_2'}(\chi_2)$$

$$\times d^{s_3}_{\lambda_3\lambda_3'}(\chi_3)\, d^{s_4}_{\lambda_4\lambda_4'}(\chi_4)\, f^{(t)}_{\lambda_4'\lambda_1'; \lambda_3'\lambda_2'}(s, t), \quad \text{(III-11)}$$

where the superscripts (s) and (t) denote the s-channel and t-channel helicity amplitudes, respectively, and the angles χ_1, χ_2, χ_3, and χ_4 are given by the following Eqs. (II-12):

$$\cos \chi_1 = -1, \qquad \sin \chi_1 = 0,$$

$$\cos \chi_2 = \frac{(s + M_F^2)(t + M_F^2 - M_{F'}^2) - 2M_F^2(M_F^2 - M_{F'}^2 + \mu^2)}{(s - M_F)[\{t - M_{F'} + M_F)\}^2 \{t - (M_{F'} - M_F)^2\}]^{1/2}}$$

$$\sin \chi_2 = \frac{2M_F[\phi(s, t)]^{1/2}}{(s - M_F)[\{t - (M_{F'} + M_F)^2\}\{t - (M_{F'} - M_F)^2\}]^{1/2}},$$

$$\cos \chi_3 = \frac{(s + \mu^2 - M_{F'}^2)(t + \mu^2) - 2\mu^2(M_F^2 - M_{F'}^2 + \mu^2)}{(t - \mu^2)[\{s - (M_{F'} + \mu)^2\}\{s - (M_{F'} - \mu)^2\}]^{1/2}},$$

$$\sin \chi_3 = \frac{-2\mu[\phi(s, t)]^{1/2}}{(t - \mu^2)[\{s - (M_{F'} +\ )^2\}\{s - (M_{F'} - \mu)^2\}]^{1/2}}, \qquad \text{(III-12)}$$

$$\cos \chi_4 = -\frac{(s + M_{F'}^2 - \mu^2)(t + M_{F'}^2 - M_F^2) + 2M_{F'}^2(M_F^2 - M_{F'}^2 + \mu^2)}{[\{t - (M_{F'} + M_F)^2\}\{t - (M_{F'} - M_F)^2\}]^{1/2}},$$
$$\times [\{s - (M_{F'} + \mu)^2\}\{s - (M_{F'} - \mu)^2\}]^{1/2}$$

$$\sin \chi_4 = \frac{-2M_{F'}[\phi(s, t)]^{1/2}}{[\{t - (M_{F'} + M_F)^2\}\{t - (M_{F'} - M_F)^2\}]^{1/2}}.$$
$$\times [\{s - (M_{F'} + \mu)^2\}\{s - (M_{F'} - \mu)^2\}]^{1/2}$$

It should be noted that if one only is interested in obtaining differential cross sections, with unpolarized initial and final states then the use of the crossing procedure is not explicitly necessary. In this case, the orthogonality of the crossing matrices can be invoked, permitting the s-channel differential cross section to be written directly in terms of the t-channel helicity amplitudes:

$$(d\sigma/dt)^{(s)} = [4\pi(s - M^2)]^{-1} \sum_{\lambda_1\lambda_2\lambda_3\lambda_4} |f_{\lambda_3\lambda_4;\, \lambda_1\lambda_2}^{(t)}(s, t)|^2. \qquad \text{(III-13)}$$

It is also possible to obtain the vector meson density matrix elements in terms of the t-channel helicity amplitudes alone if the Jackson frame is used (and all other particles are unpolarized). Following Gottfried and Jackson (1964), the density matrix elements are

$$\rho_{mm'} = \sum_{\lambda_1\lambda_2\lambda_\gamma} f_{\lambda_1\lambda_2;\, \lambda_\gamma m}^{(t)*}(s, t) f_{\lambda_1\lambda_2;\, \lambda_\gamma m'}^{(t)}, \qquad \text{(III-14)}$$

where λ_1, λ_2 are the helicities of the antinucleon and nucleon, respectively, λ_λ is the helicity of the photon, and m, m' denote the matrix elements (helicities) of the vector meson. As we have already seen, amplitudes involving negative photon helicity can be expressed in terms of the amplitudes involving photon helicity by parity conservation,

$$f_{\lambda_1\lambda_2;-1\lambda} = (-1)^{\lambda_1 - \lambda_2 + \lambda + 1} f_{-\lambda_1 -\lambda_2; 1\lambda}. \qquad \text{(III-15)}$$

Thus, the relevant density matrix elements are

$$\rho_{00}(s, t) = 2\{|f_{\frac{1}{2}\frac{1}{2};\,1\,1}(s, t)|^2 + |f_{\frac{1}{2}\,-\frac{1}{2};\,1\,0}(s, t)|^2 + |f_{-\frac{1}{2}\frac{1}{2};\,1\,0}(s, t)|^2$$
$$+ |f_{-\frac{1}{2}\,-\frac{1}{2};\,1\,0}(s, t)|^2\},$$

$$\rho_{1\,-1}(s, t) = 2\,\mathrm{Re}\{f_{\frac{1}{2}\frac{1}{2};\,1\,1}(s, t)f_{\frac{1}{2}\frac{1}{2};\,1\,-1}(s, t) + f^*_{\frac{1}{2}\,-\frac{1}{2};\,1\,1}(s, t)f_{\frac{1}{2}\,-\frac{1}{2};\,1\,-1}(s, t)$$
$$+ f^*_{-\frac{1}{2}\frac{1}{2};\,1\,1}(s, t)f_{-\frac{1}{2}\frac{1}{2};\,1\,-1}(s, t)$$
$$+ f^*_{\frac{1}{2}\,-\frac{1}{2};\,1\,1}(s, t)f_{-\frac{1}{2}\,-\frac{1}{2};\,1\,-1}(s, t)\},$$

$$\rho_{10}(s, t) = \{[f^*_{\frac{1}{2}\frac{1}{2};\,1\,1}(s, t) - f^*_{\frac{1}{2}\frac{1}{2};\,1\,-1}(s, t)]f_{\frac{1}{2}\frac{1}{2};\,1\,0}(s, t) + [f^*_{\frac{1}{2}\,-\frac{1}{2};\,1\,1}(s, t)$$
$$- f^*_{\frac{1}{2}\,-\frac{1}{2};\,1\,-1}(s, t)]f_{\frac{1}{2}\,-\frac{1}{2};\,1\,0}(s, t) + [f^*_{-\frac{1}{2}\frac{1}{2};\,1\,1}(s, t)$$
$$- f^*_{-\frac{1}{2}\frac{1}{2};\,1\,-1}(s, t)]f_{-\frac{1}{2}\frac{1}{2};\,1\,-1}(s, t) + [f^*_{-\frac{1}{2}\,-\frac{1}{2};\,1\,1}(s, t)$$
$$- f^*_{-\frac{1}{2}\,-\frac{1}{2};\,1\,-1}(s, t)]f_{-\frac{1}{2}\,-\frac{1}{2};\,1\,0}(s, t)\}. \tag{III-16}$$

The necessary kinematical factors and combinations of helicity amplitudes which are free from kinematical singularities have been obtained by Ader *et al.* (1968) for the general mass and spin case. The results for pseudoscalar and vector meson photoproduction are given below.

1. *Pseudoscalar meson case (equal mass fermions):*

$$F^1_{\frac{1}{2}\frac{1}{2};\,1\,0}(s, t) = (t - \mu^2)^{-1}\{\hat{f}_{\frac{1}{2}\frac{1}{2};\,1\,0}(s, t) - \hat{f}_{-\frac{1}{2}\,-\frac{1}{2};\,1\,0}(s, t)\} \equiv F^1_{0\,1}(s, t),$$

$$F^2_{\frac{1}{2}\frac{1}{2};\,1\,0}(s, t) = \{t/(t - 4M^2)\}^{1/2}(t - \mu^2)^{-1}\{\hat{f}_{\frac{1}{2}\frac{1}{2};\,1\,0}(s, t) + \hat{f}_{-\frac{1}{2}\,-\frac{1}{2};\,1\,0}(s, t)\}$$
$$\equiv F^2_{0\,1}(s, t),$$

$$F^1_{\frac{1}{2}\,-\frac{1}{2};\,1\,0}(s, t) = [\sqrt{t}/(t - \mu^2)]\{\hat{f}_{\frac{1}{2}\,-\frac{1}{2};\,1\,0}(s, t) - \hat{f}_{-\frac{1}{2}\frac{1}{2};\,1\,0}(s, t)\} \equiv F^1_{1\,1}(s, t),$$

$$F^2_{\frac{1}{2}\,-\frac{1}{2};\,1\,0}(s, t) = [t/(t - \mu^2)][\{t - 4M^2\}^{1/2}]^{-1}\{\hat{f}_{\frac{1}{2}\,-\frac{1}{2};\,1\,0}(s, t) + \hat{f}_{-\frac{1}{2}\frac{1}{2};\,1\,0}(s, t)\}$$
$$\equiv F^2_{1\,1}(s, t). \tag{III-17}$$

2. *Pseudoscalar meson case (unequal mass fermions):*

$$F^1_{\frac{1}{2}\frac{1}{2};\,1\,0}(s, t) = (t - \mu^2)^{-1}[t/\{t - (M - Y)^2\}]^{1/2}\{\hat{f}_{\frac{1}{2}\frac{1}{2};\,1\,0}(s, t)$$
$$- \hat{f}_{-\frac{1}{2}\,-\frac{1}{2};\,1\,0}(s, t)\} \equiv F^1_{0\,1}(s, t),$$

$$F^2_{\frac{1}{2}\frac{1}{2};\,1\,0}(s, t) = (t - \mu^2)^{-1}\{t - (M + Y)^2\}^{1/2}\{\hat{f}_{\frac{1}{2}\frac{1}{2};\,1\,0}(s, t)$$
$$+ \hat{f}_{-\frac{1}{2}\frac{1}{2};\,1\,0}(s, t)\} \equiv F^2_{0\,1}(s, t), \tag{III-18}$$

$$F^1_{\frac{1}{2}\,-\frac{1}{2};\,1\,0}(s, t) = [t/(t - \mu^2)][\{t - (M - Y)^2\}^{1/2}]^{-1}\{\hat{f}_{\frac{1}{2}\,-\frac{1}{2};\,1\,0}(s, t)$$
$$- \hat{f}_{-\frac{1}{2}\frac{1}{2};\,1\,0}(s, t)\} \equiv F^1_{1\,1}(s, t),$$

$$F^2_{\frac{1}{2}\,-\frac{1}{2};\,1\,0}(s, t) = [t/(t - \mu^2)][\{t - (M + Y)^2\}^{1/2}]^{-1}\{\hat{f}_{\frac{1}{2}\,-\frac{1}{2};\,1\,0}(s, t)$$
$$+ \hat{f}_{-\frac{1}{2}\frac{1}{2};\,1\,0}(s, t)\} \equiv F^2_{1\,1}(s, t).$$

Here we have reverted to our original notation, with M denoting the mass of the nucleon and Y the mass of the hyperon (Λ, Σ).

Vector meson case:

$$F^1_{\frac{1}{2}\frac{1}{2};\,1\,1}(s, t) = (t - \mu^2)\sqrt{t}\{\hat{f}_{\frac{1}{2}\frac{1}{2};\,1\,1}(s, t) - \hat{f}_{-\frac{1}{2}-\frac{1}{2};\,1\,1}(s, t)\},$$

$$F^2_{\frac{1}{2}\frac{1}{2};\,1\,1}(s, t) = (t - \mu^2)(t - 4M^2)^{1/2}\{\hat{f}_{\frac{1}{2}\frac{1}{2};\,1\,1}(s, t) + \hat{f}_{-\frac{1}{2}-\frac{1}{2};\,1\,1}(s, t)\},$$

$$F^1_{-\frac{1}{2}\frac{1}{2};\,1\,1}(s, t) = \{\hat{f}_{-\frac{1}{2}\frac{1}{2};\,1\,1}(s, t) - \hat{f}_{\frac{1}{2}-\frac{1}{2};\,1\,1}(s, t)\},$$

$$F^2_{-\frac{1}{2}\frac{1}{2};\,1\,1}(s, t) = \{t/(t - 4M^2)\}^{1/2}\{\hat{f}_{-\frac{1}{2}\frac{1}{2};\,1\,1}(s, t) + \hat{f}_{\frac{1}{2}-\frac{1}{2};\,1\,1}(s, t)\},$$

$$F^1_{\frac{1}{2}\frac{1}{2};\,1\,0}(s, t) = -\{\hat{f}_{\frac{1}{2}\frac{1}{2};\,1\,0}(s, t) + \hat{f}_{-\frac{1}{2}-\frac{1}{2};\,1\,0}(s, t)\},$$

$$F^2_{\frac{1}{2}\frac{1}{2};\,1\,0}(s, t) = -\{t/(t - 4M^2)\}^{1/2}\{\hat{f}_{\frac{1}{2}\frac{1}{2};\,1\,0}(s, t) - \hat{f}_{-\frac{1}{2}-\frac{1}{2};\,1\,0}(s, t)\},$$

$$F^1_{-\frac{1}{2}\frac{1}{2};\,1\,0}(s, t) = -\sqrt{t}\{\hat{f}_{-\frac{1}{2}\frac{1}{2};\,1\,0}(s, t) + \hat{f}_{-\frac{1}{2}\frac{1}{2};\,1\,0}(s, t)\}, \qquad \text{(III-19)}$$

$$F^2_{-\frac{1}{2}\frac{1}{2};\,1\,0}(s, t) = -\{t/(t - 4M^2)\}^{1/2}\{\hat{f}_{-\frac{1}{2}\frac{1}{2};\,1\,0}(s, t) - \hat{f}_{-\frac{1}{2}\frac{1}{2};\,1\,0}(s, t)\},$$

$$F^1_{\frac{1}{2}\frac{1}{2};\,1\,-1}(s, t) = [\sqrt{t/(t - \mu^2)(t - 4M^2)}]\{\hat{f}_{\frac{1}{2}\frac{1}{2};\,1\,-1}(s, t) - \hat{f}_{-\frac{1}{2}-\frac{1}{2};\,1\,-1}(s, t)\},$$

$$F^2_{\frac{1}{2}\frac{1}{2};\,1\,-1}(s, t) = [(t - \mu^2)(t - 4M^2)^{1/2}]^{-1}\{\hat{f}_{\frac{1}{2}\frac{1}{2};\,1\,-1}(s, t) + \hat{f}_{-\frac{1}{2}-\frac{1}{2};\,1\,-1}(s, t)\}.$$

$$F^1_{-\frac{1}{2}\frac{1}{2};\,1\,-1}(s, t) = [(t - \mu^2)(t - 4M^2)]^{-1}\{\hat{f}_{-\frac{1}{2}\frac{1}{2};\,1\,-\frac{1}{2}}(s, t) - \hat{f}_{\frac{1}{2}-\frac{1}{2};\,1\,-1}(s, t)\},$$

$$F^2_{-\frac{1}{2}\frac{1}{2};\,1\,-1}(s, t) = [(t - \mu^2)]\{t/(t - 4M^2)\}^{1/2}\{\hat{f}_{-\frac{1}{2}\frac{1}{2};\,1\,-1}(s, t) + \hat{f}_{\frac{1}{2}-\frac{1}{2};\,1\,-1}(s, t)\}.$$

As we have said, these amplitudes are free from any kinematical singularities and constitute an independent set of amplitudes. The only further point which need be considered is that of the constraint relations, which reduce the number of nonvanishing independent amplitudes at the thresholds and pseudo thresholds of the t-channel reactions, that is, at $t = \mu^2$ and $t = (M_F \pm M_{F'})^2$. Again following Cohen-Tannoudji *et al.* (1968), the constraints are as follows.

1. *Pseudoscalar meson case (equal mass fermions):*

  (i)   At $t = 0$: $(F^1_{1\,1} + 2MF^2_{0\,1})$ goes to zero as t;
  (ii) at $t = \mu^2$: no constraint equation;
  (iii) at $t = 4M^2$: $(2MF^1_{1\,0} - F^1_{1\,1})$ goes to zero as $(t - 4M^2)$.

2. *Pseudoscalar meson case (unequal mass fermions):*

  (i)   At $t = 0$: $[(M_F - M_{F'})F^1_{1\,1} + (M_F + M_{F'})F^2_{1\,1}]$ goes to zero as t;
  (ii) at $t = \mu^2$: no constraint equation;
  (iii) at $t = (M_{F'} - M_F)^2$: $\{(M_F - M_{F'})F^2_{0\,1} + F^1_{1\,1}\}$ goes to zero as $\{t - (M_F - M_{F'})^2\}$;

(iv) at $t = (M_{F'} + M_F)^2$: $\{(M_F + M_{F'})F_{0\,1}^1 - F_{11}^1\}$ goes to zero as $\{t - (M_F + M_{F'})^2\}$.

In most applications of the formalism, the constraint at $t = 4M^2$ is sufficiently far from the physical region to be ignored. This is not true of the constraints at $t = 0$ and μ^2, for which allowance must be made. This can be done either by an evasive mechanism, in which all the Regge exchanges contributing to a constrained amplitude have the required t dependence built into their residue functions or by conspiracy when two (or more) Regge exchanges contribute to produce the required constraint behavior, although the individual Regge exchanges do not satisfy it. For example, in pion photoproduction one can either have, for small t

$$F_{1\,1}^1(t) \sim t, \quad F_{0\,1}^2(t) \sim t \quad \text{(evasion)}$$

or

$$F_{1\,1}^1(t) \sim -2MF_{0\,1}^2(t) \quad \text{(conspiracy)}.$$

When dealing with pseudoscalar meson photoproduction, it is frequently convenient to make use of the relations between the t-channel helicity amplitudes and the usual invariant amplitudes A_1, A_2, A_3, and A_4 rather than invoking the procedure of helicity crossing. These relations, which are obtained by projecting the invariant amplitudes on to the helicity amplitudes in the t channel, are (for π, η photoproduction, with equal mass fermions)

$$F_{0\,1}^1 = A_1 - 2MA_4, \quad F_{0\,1}^2 = A_1 + tA_2,$$
$$F_{1\,1}^1 = 2MA_1 - tA_4, \quad F_{1\,1}^2 = tA_3, \tag{III-20}$$

with the inverse

$$A_1 = [t/(t - 4M^2)]\{F_{0\,1}^1 - (2M/t)\,F_{1\,1}^1\},$$
$$A_2 = t^{-1}F_{0\,1}^2 - (t - 4M^2)^{-1}\{F_{0\,1}^1 - (2M/t)F_{1\,1}^1\},$$
$$A_3 = t^{-1}F_{1\,1}^2, \tag{III-21}$$
$$A_4 = (t - 4M^2)^{-1}\{2MF_{0\,1}^1 - F_{1\,1}^1\}.$$

For K photoproduction (with unequal mass fermions), they are

$$F_{0\,1}^1 = A_1 - (M + Y)A_4,$$
$$F_{0\,1}^2 = A_1 + [t - (M - Y)^2]A_2 + (M - Y)A_3,$$
$$F_{1\,1}^1 = (M + Y)A_1 - tA_4, \tag{III-22}$$
$$F_{1\,1}^2 = (M - Y)A_1 + tA_3,$$

with the inverse

$$A_1 = (t/[t - (M + Y)^2])\{F_{0\,1}^1 - [(M + Y)/t]F_{1\,1}^1\},$$

$$A_2 = [t - (M - Y)^2]^{-1}\{F_{0\,1}^2 - [(M - Y)/t]F_{1\,1}^2$$

$$+ ([t - (M + Y)^2]/[t - (M + Y)^2])(F_{0\,1}^1 - [(M + Y)/t]F_{1\,1}^1)\},$$

$$A_3 = t^{-1}F_{1\,1}^2 - \{(M - Y)/[t - (M + Y)^2]\}\{F_{0\,1}^1 - [(M + Y)/t]F_{1\,1}^1\},$$

$$A_4 = [t - (M + Y)^2]^{-1}\{(M + Y)F_{0\,1}^1 - F_{1\,1}^1\}.$$

$$\text{(III-23)}$$

There is an important result on pion photoproduction with linearly polarized photons due to Stichel (1964), frequently known as Stichel's theorem. This states that in the high energy limit, natural parity exchanges ($P = (-1)^J$] contribute only to $\sigma_\perp$, that is, to the cross section obtained with photons polarized perpendicular to the production plane, and unnatural parity exchanges [$P = (-1)^{J+1}$] contribute only to $\sigma_\parallel$, that is, to the cross section obtained with photons polarized parallel to the production plane.

To derive this result, we start from the expression for the differential cross section for pion photoproduction with linearly polarized photons in terms of the Pauli amplitudes $\mathscr{F}_i$, which can be conveniently rewritten as

$$d\sigma/d\Omega = (q/k)\{\sin^2 \phi[|\mathscr{F}_1 - \cos \theta \mathscr{F}_2|^2 + \sin^2 \theta|\mathscr{F}_2|^2$$

$$+ \cos^2 \phi|(\mathscr{F}_1 - \cos \theta \mathscr{F}_2) + \sin^2 \theta \mathscr{F}_4|^2$$

$$+ \sin^2 \theta|(\mathscr{F}_2 + \mathscr{F}_3) + \cos \theta \mathscr{F}_4|^2]\}.$$

$$\text{(III-24)}$$

Recalling the relation between the Pauli amplitudes $\mathscr{F}_i$ and the invariant amplitudes A_i, it is straightforward to derive the asymptotic limits

$$|\sin \theta|\mathscr{F}_2 \xrightarrow[s \to \infty]{} [(s|t|^{1/2})8\pi]A_4,$$

$$(\mathscr{F}_1 - \cos \theta \mathscr{F}_2) \xrightarrow[s \to \infty]{} (\sqrt{s}/8\pi)A_1,$$

$$\text{(III-25)}$$

$$(\mathscr{F}_1 - \cos \theta \mathscr{F}_2) + \sin^2 0\mathscr{F}_4 \xrightarrow[s \to \infty]{} (\sqrt{s}/8\pi)(A_1 + tA_2),$$

$$|\sin \theta|[(\mathscr{F}_2 + \mathscr{F}_3) + \cos \theta \mathscr{F}_4] \xrightarrow[s \to \infty]{} [(s|t|)^{1/2}/8\pi]A_3.$$

Now natural parity exchanges contribute to the helicity amplitudes $F_{0\,1}^1, F_{1\,1}^1$ but not to $F_{0\,1}^2, F_{1\,1}^2$ and hence, from (III-21), for natural parity exchanges,

$$A_3 = 0, \qquad A_1 + tA_2 = 0, \qquad d\sigma_\parallel/d\Omega = 0. \qquad \text{(III-26)}$$

On the other hand, unnatural parity exchanges do not contribute to $F_{0\,1}^1, F_{1\,1}^1$, but do contribute to $F_{0\,1}^2, F_{1\,1}^2$. In this case, again using (III-21),

$$A_1 = 0, \qquad A_4 = 0, \qquad d\sigma_\perp/d\Omega = 0. \qquad \text{(III-27)}$$

The theorem is exactly true only in the high energy limit $s \to \infty$, but even at the lower end of the energy range in which Regge theory is applicable the corrections to it are only of the order of a few percent.

B.   ALLOWED EXCHANGES AND THE INCLUSION OF CUTS

The procedure for obtaining the allowed exchange contributions to the different helicity amplitudes will be illustrated by explicit consideration of pion photoproduction. The results for other reactions shall only be stated.

The first step is to obtain the allowed exchanges in general, without specifying particular helicity amplitudes, by recalling that the exchanged state, B_{ex}, must have quantum numbers such that

(i)   B, I, P, G, C, S, and Q are conserved at the NNB_{ex} vertex;
(ii)   B, P, C, S, and Q are conserved at the $\gamma \pi B_{ex}$ vertex;
(iii)   the eigenvalues of the charge conjugation operator C is a good quantum number only for neutral states.

For pion photoproduction, the exchanged state must have $B = S = 0$. The unknown even-parity states are the two vacuum trajectories P and P' and the trajectories associated with the resonances $A_1(1^+)$, $A_2(2^+)$, and $B(1^+)$. The first four of these are forbidden for π^0 photoproduction by C conservation and the first two for $\pi^{\pm}$ photoproduction by charge conservation. The known odd-parity states are $\pi(0^-)$, $\eta(0^-)$, $\rho(1^-)$, $\omega(1^-)$, and $\phi(1^-)$. Eta exchange is forbidden in π^0 photoproduction by C invariance and in $\pi^{\pm}$ photoproduction by charge conservation; π^0 exchange is forbidden in π^0 photoproduction by C invariance and ω, ϕ exchange in $\pi^{\pm}$ photoproduction by charge conservation and isotopic spin conservation. Thus, the following picture emerges

$$
\left.\begin{array}{c} \gamma p \to \pi^0 p \\ \gamma n \to \pi^0 n \end{array}\right\} \quad \rho, \omega, \phi, \text{ and B exchanges allowed,}
$$

$$
\left.\begin{array}{c} \gamma p \to \pi^+ n \\ \gamma n \to \pi^- p \end{array}\right\} \quad \pi, \rho, A_1, A_2, \text{ and B exchanges allowed.}
$$

For charged pion photoproduction, the contributions to the different helicity states can be obtained from considerations of G parity. In terms of the parity eigenstates

$$
|JM\lambda\lambda'\rangle_{\pm} = |JM\lambda\lambda'\rangle \pm |JM-\lambda-\lambda'\rangle, \tag{III-28}
$$

where λ and λ' are the nucleon helicities,

$$
G|JM\lambda\lambda\rangle_{\pm} = (-1)^{J+1}|JM\lambda\lambda\rangle_{\pm},
$$

$$
G|JM\lambda\lambda'\rangle_{\pm} = \pm(-1)^{J+1}|JM\lambda\lambda'\rangle_{\pm} \qquad (\lambda \neq \lambda'), \tag{III-29}
$$

since $p\bar{n}$ and $\bar{p}n$ are pure isovectors.

(1) π *exchange*. The π trajectory has even signature and odd parity, and since states with different nucleon helicity are forbidden by G parity, the π can thus contribute only to $g^{J^-}_{\frac{1}{2}\frac{1}{2};1\,0}$ and hence to $F^2_{0\,1}$.

(2) ρ *exchange*. Both the parity and signature of the ρ exchange are odd, it couples only to states $|JM\lambda\lambda'\rangle_+$ and therefore it contributes only to $g^{J^+}_{\frac{1}{2}\frac{1}{2};1\,0}$ and $g^{J^+}_{\frac{1}{2}-\frac{1}{2};1\,0}$ and thus to $F^1_{0\,1}$, $F^1_{1\,1}$.

(3) A_1 *exchange*. This is an odd signature, even parity exchange. G parity allows only states with opposite nucleon helicities and consequently it can contribute only to $g^{J^-}_{\frac{1}{2}-\frac{1}{2};1\,0}$, that is, to $F^2_{1\,1}$.

(4) A_2 *exchange*. This has both even signature and parity and both helicity states are allowed for the nucleons. Thus, it contributes to both $g^{J^+}_{\frac{1}{2}\frac{1}{2};1\,0}$ and $g^{J^+}_{\frac{1}{2}-\frac{1}{2};1\,0}$, that is, to $F^1_{0\,1}$ and $F^1_{1\,1}$.

(5) B *exchange*. This has odd signature and even parity and states with different nucleon helicities are forbidden, requiring that it can contribute only to $g^{J^-}_{\frac{1}{2}\frac{1}{2};1\,0}$, that is, to $F^2_{0\,1}$.

For neutral pion photoproduction, G parity is always conserved since the photon has both isoscalar and isovector character. However, in this case C has to be conserved, and

$$C|JM\lambda\lambda\rangle_\pm = (-1)^J|JM\lambda\lambda\rangle_\pm$$
$$C|JM\lambda\lambda'\rangle_\pm = \pm(-1)^J|JM\lambda\lambda'\rangle_\pm \qquad (\lambda \neq \lambda'). \tag{III-30}$$

(1) ρ, ω, ϕ *exchange*. Each of these have both odd signature and parity. Both helicity states of the nucleons are allowed, and so they contribute to both $g^{J^+}_{\frac{1}{2}\frac{1}{2};1\,0}$ and $g^{J^+}_{\frac{1}{2}-\frac{1}{2};1\,0}$, that is, to $F^1_{0\,1}$ and $F^1_{1\,1}$.

(2) B *exchange*. This time, states with different nucleon helicity are forbidden by C, so that B can contribute only to $g^{J^-}_{\frac{1}{2}\frac{1}{2};1\,0}$, that is, to $F^2_{0\,1}$.

The results are summarized in Table VII.

TABLE VII

ALLOWED EXCHANGES FOR FORWARD PION PHOTOPRODUCTION

	$\pi^\pm$					π^0	
	π	ρ	A_1	A_2	B	ρ, ω, ϕ	B
$F^2_{0\,1}$	Yes				Yes		Yes
$F^1_{0\,1}$		Yes		Yes		Yes	
$F^2_{1\,1}$			Yes				
$F^1_{1\,1}$		Yes		Yes		Yes	

The usual custom in Regge-pole exchange is to take only the leading trajectories as being the important contributors, that is, ρ and ω exchange for π^0 photoproduction and π, ρ, and A_2 exchange for $\pi^{\pm}$ photoproduction. Further, in the case of π^0 photoproduction the almost complete degeneracy of the ρ and ω masses and trajectory slopes has meant that until the recent advent of data on the reaction $\gamma + n \rightarrow n + \pi^0$, it was possible to treat these as a combined vector exchange term, usually designated as ω exchange, since $SU(3)$ would require that this be the larger of the two. In the early work on photoproduction, for example, that of Ader *et al.* (1967), when only a limited amount of data was available, B exchange was invoked to fill the dip at $t \simeq -0.6$ $(\text{GeV}/c)^2$, which is a natural consequence of the normal vector meson exchange, but subsequent data has ruled this out and the currently accepted explanation are that cut effects provide the necessary mechanism. We shall return to this point in more detail later.

The next step is to decide whether the exchanged trajectories choose an evasive or conspiring mechanism to satisfy the constraint equation at $t = 0$. The natural mechanism in all cases (see, for example Ader *et al.*, 1967) is one of evasion and with this choice typical Regge-pole parametrizations of the amplitudes are as follows.

1. π^0 *Photoproduction*

$$F_{0\,1}^1 = g_{\omega 0}(\alpha_\omega - 1)^{l_\omega}\alpha_\omega^{m_\omega}\,\Pi_{\alpha_\omega}\,\xi_{\alpha_\omega}(s/s_0)^{\alpha_\omega - 1}; \qquad F_{0\,1}^2 = 0,$$

$$F_{1\,1}^1 = tg_{\omega 1}(\alpha_\omega - 1)^{l_\omega}\alpha_\omega^{n_\omega}\Pi_{\alpha_\omega}\,\xi_{\alpha_\omega}(s/s_0)^{\alpha_\omega - 1}, \qquad F_{1\,1}^2 = 0,$$

$$\text{(III-31)}$$

with

$$\Pi_{\alpha_\omega} = (\alpha_\omega + 1)(\alpha_\omega + 2) \qquad \text{or} \qquad (\alpha_\omega + 1)(\alpha_\omega + 2)(\alpha_\omega + 3) \quad \text{(III-32)}$$

and

$$\xi_{\alpha_\omega} = [1 - \exp(-i\pi\alpha_\omega)]/\sin(\pi\alpha_\omega). \qquad \text{(III-33)}$$

TABLE VIII

POWERS OF α REQUIRED FOR VARIOUS CHOOSING MECHANISMS
NEAR THE SENSE/NONSENSE POINT $\alpha = 0$

	Sense-sense	Sense-nonsense	Nonsense-nonsense
Mechanism	α^l	α^m	α^n
Sense	1	α	α^2
Nonsense	α	α	α
Chew	α	α^2	α^3
No compensating	α^2	α^2	α^2

Here $g_{\omega 0}$, $g_{\omega 1}$, and s_0 are constants. The indices l_ω, m_ω, and n_ω depend on the choosing mechanism of the trajectory at its sense-nonsense points, and are given in Table VIII.

2. $\pi^\pm$ Photoproduction

$$F_{0\,1}^1 = g_{\rho 0}(\alpha_\rho - 1)^{l_\rho}\alpha_\rho{}^{m_\rho} \Pi_{\alpha_\rho} \xi_{\alpha_\rho} (s/s_0)^{\alpha_\rho - 1}$$
$$+ g_{A_2 1}(\alpha_{A_2} - 1)^{l_{A_2}}\alpha_{A_2}^{m_{A_2}}\Pi_{\alpha_{A_2}}\xi_{\alpha_{A_2}} (s/s_0)^{\alpha_{A_2} - 1},$$

$$F_{0\,1}^2 = [t/(t - \mu^2)]g_\pi(\alpha_\pi - 1)^{l_\pi}\alpha_\pi^{m_\pi}\Pi_{\alpha_\pi} \xi_{\alpha_\pi} (s/s_0)^{\alpha_\pi - 1},$$

$$F_{1\,1}^1 = t\{g_{\rho_1}(\alpha_\rho - 1)^{l_\rho}\alpha_\rho^{m_\rho}\Pi_{\alpha_\rho} \xi_{\alpha_\rho} (s/s_0)^{\alpha_\rho - 1}$$
$$+ g_{A_2 1}(\alpha_{A_2} - 1)^{l_{A_2}}\alpha_{A_2}^{n_{A_2}}\Pi_{\alpha_{A_2}} \xi_{\alpha_{A_2}} (s/s_0)^{\alpha_{A_2} - 1}\},$$

$$F_{1\,1}^2 = 0,$$

(III-34)

with

$$\Pi_{\alpha_\rho} = (\alpha_\rho + 1)(\alpha_\rho + 2) \quad \text{or} \quad (\alpha_\rho + 1)(\alpha_\rho + 2)(\alpha_\rho + 3),$$
$$\Pi_{\alpha_{A_2}} = (\alpha_{A_2} + 1)(\alpha_{A_2} + 2) \quad \text{or} \quad (\alpha_{A_2} + 1)(\alpha_{A_2} + 2)(\alpha_{A_2} + 3),$$
$$\Pi_{\alpha_\pi} = (\alpha_\pi + 1)(\alpha_\pi + 2) \quad \text{or} \quad (\alpha_\pi + 1)(\alpha_\pi + 2)(\alpha_\pi + 3)$$

(III-35)

and

$$\xi_{\alpha_\rho} = [1 - \exp(-i\pi\alpha_\rho)]/\sin(\pi\alpha_\rho),$$
$$\xi_\alpha = [1 + \exp(-i\pi\alpha_{A_2})]/\sin(\pi\alpha_\rho),$$
$$\xi_{\alpha_\pi} = [1 + \exp(-i\pi\alpha_\pi)]/\sin(\pi\alpha_\pi).$$

(III-36)

In this case $g_{\rho 0}$, $g_{\rho 1}$, $g_{A_2 0}$, $g_{A_2 1}$, g_π, and s_0 are constant, and the indices l_ρ, m_ρ, n_ρ, l_{A_2}, m_{A_2}, n_{A_2}, l_π, m_π are given in Table VIII.

A feature of this "all-evasive" approach to pion photoproduction is that it produces a differential cross-section which dips in the forward direction. There is nothing objectionable to this in π^0 photoproduction, but it is completely inconsistent with the experimental results for $\pi^\pm$ photoproduction, which exhibit a strong forward peak. This peak is very narrow, and it is natural to associate it with the pion pole. To obtain the peak within a simple Regge-pole exchange formalism it is necessary to call upon a conspiring mechanism. The difficulty here is that none of the exchanges thus far written down is a suitable conspirator, and therefore one must be created.

Two approaches are possible. In the first, one conspirator π_c is postulated, the pion term itself is no longer evasive and the residue functions of the terms are such that the constraint relation is satisfied. For example, one could write

$$F_{01}^2 = [g_\pi/(t - \mu^2)]\alpha_\pi(\alpha_\pi + 1)(\alpha_\pi + 2)\xi_{\alpha_\pi}(s/s_0)^{\alpha_\pi - 1}$$

(III-37)

and insert in $F_{1\,1}^1$ an additional term

$$F_{(1\,1)_c}^1 = g_{\pi_c}\alpha_{\pi_c}(\alpha_{\pi_c} + 1)(\alpha_{\pi_c} + 2)\xi_{\alpha_{\pi_c}} s/s_0)^{\alpha_{\pi_c} - 1},$$

(III-38)

which must be such that $\alpha_{\pi_c}(0) = \alpha_\pi(0)$, that is, the pion and its conspirator must have degenerate trajectories at $t = 0$, the conspirator must have opposite parity to the pion, but all other quantum numbers the same, and $g_{\pi_c}(0) = (2M/\mu^2)g_\pi(0)$. To obtain a good description of the forward peak with this mechanism and at the same time reconcile the height of the forward peak with the known value of the pion-nucleon coupling constant g^2, it is necessary to introduce a variation of the pion residue function with t. For example, Ball *et al.* (1968) used a linear form

$$g^2 \rightarrow g^2\{1 + \lambda(t - \mu^2)/\mu^2\}, \tag{III-39}$$

and succeeded in fitting the forward peak with only these two terms, with g_{π_c} taken to be a constant.

The second possibility is to keep the pion evasive, but to invoke a conspiring parity doublet π_c, $\pi_{c'}$ in addition, one of which contributes to $F_{0\,1}^2$ and the other to $F_{1\,1}^1$. Since they are conspirators, they must have degenerate trajectories at $t = 0$ and their residue functions must be suitably related at $t = 0$ to yield the constraint condition. The forward peak in the data is now obtained by coherent interference of the evasive pion with the term contributing to $F_{0\,1}^2$.

Since single-pion exchange occurs in many reactions and since factorization can be applied to simple Regge exchanges, it is possible to attempt a consistency check of the conspiring π, π_c contributions. This check does not work out, and conspiracy, at least in this simple form, is no longer looked on as being the requisite mechanism but, as in π^0 photoproduction, reliance is placed on cuts. Another possibility does present itself here, namely, that the pion pole is *not* a Regge pole but is a fixed pole—it is, in fact, the usual gauge-invariant combination of the t-channel pion pole and the electric part of the u-channel nucleon pole.

3. K *Photoproduction*

The allowed exchanges are $K(0^-)$, $K^*(1^-)$, and possibly $K^{**}(2^-)$ and they will contribute similarly to π, ρ, and A_2 in charged pion photoproduction:

$$
\begin{aligned}
F_{0\,1}^1 &= g_{K^*0}(\alpha_{K^*} - 1)^{l_{K^*}}\alpha_{K^*}^{m_{K^*}}\Pi_{\alpha_{K^*}}(s/s_0)^{\alpha_{K^*}-1} \\
&\quad + g_{K^{**}0}(\alpha_{K^{**}} - 1)^{l_{K^{**}}}\alpha_{K^{**}}^{m_{K^{**}}}\Pi_{\alpha_{K^{**}}}\,\xi_{\alpha_{K^{**}}}(s/s_0)^{\alpha_{K^{**}}-1} \\
F_{0\,1}^2 &= [(t - \Delta^2)/(t - \mu^2)]g_K(\alpha_K - 1)^{l_K}\alpha_K^{m_K}\Pi_{\alpha_K}\,\xi_{\alpha_K}(s/s_0)^{\alpha_K-1}, \\
F_{1\,1}^1 &= t\{g_{K^*1}(\alpha_{K^*} - 1)^{l_{K^*}}\alpha_{K^*}^{n_{K^*}}\Pi_{\alpha_{K^*}}\,\xi_{\alpha_{K^*}}(s/s_0)^{\alpha_{K^*}-1} \\
&\quad + g_{K^{**}1}(\alpha_{K^{**}} - 1)^{l_{K^{**}}}\alpha_{K^{**}}^{n_{K^{**}}}\Pi_{\alpha_{K^{**}}}\,\xi_{\alpha_{K^{**}}}(s/s_0)^{\alpha_{K^{**}}-1}\}, \\
F_{1\,1}^2 &= 0.
\end{aligned}
\tag{III-40}
$$

Here $\Delta = (M - Y)$.

We have again written down the evasive solution as an explicit example, although, of course, a conspiring solution can also be chosen, invoking a kaon conspirator K_c contributing to $F_{1\ 1}^2$ and/or a K* (or even K**) conspirator also contributing to $F_{1\ 1}^2$ to satisfy the constraint equations.

The question of conspiracy or evasion in $K^\pm$ photoproduction does not appear to be such a crucial one as in $\pi^\pm$ photoproduction since the constraint equations are being applied further away from the physical region because of the increased meson mass, the kaon pole is also much further from the physical region and the data for both the reactions $\gamma + p \to K^+ + \Lambda^0$, $\gamma + p \to K^+ + \Sigma^0$ show a dip in the forward direction. These reactions also appear to require cuts. Although it is possible to fit each reaction separately in terms of K, K*, K** Regge exchange, it is not possible to fit them simultaneously with the same trajectory parameters, a strong indication that cut effects are present.

4. η^0 Photoproduction

This is precisely the same as π^0 photoproduction, except that the data does not exclude B exchange:

$$F_{0\ 1} = g_{\omega_0}(\alpha_\omega - 1)^{l_\omega}\alpha_\omega^{m_\omega}\Pi_{\alpha_\omega}\xi_{\alpha_\omega}(s/s_0)^{\alpha_\omega - 1},$$

$$F_{0\ 1} = tg_{\rm B}(\alpha_{\rm B} - 1)^{l_{\rm B}}\alpha_{\rm B}^{m_{\rm B}}\Pi_{\alpha_{\rm B}}\xi_{\alpha_{\rm B}}(s/s_0)^{\alpha_{\rm B} - 1},$$

$$F_{1\ 1} = tg_{\omega_1}(\alpha_\omega - 1)^{l_\omega}\alpha_\omega^{m_\omega}\Pi_{\alpha_\omega}\xi_{\alpha_\omega}(s/s_0)^{\alpha_\omega - 1},$$

$$F_{1\ 1} = 0,$$

(III-41)

with

$$\xi_{\rm B} = [1 - \exp(-i\pi\alpha_{\rm B})]/\sin(\pi\alpha_{\rm B}).$$

As in π^0 photoproduction, ρ exchange can contribute in an identical fashion to ω exchange and there is no reason for not choosing an evasive mechanism.

5. ρ^0, ω, ϕ Photoproduction

The allowed exchanges are (1) π, which contributes to $F_{\frac{1}{2}\frac{1}{2};1\ 1}$, $F_{\frac{1}{2}\frac{1}{2};1\ 0}$, and $F_{\frac{1}{2}\frac{1}{2};1\ -1}$; (2) A_1, which contributes to $F_{-\frac{1}{2}\frac{1}{2};1\ 1}$, $F_{-\frac{1}{2}\frac{1}{2};1\ 0}$, and $F_{-\frac{1}{2}\frac{1}{2};1\ -1}$; and (3) P, P′, and A_2, which contribute to the remaining six amplitudes. The dominant mechanism here is pomeron exchange and this, with some contribution from π exchange, is sufficient to explain all the existing data without any need for the lower-lying P′, A_2, and A_1 trajectories. Again choosing an evasive mechanism, one obtains Eqs. (III-42):

$$F_{\frac{1}{2}\frac{1}{2};1\ 1} = t(t - \mu^2)^2 g_{0\ 0\ 3}\alpha_\pi^{l_\pi}\Pi_{\alpha_\pi}\xi_{\alpha_\pi}(s/s_0)^{\alpha_\pi},$$

$$F_{\frac{1}{2}\frac{1}{2};1\ 1} = (t - \mu^2)^2 g_{0\ 0\ 0}\alpha_{\rm P}^{l_{\rm P}}\Pi_{\alpha_{\rm P}}\xi_{\alpha_{\rm P}}(s/s_0)_{\alpha_{\rm P}},$$

(III-42)

$$F_{\frac{1}{2}\frac{1}{2};1\ 0} = (t - \mu^2)^2 g_{0\ 1\ 3}\alpha_{\rm P}^{m_{\rm P}}(\alpha_{\rm P} - 1)^{l_{\rm P}}\Pi_{\alpha_{\rm P}}\xi_{\alpha_{\rm P}}(s/s_0)^{\alpha_{\rm P} - 1},$$

$$F_{\frac{1}{2}\frac{1}{2};1\,0} = t(t - \mu^2)^2 g_{0\,1\,0} \alpha_\pi^{m_\pi}(\alpha_\pi - 1)^{l_\pi} \Pi_{\alpha_\pi} \xi_{\alpha_\pi}(s/s_0)^{\alpha_\pi - 1},$$

$$F_{\frac{1}{2}\frac{1}{2};1\,-1} = t(t - \mu^2)^2 g_{0\,2\,3} \alpha_\pi^{m_\pi}(\alpha_\pi - 1)^{m_\pi}(\alpha_\pi - 2)^{l_\pi} \Pi_{\alpha_\pi} \xi_{\alpha_\pi}(s/s_0)^{\alpha_\pi - 2},$$

$$F_{\frac{1}{2}\frac{1}{2};1\,-1} = (t - \mu^2)^2 g_{0\,2\,0} \alpha_P^{m_P}(\alpha_P - 1)^{m_P}(\alpha_P - 2)^{l_P} \Pi_{\alpha_P} \xi_{\alpha_P}(s/s_0)^{\alpha_P - 2}$$

$$F_{-\frac{1}{2}\frac{1}{2};1\,1} = 0$$

$$F_{-\frac{1}{2}\frac{1}{2};1\,1} = t(t - \mu^2)^2 g_{1\,0\,0} \alpha_P^{m_P}(\alpha_P - 1)^{l_P} \Pi_{\alpha_P} \xi_{\alpha_P}(s/s_0)^{\alpha_P - 1},$$

$$F_{-\frac{1}{2}\frac{1}{2};1\,0} = t(t - \mu^2)^2 g_{1\,1\,0} \alpha_P^{m_P}(\alpha_P - 1)^{l_P} \Pi_{\alpha_P} \xi_{\alpha_P}(s/s_0)^{\alpha_P - 1},$$

$$F_{-\frac{1}{2}\frac{1}{2};1\,0} = 0$$

$$F_{-\frac{1}{2}\frac{1}{2};1\,-1} = 0$$

$$F_{-\frac{1}{2}\frac{1}{2};1\,-1} = t(t - \mu^2) g_{1\,2\,0} \alpha_P^{m_P}(\alpha_P - 1)^{m_P}(\alpha_P - 2)^{l_P} \Pi_{\alpha_P} \xi_{\alpha_P}(s/s_0)^{\alpha_P - 2},$$

(III-42,
contd.)

with

$$\xi_{\alpha_P} = [1 + \exp(-i\pi\alpha_P)]/\sin(\pi\alpha_P). \qquad \text{(III-43)}$$

6. Primakoff Effect

An additional effect which should be taken into account for π^0 and η^0 photoproduction is the Primakoff effect. This contributes only to $F_{0\,1}$ and has the form

$$F_{0\,1}(\text{Primakoff}) \propto e f_\pi [(t_0 - t)^{1/2}/t](1 - t/m_\rho^2)^{-1}, \qquad \text{(III-44)}$$

where $t_0 = t\,(= 0)$ and f_π is the $\pi^0\gamma\gamma(\eta^0\gamma\gamma)$ coupling constant. Strictly speaking, the form factor term should not be present. It is merely introduced in phenomenology to prevent the interference between $F_{0\,1}^1$ (Primakoff) and the rest of the amplitude becoming dominant at very high s and large $|t|$. It has no effect near $t = t_0$, which is the only region in which the Primakoff effect is physically important.

7. Cut Effects

At various points in the above discussion on Regge-pole parametrization we have indicated that cut effects may be important. A cut contribution is typically of the general form

$$F_{ij}(\text{cut}) = h_{ij}(t) \exp(i\phi_{ij}) \frac{1 - \exp[-i\pi_{\alpha_c}(t)]}{\sin \pi_{\alpha_c}(t)} \frac{(s/s_0)^{\alpha_c(t)-1}}{\ln(s/s_0) + d_{ij}}, \qquad \text{(III-45)}$$

where $h_{ij}(t)$ if to be interpreted as a discontinuity function and is usually parametrized by

$$h_{ij}(t) = h_{ij} \exp(a_{ij} t), \qquad \text{(III-46)}$$

with h_{ij} and a_{ij} constants (and free parameters). The terms ϕ_{ij} and d_{ij} are possible additional parameters and $\alpha_c(t) = \alpha_c(0) + \alpha_c'(t)$ is the cut branch point.

$\alpha_c(0)$ and α_c' can also be treated as free parameters or can be specified if a particular cut mechanism is postulated. The usual mechanism postulated is Regge exchange plus absorption, the latter being given by pomeron exchange. In the case of linear trajectories for the Regge and pomeron exchange.

$$\alpha_R = \alpha_R(0) + \alpha_{R'}t, \qquad \alpha_P = \alpha_P(0) + \alpha_{P'}t, \qquad \text{(III-47)}$$

then

$$\alpha_c(t) = \alpha_R(0) + \alpha_P(0) - 1 + [\alpha_{R'}\alpha_{P'}/(\alpha_{R'} + \alpha_{P'})]t. \qquad \text{(III-48)}$$

Just as can the Regge pole exchange contributions choose evasion or conspiracy to satisfy the constraint equations, so too can the cut contributions. For example, a parametrization of π^0 photoproduction with evasive ω exchange and conspiring ω-P cut would be

$$F_{0\,1}^1 = g_{\omega 0}\,\Pi_{\alpha_\omega}\,\xi_{\alpha_\omega}(s/s_0)^{\alpha_\omega - 1} + g_{c0}(t)\,\xi_{\alpha_c}(s/s_0)^{\alpha_c - 1}[\ln(s/s_0)]^{-1},$$

$$F_{0\,1}^2 = g_{c1}(t)\xi_{\alpha_c}(s/s_0)^{\alpha_c - 1}[\ln(s/s_0)]^{-1},$$

$$F_{1\,1}^1 = tg_{\omega 1}\,\Pi_{\alpha_\omega}\,\xi_{\alpha_\omega}(s/s_0)^{\alpha_\omega - 1} - 2Mg_{c1}(t)\xi_{\alpha_c}(s/s_0)^{\alpha_c - 1}[\ln(s/s_0)]^{-1},$$

$$F_{1\,1}^2 = 0. \qquad \text{(III-49)}$$

This is the most economical pole and cut parametrization possible, and with $g_{ci}(t) = g_{ci}\exp(a_i\,t)$ $(i = 0, 1)$ contains six parameters.

Calculations of the cut contribution can be made using the eikonal approximation or (equivalently to first order) the K matrix approximation. However, it is well known that these calculations, while giving the correct sign and order of magnitude of the cut certainly underestimate its strength and it is necessary to introduce an arbitrary multiplicative parameter. Consequently it is probably much more useful from a purely phenomenological point of view to use these calculations only as a guide to the estimated strength of the cut and its phase with respect to the generating pole.

C. FORWARD PSEUDOSCALAR MESON PHOTOPRODUCTION

1. π^0 Photoproduction

A considerable amount of data is now available on the differential cross section for the reaction

$$\gamma + p \rightarrow p + \pi^0,$$

that below 6.0 GeV having been obtained by Bolon et al. (1967) and by Braunschweig et al. (1966, 1968), and that above 6.0 GeV by Anderson et al. (1968a, 1969a). In addition there is information at 3 GeV on the asymmetry with polarized photons obtained by Bellenger *et al.* (1968) and on the ratio

$$R = \frac{(d\sigma/dt)(\pi^0 n)}{(d\sigma/dt)(\pi^0 p)} \tag{III-50}$$

at 4 GeV, obtained by Bolon *et al.* (1971).

Excluding the Primakoff region, the general features of the differential cross-sections on protons are a forward dip, the cross section peaking at $t \simeq -0.1$ $(GeV/c)^2$, then decreasing to a minimum at $t \simeq -0.6$ $(GeV/c)^2$, rising to a small secondary maximum at $t \simeq -0.9$ $(GeV/c)^2$ and then decreasing smoothly out to the furthest limit of the experiments, $t \simeq -1.8$ $(GeV/c)^2$. This secondary structure in the cross section gradually diminishes with increasing energy.

Since ω exchange is expected to be the dominant mechanism, it is natural to ascribe this dip to the nonsense zero of this Regge trajectory at the wrong-signature point $\alpha_\omega = 0$, which occurs precisely in the dip region. However, since this contribution by itself would go exactly to zero at this point, some dip filling mechanism must be proposed. The initial choice fell on B exchange, and with evasive ω and B exchange Ader *et al.* (1967) obtained a good fit to the data up to 5 GeV, which was the highest energy then available.

Further developments have ruled out this possibility. First, this model predicts an increase in the depth of the dip with increasing energy since the B trajectory is lower lying than that of ω, and consequently its contribution to the cross section will fall off more rapidly than that of ω. This is in direct contradiction to what is experimentally observed. Second, this model predicts that the asymmetry with polarized photons be close to -1 in the dip region, since it is completely dominated there by unnatural parity B exchange. Experimentally it is found to be positive, of the order of 0.6. Now although the energy at which this data was obtained, namely 3 GeV, is rather low and the Regge pole models have some difficulty in giving a precise quantitative description of the normal differential cross section, allowing one to postulate that there may be residual effects from the resonance region, these effects should be small and would certainly not be sufficient to change completely the character of the asymmetry. Thus, the model is ruled out again.

There appears to be no alternative but to require that cut effects are important, and on this premise, three calculations with different input have succeeded in giving a reasonable description of most of the data.

Capella and Tran Thanh Van (1969) used evasive ω exchange with a conspiring ω-P cut, the latter contributing to $F_{1\,0}^{(1,\,2)}$ and $F_{1\,1}^{(1)}$. Assuming $\alpha_\omega(t) = 0.45 + 0.90t$ and $\alpha_\rho(t) \simeq 1.0$, and by taking the exponent of the exponential terms in the discontinuity functions of the cuts to be the same for all the cut contributions, the number of free parameters was reduced to five. A good description of all the data was obtained, with the exception of the $(\pi^0 n)/(\pi^0 p)$

ratio, which is perforce unity since with only ω exchange included only the isovector pion photoproduction amplitude is present.

This latter situation was remedied in the calculation of Frøyland (1969), who considered evasive ω and ρ exchange, with conspiring ωP and ρP cuts. It was assumed that the ω and ρ trajectories are degenerate and in addition that the ωP and ρP cuts are degenerate. This assumption effectively decouples the $(\pi^0 p)$ from the $(\pi^0 n)$ reaction as far as data fitting is concerned. The specific parametrization used was more complicated than that of Capella and Tran Thanh Van (1969), since not only did it have the addition of ρ exchange, but it allowed for different t dependence of each cut contribution as well as including the extra phase ϕ and the scaling factor d and treating α_c' as a parameter, giving fourteen possible parameters in all. In practice, this was reduced somewhat by setting d to a value found previously for charged pion photoproduction (Frøyland and Gordon, 1969) and the discovery that α_c' was much as would be expected on theoretical grounds anyway. Further, because of the paucity of data on $\gamma + n \rightarrow \pi^0 + n$, the fit was not sensitive to the parameters associated with ρ exchange (ω exchange was assumed to be the dominant vector meson exchange).

A good fit to the π^0 data has also been obtained by Kellett (1970), as part of an overall fit to the π^0, π^+, and π^- photoproduction data. The model assumed the dominance of the electric Born term in charged pion photoproduction and the dominance of ω exchange in the neutral case. The ρ, ω, and A_2 Regge poles were taken to be evasive, and phenomenological evasive cuts were included as well. The Regge pole parametrization was conventional, with nondegenerate trajectories, and a fully crossing-symmetric form for the cuts was used and product signature was assumed. The parametrization was appreciably more restrictive than that of Frøyland and Gordon (1969).

A comparison of the fit of Kellett (1970) with the experimental data is made in Figs. 58–60. The results are good for energies of 4.0 GeV and above, acceptable at 3.0 GeV, but decidedly in error at 2.0 GeV.

2. η^0 Photoproduction

Since high energy η^0 photoproduction can proceed by precisely the same exchanges as π^0 photoproduction, one could perhaps expect a strong similarity in their cross sections. However, while having the forward dip and peaking at $t \simeq -0.2 \ (\text{GeV}/c)^2$, the data does not exhibit the minimum at $t \simeq -0.6 \ (\text{GeV}/c)^2$, the differential cross section decreasing smoothly out to the largest momentum transfers measured, $t \simeq -1.4 \ (\text{GeV}/c)^2$. At all momentum transfers the energy dependence is consistent with s^{-2} from 4.0 to 9.0 GeV. The collected data are shown in Fig. 61, being taken from Bellenger et al. (1968), Anderson et al. (1968a, b, 1969), and Braunschweig et al. (1969). On this

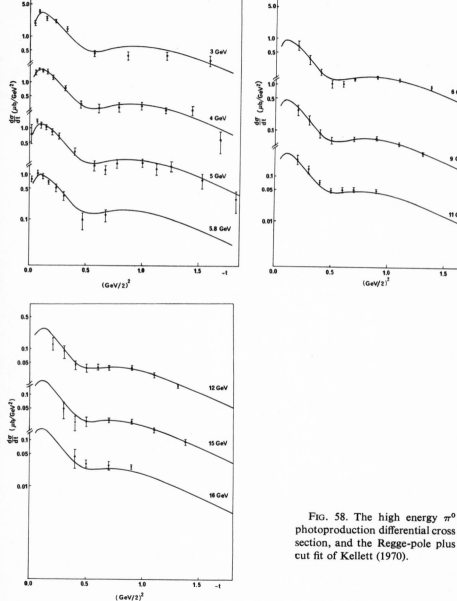

FIG. 58. The high energy π^0 photoproduction differential cross section, and the Regge-pole plus cut fit of Kellett (1970).

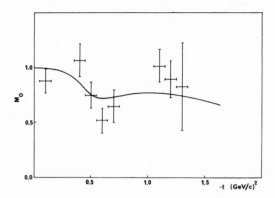

FIG. 59. The polarized photon asymmetry ratio in high energy π^0 photoproduction and the Regge-pole plus cut fit of Kellett (1970).

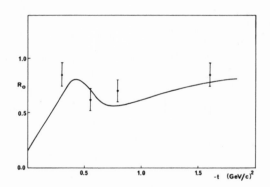

FIG. 60. The ratio R_0 of the reactions $(\gamma + n \rightarrow n + \pi^0)/(\gamma + p \rightarrow p + \pi^0)$ in high energy photoproduction, and the Regge-pole plus cut fit of Kellett (1970).

comparison, the data of Bellenger *et al.* (1968) appear to be somewhat lower than the rest, a feature which is borne out by Regge-pole fits to the data.

Because of the lack of structure in the data and the rather large errors, obtaining a Regge-pole fit is easy but obtaining a reasonably unique description is impossible. For example, Gorczyka and Hayashi (1968) considered ρ, ω, and B exchange, and Bajpai and Donnachie (1969b) obtained a perfectly reasonable fit both with $\rho + B$ exchange and with nondegenerate $\rho + \omega$ exchange. An interesting feature of these fits is that they can give a good description of the data at energies below 2.0 GeV, as we have seen already in Section II (Fig. 49).

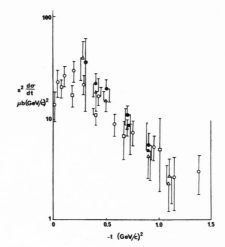

FIG. 61. Collected data for the differential cross section in high energy η photoproduction: (O, □) 4.0 GeV; (●) 5.5 GeV; (△) 6.5 GeV; (■) 9.0 GeV. (Data from Bellenger *et al.*, 1968.)

3. $\pi^{\pm}$ *Photoproduction*

A considerable amount of data is now available on π^+ photoproduction in the energy range from 3 to 16 GeV and for momentum transfers up to $2\,(\text{GeV}/c)^2$. The data are from Buschorn *et al.* (1967a, b), Joseph *et al.* (1967), Bar-Yam *et al.* (1967), Heide *et al.* (1968), and Boyarski *et al.* (1968a, b). The general features of these data are a very sharp forward peak for momentum transfers of $|t| \lesssim m_\pi^2$, a t dependence of $\simeq e^{2.5t}$ up to momentum transfers of $|t| \simeq 0.6$ GeV^2 and then a fairly abrupt change of slope to fall off as $\simeq e^{3.0t}$ up to the maximum momentum transfers measured.

The data from π^- production are obtained from photoproduction off deuterium. To obtain the cross section off free neutrons, it is assumed that nuclear physics problems, in particular those associated with the Pauli exclusion principle and Glauber corrections, will be the same for π^- and π^+ off deuterium, so that by investigating both the reactions

$$\gamma + D \to \pi^+ + n + n, \qquad \gamma + D \to \pi^- + p + p,$$

the cross section for π^- photoproduction from free neutrons will be given by

$$(d\sigma/dt)(n\pi^-) = \frac{(d\sigma/dt)(D\pi^-)}{(d\sigma/dt)(D\pi^+)}\,(d\sigma/dt)(p\pi^+) = R(d\sigma/dt)(p\pi^+). \qquad \text{(III-51)}$$

The ratio R varies little over a wide range of energies for a given t, but changes rapidly with t at a given energy, being close to unity in the forward

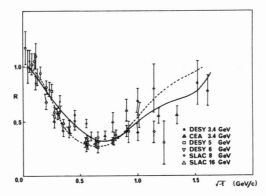

FIG. 62. The angular dependence of the ratio of the reactions $(\gamma + n \rightarrow \pi^- + p)/(\gamma + p \rightarrow \pi^+ + n)$ in high energy photoproduction, and the Regge-pole plus cut fit of Kellett (1970).

direction, decreasing to about 0.3 at $t \simeq 0.4$ $(\text{GeV}/c)^2$ and then increasing again. This behavior is illustrated specifically in Fig. 62. The value close to unity in the forward direction implies that the π^- differential cross section also has the very sharp peak observed in π^+ photoproduction, and the pronounced minimum requires that there be a strong interference between the isoscalar and isovector parts of the amplitudes.

Data on both reactions with linearly polarized photons is more extensive than that on π^0 photoproduction, and has been obtained by Geweniger *et al.* (1968, 1969), Burfeindt *et al.* (1969, 1970), and Bar-Yam *et al.* (1970a, b). Not only have the asymmetry ratios $A^\pm$ for $\pi^\pm$ photoproduction been obtained but it has also been possible to obtain explicitly the separate cross sections for $\pi^\pm$ production by transversely polarized photons, $d\sigma_\perp/dt$, and for photons polarized parallel to the production plane $d\sigma_\parallel/dt$, enabling a detailed study to be made of the separate roles of natural and unnatural parity exchanges in these reactions. Quite apart from any other features of the data, the sharp forward peak, which is naturally associated with the nearby pion pole, excludes a purely evasive solution since this would require the cross section to dip in the forward direction, as exhibited explicitly by Ader *et al.* (1967).

The initial solution in terms of Regge poles was to postulate a conspiring pion and to introduce a pion conspirator π_c, as was done by Frautschi and Jones (1967), Henyey (1968), and Ball *et al.* (1968), enabling the constraint equation to be satisfied, the peak to be obtained, and a reasonable description of the near forward data $[|t| \lesssim 0.1 \ (\text{GeV}/c)^2]$ to be given. This approach probably reached its zenith in the work of Brower and Dash (1968), who, by adding evasive ρ, A_2, and B exchange to the conspiring π and π_c trajectories, achieved success with the π^+ differential cross sections up to $|t| \simeq 0.5 \ (\text{GeV}/c)^2$

(that is, approximately up to the change in slope), but their calculation is less successful at higher momentum transfers. Further, the agreement with the polarized photon data is at best qualitative and the predicted π^-/π^+ ratio is considerably greater than that found experimentally, particularly in the vicinity of the dip. There are more fundamental objections to the conspiring pion solution. First, a strong t dependence of the pion residue function is required simultaneously to explain the peak and to give the expected value of the residue on the pion mass shell, which is an unsatisfactory feature. Second, and much more importantly, since all the exchanges are conventional Regge-pole exchanges factorization applies, π_c must contribute to other reactions and consistency tests can be applied. These tests fail. For example, forward peaking in the reaction $\pi p \to \rho \Delta$ is naturally associated with pion exchange and this, together with the fact that the ρ is produced dominantly with zero helicity near the forward direction, cannot be explained by a conspiring pion, a fact first noted by Le Bellac (1967).

Given that the pion must evade, a Regge pole solution can still be obtained if a conspiring parity doublet (c, c′) is postulated in addition. Arguing that since this mechanism cannot operate for π^0 photoproduction (it shows a dip in the forward direction), Dietz and Korth (1968) chose the isospin and G parity of this pair to be $+1$ and -1, respectively, and with the addition of isoscalar B exchange obtained a reasonable fit to the π^+ differential cross section up to $|t| \simeq 0.5$ (GeV/c)^2. This approach has been extended considerably by Korth (1969), who used evasive π and ρ exchange with fixed poles at the wrong signature points $\alpha_\pi = -1$ and $\alpha_\rho = 0$ plus a conspiring parity doublet. The results of this calculation are quoted by Lübelsmeyer (1969).

Another possibility is to follow the route taken for π^0 photoproduction and introduce cuts, some of which must conspire. The first calculation along these lines was by Amati et al. (1968), who used an evasive pion with evasive B exchange, conspiring A_1 and A_2 exchange and a conspiring A_1 and A_2 "background" (cuts) and obtained good agreement with the π^+ differential cross sections over the whole energy and momentum transfer range available. Further developments have been by Frøyland and Gordon (1969), Blackmon et al. (1969), and Kane (1969).

In particular, Frøyland and Gordon (1969) considered an evading π trajectory and an evading ρ trajectory (contributing only to $F^1_{1\,1}$) together with a πP cut, containing a conspiring and evading part, and a ρP cut with only an evading part, both cuts contributing to $F^2_{1\,0}$ and $F^1_{1\,1}$.

The final possibility is to assume that the pion pole is not a Regge pole, but is given by the normal gauge-invariant pion pole term, with a suitable form factor. This was the approach adopted by Kellett (1970). The details of this fit were outlined in the discussion on π^0 photoproduction, and the comparison of this fit with the π^+ and π^- data is given in Figs. 63–65.

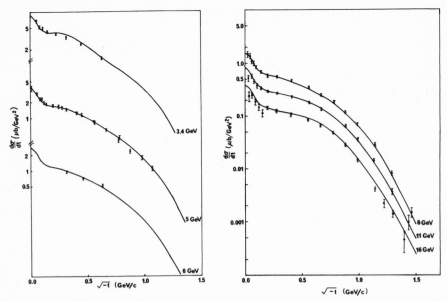

FIG. 63. The high energy π^+ photoproduction differential cross section and the Regge-pole plus cut fit of Kellett (1970).

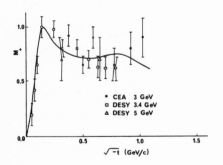

FIG. 64. The polarized photon asymmetry ratio in high energy π^+ photoproduction and the Regge-pole plus cut fit of Kellett (1970).

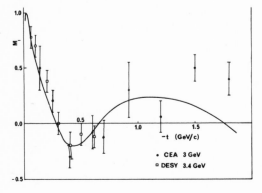

FIG. 65. The polarized photon asymmetry ratio in high energy π^- photoproduction and the Regge-pole plus cut fit of Kellett (1970).

The difficulty of discriminating between the different models of forward $\pi^{\pm}$ photoproduction at high energies has been exhibited explicitly by Jackson and Quigg (1969), who first of all showed that it was possible to construct both evasive and conspiring pion fits to the data which simultaneously satisfied continuous moment sum rules. Making the additional assumption that the high energy amplitudes are predominantly real [the power law behavior of the data is consistent with $(s - M^2)^{-2}$, that is, with an effective trajectory $\alpha_{\text{eff}} \simeq 0$, and since the phase is given by the signature factors $(1 \pm e^{-i\pi\alpha})$, the reality of the amplitudes follows], Jackson and Quigg (1969) constructed a pseudo-model in which the high energy forward $\pi^{\pm}$ cross sections (including those with linearly polarized photons) are given in terms of the finite energy sum rules alone, without the intervention of a specific high energy model.

4. Charged $\pi\Delta(1236)$ Photoproduction

The reaction

$$\gamma + p \rightarrow \pi^- \Delta^{++}(1236)$$

has been investigated by Boyarski et al. (1969a) at 5, 8, 11 and 16 GeV from the near forward direction to momentum transfers of between 1 and 2 $(\text{GeV}/c)^2$ and the reactions

$$\gamma + p \rightarrow \begin{cases} \pi^- \Delta^{++}, \\ \pi^+ \Delta^0, \end{cases} \qquad \gamma + n \rightarrow \begin{cases} \pi^- \Delta^+ \\ \pi^+ \Delta^- \end{cases}$$

by Boyarski et al. (1970) at 16 GeV, and for momentum transfers up to 2 $(\text{GeV}/c)^2$. The neutron cross sections were obtained by subtracting the (γp) crosssections from the (γd) cross sections, assuming absorption effects in production from the deuteron to be small. These data are shown in Figs. 66 and 67. The energy dependence for the reaction $\gamma + p \rightarrow \pi^- \Delta^{++}$ is again given more or less by $(s - M^2)^{-2}$ over the whole momentum transfer range and the momentum transfer dependence is almost identical to that of the reaction $\gamma + p \rightarrow \pi^+ + n$, the sharp forward peak ($\simeq e^{12t}$) then the two regions, $e^{2.5t}$ out to $|t| \simeq 0.6$ $(\text{GeV}/c)^2$ and then $e^{3.0t}$ thereafter. The one difference is that the forward peak does not continue right to $0°$, but dips to a finite (nonzero) value, the maximum in the cross section occurring at $\sqrt{|t|} \simeq 0.1$ $(\text{GeV}/c)^2$. This dip is a natural consequence of having unequal mass baryons, and the peak is still to be associated with the pion pole. The other reactions have a qualitatively similar behavior in the near forward direction, although the peak is less marked than in $\gamma + p \rightarrow \pi^- \Delta^{++}$, and at higher momentum transfers the two distinct regions are again observed. The extent of the intermediate region varies from one reaction to the other, but in each case the change to the slope of $\simeq e^{3.0}$ occurs at $|t| \simeq 0.6$ $(\text{GeV}/c)^2$.

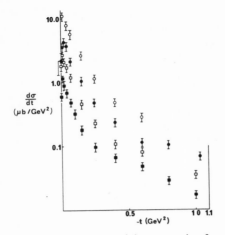

FIG. 66. The differential cross-section for the reaction $\gamma + p \to \pi^- \Delta^{++}$. The data are successively (in increasing cross section) at 5.0, 8.0, 11.0, and 16.0 GeV.

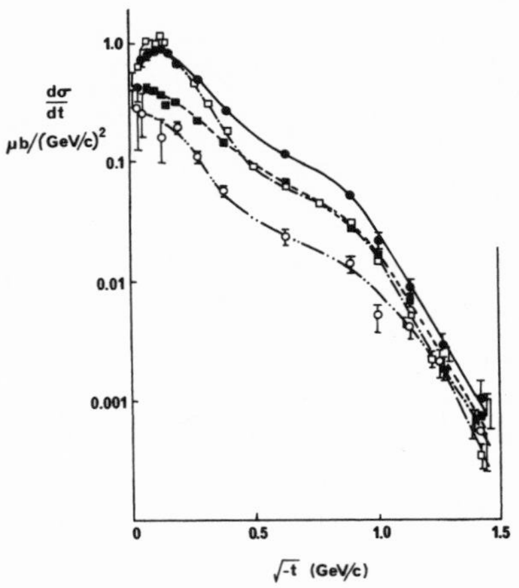

FIG. 67. The differential cross section for the reactions $\gamma + p \to \pi^- + \Delta^{++}$ ($\square$), $\gamma + p \to \pi^+ + \Delta^0$ ($\blacksquare$), $\gamma + n \to \pi^- + \Delta^+$ ($\bigcirc$), and $\gamma + n \to \pi^+ + \Delta^-$ ($\bullet$) at 16.0 GeV.

The preliminary results of Boyarski *et al.* (1969b) for the ratios

$$R_1 = \frac{\sigma(\gamma n \to \pi^- \Delta^+)}{\sigma(\gamma p \to \pi^+ \Delta^0)}, \qquad R_2 = \frac{\sigma(\gamma p \to \pi^- \Delta^{++})}{\sigma(\gamma n \to \pi^+ \Delta^-)} \qquad \text{(III-52)}$$

are qualitatively similar to the ratio $(\pi^- p)/(\pi^+ n)$ in charged pion production, being close to unity in the forward direction, then decreasing with a possible increase beyond $|t| \simeq 0.5 \, (\text{GeV}/c)^2$. These results are shown in Figs. 68 and 69. The implications of all this are that the mechanism for these reactions is very similar to that of $\gamma + p \to \pi^+ + n$ and $\gamma + n \to \pi^- + p$.

However, the preliminary results of Boyarski *et al.* (1970) indicate that the situation may well be more complicated than this. Since unit charge is being exchanged, the normal assumption is that the production proceeds by isospin one exchange in the t-channel. Isospin invariance then requires the following ratios to hold:

$$\frac{D_2}{H_2}\bigg|_{\pi^-} = \frac{\sigma(\gamma p \to \pi^- \Delta^{++}) + \sigma(\gamma n \to \pi^- \Delta^+)}{\sigma(\gamma p \to \pi^- \Delta^{++})} = \frac{4}{3},$$

$$\frac{D_2}{H_2}\bigg|_{\pi^+} = \frac{\sigma(\gamma p \to \pi^+ \Delta^0) + \sigma(\gamma n \to \pi^+ \Delta^-)}{\sigma(\gamma p \to \pi^+ \Delta^0)} = 4. \qquad \text{(III-53)}$$

These ratios are shown in Fig. 70 and in the latter case clearly deviate appreciably from the expected value, indicating that some isospin two exchange is present.

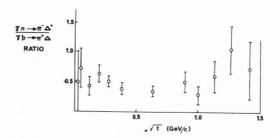

FIG. 68. The angular distribution of the ratio of the reactions $(\gamma + n \to \pi^- \Delta^+)/(\gamma + p \to \pi^+ \Delta^0)$ at 16.0 GeV.

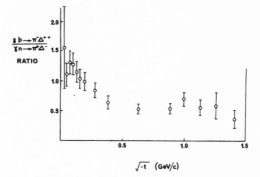

FIG. 69. The angular distribution of the ratio of the reactions $(\gamma + p \to \pi^- \Delta^{++})/(\gamma + n \to \pi^+ + \Delta^-)$ at 16.0 GeV.

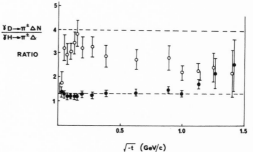

FIG. 70. The ratio of the reactions $(\gamma D \to \pi^\pm \Delta N_s)/(\gamma H \to \pi^\pm \Delta)$ at 16.0 GeV.

5. The Born Approximation in Charged Pion Photoproduction

Dombey (1969) has argued that the natural way to explain forward π^+ and π^- photoproduction is just to use the Born approximation, the reasons for this being based on the fact that the Coulomb coupling of a particle is a very special and nonanalytic coupling and only the Born approximation is consistent with the constraints imposed by that coupling. At high energies, what is meant by the Born approximation is the pion-pole term plus the electric part of the crossed (u-channel) nucleon-pole term, that is, that part required for gauge invariance. This gives an absolute prediction for the forward cross sections, since all couplings are known, and also predicts that the energy dependence of the forward cross sections will be $(s - M^2)^{-2}$, in excellent agreement with what is observed. The predictions of the cross section for the reactions $\gamma + p \rightarrow \pi^+ + n$ and $\gamma + p \rightarrow \pi^- + \Delta^{++}$ are shown in Figs. 71 and 72, and are obviously good for $\sqrt{|t|} \lesssim 2m_\pi$, beyond which they become increasingly too large. This can be remedied by introducing a form factor effect $e^{\alpha t}$ with α of the order of 2–3, justifying this on the grounds that it is an off mass-shell pole. One of the implications of this model is that non-Regge behavior should be observed in backward charged pion photoproduction because of the fixed nucleon pole, but due to the form-factor effect one would not expect it to become important until energies much higher than those presently available. It this hypothesis is accepted, the rest of the cross section can be explained in conventional Regge-pole terms, with minimal cuts.

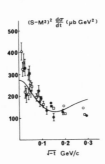

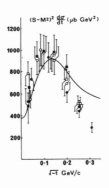

Fig. 71. Comparison of the near-forward differential cross section for $\gamma + p \rightarrow n + \pi^+$ with the gauge-invariant Born approximation.

Fig. 72. Comparison of the near-forward differential cross section for $\gamma + p \rightarrow \pi^- + \Delta^{++}$ with the gauge-invariant Born approximation.

6. $K^{\pm}$ Photoproduction

Some low energy data on the reactions $\gamma + p \rightarrow K^+ + \Lambda^0$, $\gamma + p \rightarrow K^+ + \Sigma^0$ was obtained by Elings et al. (1966) but the most extensive data comes from Boyarski et al. (1969b), covering the energy range 5–16 GeV and the momentum transfer range from near threshold to $|t| \simeq 1.4 \ (\text{GeV}/c)^2$.

The characteristic features of the data are first, that both the forward differential cross sections dip as $|t|$ goes to zero and do not exhibit the sharp forward peak observed in charged pion photoproduction, second, that the differential cross section decreases smoothly with increasing $|t|$ and shows no structure, and third, that the ratio of the $K^+\Lambda^0$ and $K^+\Sigma^0$ cross section is consistent with unity at all photon energies and at all momentum transfers except very near the forward direction, where it decreases. This latter fact rules out any model for both $K^+\Lambda$ and $K^+\Sigma$ reactions which is based on dominant K exchange, since $g^2_{K\Lambda N} \ll g^2_{K\Sigma N}$ and one would expect the $K\Lambda$ reaction to be appreciably greater than the $K\Sigma$ reaction. If K exchange is assumed to be important for the $K\Lambda$ reaction, then the forward dip indicates that it is almost certainly evasive.

This conclusion was the one drawn by Ader *et al.* (1967), Ball *et al.* (1968), and Henyey (1968) in fits to the data in the near-forward direction in terms of simple Regge exchanges. However, when the complete data set is tackled, it turns out that while it is possible to fit the reactions separately in terms of evasive K* and K** exchange, the trajectory parameters required in the two cases are different and they cannot be fitted simultaneously.

The situation is again set for the inclusion of cuts, and this has been done by Borgese and Colocci (1969) who used a simple model based on K* exchange plus the corresponding K*–P cut. The K* was assumed to choose an evasive mechanism and both evasion and conspiracy were considered for the cut contributions, which were taken to contribute to all four helicity amplitudes.

The model contained eight free parameters in the evasive case and six in the conspiring case. It was found impossible to obtain an acceptable solution for evasive cut contributions, the best χ^2 being 560 for 54 data points. In contrast, the agreement for the conspiratorial solution was good, the best fit yielding $\chi^2 = 6.5$. This fit, and the data, are given in Figs. 73 and 74.

As for $\pi\Delta$ production on neutrons, preliminary data from Boyarski *et al.* (1969c) on the reactions

$$\gamma + p \to K^+ + \Lambda^0, \qquad \gamma + p \to K^+ + \Sigma^0, \qquad \gamma + n \to K^+ + \Sigma^-$$

in deuterium are again strongly indicative that this comparatively simple picture is not the whole story. The D_2/H_2 ratios for the $K^+\Lambda$ and $K^+\Sigma$ reactions are shown in Fig. 75. Since Λ^0 has zero isospin, the $K\Lambda$ reaction can proceed only via pure $I = \frac{1}{2}$ exchange, requiring the D_2/H_2 ratio to be unity. This is indeed found (explicitly it is $1.02 \pm .04$) and indicates that absorption in the deuteron nucleus is negligible and consequently it is again reasonable to subtract hydrogen cross sections from deuterium cross sections to get the free neutron cross section. Now if the $K^+\Lambda^0$ reactions proceed entirely by pure $I = \frac{1}{2}$ exchange (and this is the usual assumption), then

$$\frac{(d\sigma/dt)(\gamma n \to K^+\Sigma^-)}{(d\sigma/dt)(\gamma p \to K^+\Sigma^0)} = 2, \tag{III-54}$$

or the D_2/H_2 ratio for $(K + \Sigma)$ should be 3. As can be seen from Fig. 75, it is much less than this, in fact 2.37 ± 0.11, implying that at least 10% of the amplitude is $I = \frac{3}{2}$ exchange.

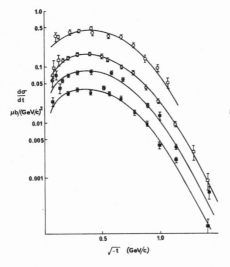

FIG. 73. The differential cross section for the reaction $\gamma + p \rightarrow K^+ + \Lambda^0$ and the Regge-pole plus cut fit of Borgese and Colocci (1969). The data are successively at 5.0, 8.0, 11.0, and 16.0 GeV.

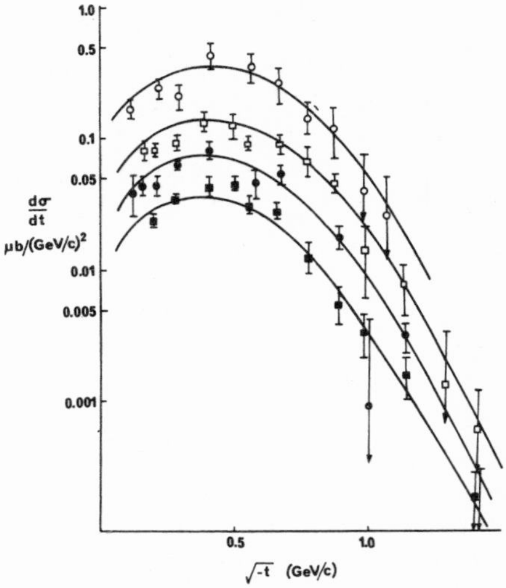

FIG. 74. The differential cross-section for the reaction $\gamma + p \rightarrow K^+ + \Sigma^0$ and the Regge-pole plus cut fit of Borgese and Colocci (1969). The data are successively at 5.0, 8.0, 11.0, and 16.0 GeV.

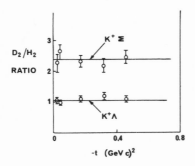

FIG. 75. The ratios of K^+ photoproduction from deuterium and hydrogen at 11.0 GeV.

7. π, K *Photoproduction on Complex Nuclei*

Differential cross sections for π^+ photoproduction on Be, C, Al, Cu, Ag, and Pb have been measured by Boyarski *et al.* (1969d) at 8 and 16 GeV for momentum transfers $|t| \lesssim 0.5$ $(GeV/c)^2$, as well as the cross sections for π^- photoproduction on the same nuclei at 16 GeV and $t = -0.15$ $(GeV/c)^2$. Pi-zero photoproduction has been measured on C, Zn, Ag and Pb at 1.5 and 2.0 GeV by Belletini *et al.* (1969), this experiment being designed primarily to study the Primakoff effect.

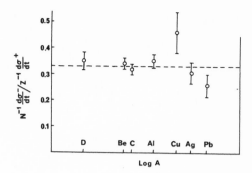

FIG. 76. The dependence of the ratio $Z_{eff} = (d\sigma/dt)(\gamma A \to \pi^+ A^*)/[(d\sigma/dt)(\gamma p \to \pi^+ n)]$ on momentum transfer and nuclear charge. The curves are from the calculation of Gottfried and Yennie (1969) and are normalized to carbon.

For charged pion photoproduction it is expected that the total amplitude will be obtained by the completely incoherent addition of the individual nucleon amplitudes. The results of Boyarski *et al.* (1969d) for π^+ photoproduction, are shown in Fig. 76 where the quantity

$$Z_{eff} = \frac{(d\sigma/dt)(\gamma + A \to \pi^+ + A^*)}{(d\sigma/dt)(\gamma + p \to \pi^+ + n)} \qquad (III\text{-}55)$$

is plotted against $\log Z$. Both the 8- and 16-GeV results are consistent with no energy dependence of Z_{eff} and the effect of the Pauli exclusion principle can clearly be seen in the steady decrease of Z_{eff} from the highest to the lowest momentum transfers for any given nucleus.

Classical calculations with the fermi gas model do not agree particularly well with the data. However, a more complete theory has been developed by Gottfried and Yennie and their results, normalized to the carbon data at each momentum transfer are shown in Fig. 76. They are clearly satisfactory at 8 GeV but work less well for large Z at 16 GeV.

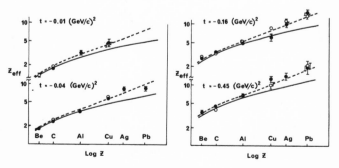

FIG. 77. The mass-number dependence of the ratio $[N^{-1}(d\sigma/dt)(\gamma A \to \pi^- A^*)]/[Z^{-1}(d\sigma/dt) \times (\gamma A \to \pi^+ A^*)]$, where N and Z are the numbers of neutrons and protons in the nuclei, respectively.

The results of Boyarski *et al.* (1969d) for π^- photoproduction are shown in Fig. 77, where the ratio

$$\frac{N^{-1}(d\sigma/dt)(\gamma A \to \pi^- A^*)}{Z^{-1}(d\sigma/dt)(\gamma A \to \pi^+ A^*)}, \qquad \text{(III-56)}$$

is plotted against the atomic number of the nucleus (N is the number of neutrons; Z is the number of protons). The data are all consistent with the ratio found in deuterium, which is also shown, and which is about 0.35 ± 0.03. The weighted average over all nuclei is $\simeq 0.33$.

By contrast, π^0 photoproduction is predominantly coherent in the forward direction. The main contribution in this angular region is the Primakoff effect, the cross section being proportional to $1/\tau$, where τ is the π^0 lifetime. The complete cross section for the reaction can be written

$$d\sigma/d\Omega = d\sigma_P/d\Omega + d\sigma_c/d\Omega + d\sigma_{inc}/d\Omega + 2[(d\sigma_P/d\Omega)(d\sigma_c/d\Omega)] \cos \phi, \quad \text{(III-57)}$$

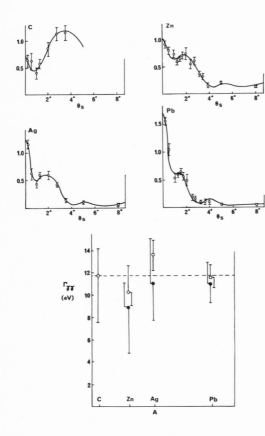

FIG. 78. Forward π^0 photoproduction at 2.0 GeV from complex nuclei. The curves are fits to the data including the Primakoff effect, coherent and incoherent production and their interference.

FIG. 79. The values of the width of the decay $\pi^0 \to \gamma\gamma$ deduced from the Primakoff effect.

where the terms are respectively Primakoff, nuclear coherent, nuclear incoherent, and the Primakoff–nuclear coherent interference. Folding this cross section into the detection efficiency and fitting to the data with the free parameters available in these cross sections [five in the case of Belletini *et al.* (1969)], it is possible to reproduce the data rather well. The results of this fit to the 2.0 GeV data are shown in Fig. 78. The characteristic Primakoff behavior is clearly seen and becomes more and more pronounced in the higher Z elements. The results for the π^0 decay width into two photons, $\Gamma_{\gamma\gamma}$ is shown in Fig. 79. The mean value, $\Gamma_{\gamma\gamma} = (11.7 \pm 1.1)$ eV corresponds to a π^0 lifetime of $(0.56 \pm 0.05) \times 10^{-16}$ sec.

D. BACKWARD PSEUDOSCALAR MESON PHOTOPRODUCTION

Just as forward photoproduction is dominated by reggeized meson exchange, so backward photoproduction is dominated by reggeized baryon exchange. Unlike reggeized meson exchange, however, it is not possible in baryon exchange to factor out all the kinematical singularities and consequently it is not clear how to introduce the Regge pole parametrization.

Let us consider photopion production explicitly. One possibility, which was adopted by Paschos (1968) and by Beaupre and Paschos (1969) is to parametrize the amplitudes $F_1 \ldots, F_4$ introduced initially by Chew *et al.* (1956). In terms of the Pauli amplitudes $\mathscr{F}_1, \ldots, \mathscr{F}_4$ and the invariant amplitudes $A_1, \ldots, A_4$, these are given by

$$F_1 = \frac{8\pi W}{(W-M)} \frac{\mathscr{F}_1}{[(E_1 + M)(E_2 + M)]^{1/2}}$$

$$= A_1 - \frac{Q \cdot K}{(W-M)} A_3 + \left\{ (W-M) + \frac{Q \cdot K}{(W-M)} \right\} A_4,$$

$$F_2 = \frac{8\pi W}{(W-M)} \left(\frac{E_2 + M}{E_1 + M} \right)^{1/2} \frac{\mathscr{F}_2}{q}$$

$$= -A_1 - \frac{Q \cdot K}{(W+M)} A_3 + \left\{ (W+M) + \frac{Q \cdot K}{(W+M)} \right\} A_4, \quad \text{(III-58)}$$

$$F_3 = \frac{8\pi W}{(W-M)} \frac{\mathscr{F}_3}{q[(E_1 + M)(E_2 + M)]^{1/2}}$$

$$= (W-M)A_2 + A_3 - A_4,$$

$$F_4 = \frac{8\pi W}{(W-M)} \left(\frac{E_2 + M}{E_1 + M} \right)^{1/2} \frac{\mathscr{F}_4}{q}$$

$$= -(W+M)A_2 + A_3 - A_4.$$

These amplitudes are related by McDowell symmetry, which requires that

$$F_1(W) = F_2(-W), \qquad F_3(W) = F_4(-W), \qquad \text{(III-59)}$$

a conclusion that is obvious in view of their expansion in terms of the invariant amplitudes.

Although this approach appears superficially simple, the effects of the various kinematical requirements are obscure, and it is not clear how best to incorporate the various factors conventionally associated with the choosing mechanisms of the trajectories. This is more transparent in terms of the parity conserving u-channel helicity amplitudes, with the appropriate kinematical singularity factors removed. Again following Ader *et al.* (1968), the required amplitudes are

$$F^1_{\frac{1}{2}\frac{1}{2}} = -(\sqrt{u}/M)\{\hat{f}_{\frac{1}{2}\frac{1}{2}} + \hat{f}_{\frac{1}{2}-\frac{1}{2}}\},$$

$$F^2_{\frac{1}{2}\frac{1}{2}} = \frac{\sqrt{u}}{M} \frac{(M^2 - \mu^2)}{[\{u - (M + \mu)^2\}\{u - (M - \mu)^2\}]^{1/2}} \{\hat{f}_{\frac{1}{2}\frac{1}{2}} - \hat{f}_{\frac{1}{2}-\frac{1}{2}}\},$$

$$\text{(III-60)}$$

$$F^1_{\frac{3}{2}\frac{1}{2}} = \frac{u}{(u - M^2)} \frac{(M^2 - \mu^2)^2}{\{u - (M + \mu)^2\}\{u - (M - \mu)^2\}} \{\hat{f}_{\frac{3}{2}\frac{1}{2}} - \hat{f}_{\frac{3}{2}-\frac{1}{2}}\},$$

$$F^2_{\frac{3}{2}\frac{1}{2}} = \frac{-u}{(u - M^2)} \frac{(M^2 - \mu^2)^2}{[\{u - (M + \mu)^2\}\{u - (M - \mu)^2\}]^{1/2}} \{\hat{f}_{\frac{3}{2}\frac{1}{2}} + \hat{f}_{\frac{3}{2}\frac{1}{2}}\},$$

with

$$\hat{f}_{\lambda\mu} = \cos(\tfrac{1}{2} - \theta_u)^{-|\lambda+\mu|} \sin(\tfrac{1}{2}\theta_u)^{-|\lambda-\mu|}\hat{f}_{\lambda\mu},$$

where $f_{\lambda\mu}$ are the conventional u-channel helicity amplitudes.

The amplitudes $F^i_{\lambda\mu}$ in (III-60) are free from all kinematical singularities other than a square-root branch point at $u = 0$, which cannot be removed.

These amplitudes are related by

$$F^1_{\frac{1}{2}\frac{1}{2}}(-\sqrt{u}) = F^2_{\frac{1}{2}\frac{1}{2}}(\sqrt{u}) \qquad F^1_{\frac{1}{2}\frac{1}{2}}(-\sqrt{u}) = F^2_{\frac{3}{2}\frac{1}{2}}(\sqrt{u}), \qquad \text{(III-61)}$$

and there are additional constraints at the thresholds and pseudothresholds. These are readily obtained from Ader *et al.* (1968). There are no constraints at $u = (M \pm \mu)^2$, but at $u = M^2$ the amplitudes are required to obey

$$F^2_{\frac{1}{2}\frac{1}{2}} + [(s - M^2)/(M^2 - \mu^2)]F^1_{\frac{3}{2}\frac{1}{2}} = o(u - M^2),$$
$$F^1_{\frac{1}{2}\frac{1}{2}} + [(s - M^2)/(M^2 - \mu^2)]F^2_{\frac{3}{2}\frac{1}{2}} = o(u - M^2). \qquad \text{(III-62)}$$

This approach has been followed by Barger and Weiler (1969) and by Bajpai and Donnachie (1969).

Regge poles with natural parity contribute to $F^2_{\frac{1}{2}\frac{1}{2}}$ and $F^1_{\frac{3}{2}\frac{1}{2}}$ and Regge poles with unnatural parity contribute to $F^1_{\frac{1}{2}\frac{1}{2}}$ and $F^2_{\frac{3}{2}\frac{1}{2}}$. The obvious condidates are N_α and N_γ for the natural parity trajectories and Δ_δ for the unnatural parity trajectories, but in terms of only those three trajectories the requirements of (III-61) cannot be satisfied because different trajectories contribute to the left and right hand sides of the equation.

One way out of this dilemma is to postulate the existence of parity doublets and introduce additional trajectories $N_{\alpha p}$, $N_{\gamma p}$, and $\Delta_{\delta p}$. There is at least superficial evidence for this in the N* resonances, although it is not yet clear whether the observed pairs of nearly equal mass, same total angular momentum but opposite parity are genuine parity doublets or not. The introduction of these extra trajectories immediately poses another problem, since the lowest-lying nucleon state $N_\alpha(938)$, $N_\gamma(1512)$, and $\Delta_\delta(1236)$ certainly do not have a parity doublet. These latter states are conveniently described as "extinguished" and they can be eliminated by requiring factors of $(1 + \sqrt{u}/M_{N*})$ and $(1 - \sqrt{u}/M_{N*})$ in the residue functions of the trajectory and of its missing twin, where M_{N*} is the mass of the leading particle of the trajectory. This then leads to the parametrization

$$F^1_{\frac{1}{2}\frac{1}{2}} = \sum_{N_{\alpha p}, \Delta_\delta, N_{\gamma p}} (\alpha - \tfrac{1}{2})^m g_1(\alpha, \sqrt{u})(1 \pm \sqrt{u}/M_{N*})R(\alpha, s)(s/s_0)^{\alpha - 1/2},$$

$$F^1_{\frac{3}{2}\frac{1}{2}} = \sum_{N_\alpha, \Delta_{\delta p}, N_\gamma} (\alpha - \tfrac{1}{2})^m g_3(\alpha, \sqrt{u})(1 \pm \sqrt{u}/M_{N*})R(\alpha, s)(s/s_0)^{\alpha - 3/2}$$

$$(\text{III-63})$$

with $F^2_{\frac{1}{2}\frac{1}{2}}$ and $F^2_{\frac{3}{2}\frac{1}{2}}$ specified by (eq. III-61), $g_1(\alpha\sqrt{u})$ and $g_3(\alpha, \sqrt{u})$ are the residue functions, m and n depend on the choosing mechanisms of the trajectory and $R(\alpha, s)$ contains the signature and can be specified by

$$R(\alpha, s) = (1 + \xi \exp[-i\pi(\alpha - \tfrac{1}{2})])/(\cos \pi\alpha)[\Gamma(\alpha + \tfrac{1}{2})]^{-1}, \quad (\text{III-64})$$

which is the choice of Barger and Weiler (1969), or by

$$R(\alpha, s) = (1 + \xi \exp[-i\pi(\alpha - \tfrac{1}{2})])\Gamma(\tfrac{1}{2} - \alpha), \quad (\text{III-65})$$

which is the choice of Bajpai and Donnachie (1969).

The equivalent parametrization of the amplitudes $F_1, \ldots, F_4$ is

$$F_2 = [(u - M^2)/u][(W - M)^2 - \mu^2]^{1/2} \sum g_2(\alpha, \sqrt{u})R(\alpha, s)(s/s_0)^{\alpha - 1/2},$$

$$F_4 = q \cdot [(u - M^2)/u][(W - M)^2 - \mu^2]^{1/2} \sum g_4(\alpha, \sqrt{u})R(\alpha, s)(s/s_0)^{\alpha - 1/2}$$

$$(\text{III-66})$$

with F_1 and F_3 specified by (III-59). This was the choice of Beaupre and Paschos (1969), with $R(\alpha, s)$ given by (III-65).

The final step is to take account of the constraint (III-62). This can be done in one of two ways. First, there is the straightforward approach of introducing the appropriate factor into the residue functions

$$g_1(\alpha, \sqrt{u}) = (u - M^2)\tilde{g}_1(\alpha, \sqrt{u}), \qquad g_3(\alpha, \sqrt{u}) = (u - M^2)\tilde{g}_3(\alpha, \sqrt{u}).$$

$$(\text{III-67})$$

Second, since the same trajectories contribute to both the amplitudes involved in the constraint relation, it is possible to satisfy (III-62) by writing, for any particular trajectory

$$[\alpha(M) - \tfrac{1}{2}]^m g_1(\alpha, M) + [(s - M^2)/(M^2 - \mu^2)][\alpha(M) - \tfrac{1}{2}]^n [\alpha(M) - \tfrac{3}{2}]^n g_3(\alpha, M)/s$$
$$= C(u - M^2). \tag{III-68}$$

The latter approach was the one adopted by Barger and Weiler (1969) and by Bajpai and Donnachie (1969).

1. π Photoproduction

Data on backward π^+ photoproduction have been obtained by Anderson et al. (1968, 1969b) and on backward π^0 photoproduction by Tompkins et al. (1969). The π^+ data covers the energy range 4.1–14.8 GeV and the momentum transfer range $-1.8 < u < 0.05$ $(\text{GeV}/c)^2$; the π^0 data covers the energy range 6–18 GeV and the momentum transfer range $-1.2 < u \leqslant 0.0$ $(\text{GeV}/c)^2$.

The most notable feature of the data is the lack of any dip, unlike backward $\pi^+ p$ scattering where there is a pronounced dip at $u \simeq -0.15$ $(\text{GeV}/c)^2$ due to the nonsense wrong signature zero of the N_α trajectory. This eliminates any possibility of N_α dominance in photoproduction, and the fact that the two cross sections are almost equal in the vicinity of $u = -0.15$ $(\text{GeV}/c)^2$ rules out the possiblity of Δ_δ exchange being the only other contributor, since this would require

$$(d\sigma/du)_{\gamma p \to p\pi^0} = 2(d\sigma/du)_{\gamma p \to \pi^+ n} \tag{III-69}$$

since any Δ exchange contribution involves only the isovector part of the photon. This necessarily requires an extra $I = \tfrac{1}{2}$ contribution which does not vanish at $u \simeq -0.15$ $(\text{GeV}/c)^2$. The obvious choice is N_γ, since the lowest-lying member of this trajectory, $D_{13}(1512)$, is strongly photoproduced and its residue at the pole (obtained from the various fits to the low and intermediate energy photoproduction data) is considerably larger than either the N_α or Δ_δ residues. The energy dependence of the data is consistent with k^{-3}. Good fits to the data, based on N_α, N_γ, and Δ_δ exchange plus the McDowell symmetry requirements have been obtained by Beaupre and Paschos (1969), Barger and Weiler (1969), and Bajpai and Donnachie (1969), although their parametrizations are somewhat different.

Beaupre and Paschos (1969) used nonlinear trajectories, basing this on the results obtained in backward pion-nucleon scattering, allowed the residue functions to have momentum transfer dependence (g_2 linearly, g_4 exponentially) and allowed the scaling factor s_0 to vary, having in all 16 parameters. Barger and Weiler (1969) used linear trajectories and ultimately took the N_α and N_γ trajectories to be completely degenerate (but different from the Δ_δ

trajectory), allowed the residue functions to have exponential momentum transfer dependence (but the same for the two N trajectories) and fixed the scaling factor s_0 at 1 $(GeV/c)^2$, leaving themselves only eight parameters. Bajpai and Donnachie (1969) also took linear trajectories, but constant residues, the price paid for this being that the trajectories were nondegenerate. Their fit contained nine parameters. All these fits give a good description of the data. The π^+ data is shown in Fig. 80 and the π^0 data in Fig. 81, together with the fits of Bajpai and Donnachie (1969). The fits of Beaupre and Paschos (1969) and of Barger and Weiler (1969) do not differ significantly from these.

One feature which all these fits have in common, apart from the importance of the N_γ exchange, is a considerable amount of Δ exchange. Kane (1969) has argued on the basis of the total backward $\pi^- p \to p\rho^-$ cross section and vector dominance that $I = \frac{3}{2}$ exchange can contribute at most 20% of the total backward π^+ photoproduction cross section. However, against this, Beaupre and Paschos (1969) claim for the reaction $\pi^- p \to p\rho^-$, on the basis of the backward differential cross sections (and vector dominance again), that the contribution of the Δ trajectory could be large enough to account for almost all of the observed π^0 cross section for -0.5 $(GeV/c)^2 < u$, while its maximum contribution to π^+ photoproduction could give up to half the observed cross section.

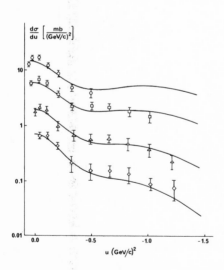
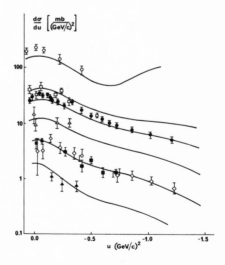

FIG. 80. The differential cross section for backward π^+ photoproduction, with the Regge-pole fits of Bajpai and Donnachie (1970). The data are successively (with increasing cross section) at 3.0, 4.5, 5.0, 9.5, and 15.0 GeV.

FIG. 81. The differential cross section for backward π^0 photoproduction, with the Regge-pole fits of Bajpai and Donnachie (1970). The data are successively (with increasing cross section) at 6.0, 8.0, 12.0, and 18.0 GeV.

2. η Production

Data on this reaction have been obtained by Tompkins *et al.* (1969), and they are shown in Fig. 82. Obviously, they are too few to allow any conclusions to be drawn, it being quite possible to fit them with N_α dominance, N_γ dominance, or any mixture of the two. On the basis of the analysis of η photoproduction in the resonance region one would not expect N_α dominance, since there is no evidence for any contribution from $F_{15}(1690)$. Nothing can be said about N_γ. Of the two known states on this trajectory, $D_{13}(1512)$ is below threshold and $F_{17}(2170)$ is above the limit of the data, so that their couplings are completely unknown. In the event of N_γ being dominant, the cross section should show a pronounced dip in the region of $u \simeq -0.45$ $(GeV/c)^2$, as indicated in Fig. 82.

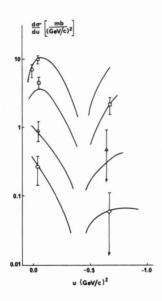

FIG. 82. The differential cross section for backward η^0 photoproduction. The data are successively (with increasing cross section) at 6.0, 8.0, 12.0, and 18.0 GeV.

3. K Production

Backward $K^+\Lambda$ and $K^+\Sigma^0$ production have been measured by Anderson *et al.* (1969b) for an incident photon energy of 4.3 GeV and for momentum transfers $-0.7 < u < 0.2$ $(GeV/c)^2$. The cross sections are shown in Fig. 83. The behavior is very similar to that of π^+ photoproduction, the cross section decreasing smoothly with increasing $|u|$. The Σ^0/Λ ratio is consistent with a mean value of (1.7 ± 0.15) which rules out any large contribution from the exchange of a Σ trajectory (see Levinson *et al.*, 1963).

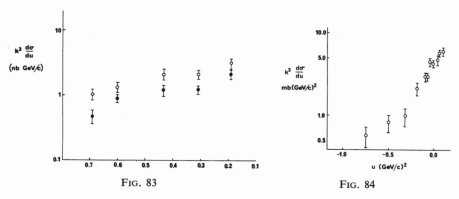

FIG. 83

FIG. 84

FIG. 83. The differential cross section at 4.3 GeV for the reactions $\gamma + p \rightarrow K^+ + \Lambda^0$ (○) and $\gamma + p \rightarrow K^+ + \Sigma^0$ (●).

FIG. 84. The differential cross section at 4.5 to 5.3 GeV for the reaction $\gamma + p \rightarrow \pi^- + \Delta^{++}$.

4. Reaction $\gamma p \rightarrow \pi^- \Delta^{++}$

This has been studied at 4.5 and 5.3 GeV by Anderson et al. (1969) for $-0.7 < u < 0.1$ $(\text{GeV}/c)^2$, and the results are shown in Fig. 84. This can proceed only via Δ exchange, and since it drops off much faster than either the $(\pi^+ n)$ or $(\pi^0 p)$ cross sections would appear to require that for large $|u|$, the latter two should be dominated by N_α and N_γ, with little contribution from Δ_δ. The rate of decrease of cross section is comparable with that in the $(\pi^0 p)$ cross section in the range $-0.3 < u < 0.0$ $(\text{GeV}/c)^2$, and so does not appear to exclude an appreciable amount of Δ exchange in this region for the $(\pi^0 p)$ [and consequently the $(\pi^+ n)$] reaction.

E. VECTOR MESON PHOTOPRODUCTION

1. ρ, ω, ϕ Photoproduction on Protons

Of the various t-channel exchanges possible in these reactions, the data presently available requires only two to be included, namely, π and P. Even then π exchange is really only important for ω photoproduction, where it dominates the lower energy range (threshold to ~ 6.0 GeV). This can be understood on the basis of $SU(3)$ (although it is never very clear how meaningful the application of $SU(3)$ is in these circumstances) which requires $g_{\gamma\pi\omega} = 3g_{\gamma\pi\rho}$ and $g_{\gamma\pi\phi} \simeq 0$, numbers which are not inconsistent with experimental observation.

The different character of ρ and ω photoproduction is clearly shown by their excitation curves which are given in Figs. 85 and 86. The initial structure observed in ρ photoproduction near threshold can be attributed to the

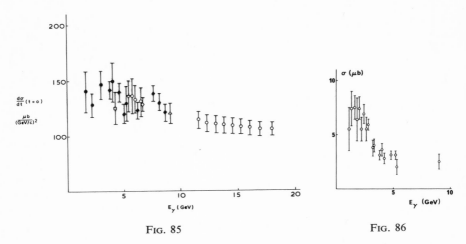

FIG. 85   FIG. 86

FIG. 85. The zero angle differential cross section for ρ photoproduction.
FIG. 86. The total cross section for ω photoproduction.

$F_{37}(1940)$ resonance. Thereafter, the cross section becomes constant, or perhaps very slowly decreasing. There is perhaps a secondary structure in the 3.0-GeV region; however, for photon energies less than 4.0 GeV, the value of the total cross section depends sensitively on the method of analysis used, and at the moment this structure cannot be considered to be significant. The total cross section for ω photoproduction is quite different. After the sharp threshold rise, it decreases steadily and smoothly until about 6.0 GeV, a characteristic feature of single-pion exchange.

Because of the simple nature of the reaction mechanism, it is possible to give a reasonable description of the data without having recourse to Regge theory. For example, ρ photoproduction can be explained in terms of an optical model (Eisenberg *et al.*, 1966a, b) and ω photoproduction in terms of one pion exchange modified by a final state absorption (Schilling and Storim, 1968). Further information on the diffractive (or nondiffractive) nature of

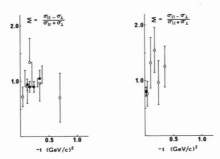

FIG. 87. The polarized photon asymmetry ratio for ρ photoproduction at (a) 2.3–2.9 GeV and (b) 3.5–4.7 GeV.

these reactions can be obtained by photoproduction with polarized photons and by photoproduction off deuterons.

Data on ρ photoproduction by polarized photons has been obtained by Criegee and Timm, and quoted by Lohrmann (1969) at 2.3 GeV, by Diambrini-Palazzi et al. (1969) at 3.5 GeV, and by Bingham et al. (1969) at 2.8 and 4.7 GeV. The data for the asymmetry ratio

$$\Sigma = (\sigma_\| - \sigma_\perp)/(\sigma_\| + \sigma_\perp) \tag{III-70}$$

are shown in Fig. 87. For purely diffractive production, $\Sigma = 1$, and, within the errors, the results are consistent with this except for that of Diambrini-Pallazzi et al. (1969), who obtain $\Sigma = 0.8 \pm 0.06$. In strong contrast, Ballam et al. (1970) find for ω photoproduction at 4.7 GeV a value of $\Sigma \simeq 0.1$. Further evidence that ω photoproduction is not predominantly diffractive, even at 7.5 GeV, is found in photoproduction off complex nuclei, a point which will be discussed in more detail.

Cross sections for the photoproduction of ρ mesons off deuterium have been obtained by McClellan et al. (1969a), Bulos et al. (1969a), and Hilpert et al. (1969). For pure diffraction, the ratio

$$R = \frac{(d\sigma/dt)\,(D,\, t = 0)}{(d\sigma/dt)\,(H,\, t = 0)} = 3.64. \tag{III-71}$$

This is not 4, as might be expected, because of the Glauber correction. McClellan et al. (1969) found $R = 3.26 \pm 0.10$ and Bulos et al. (1969) found $R = 3.5 \pm 0.3$. The former result implies some degree of nondiffractive production, but the latter is consistent both with the results of McClellan et al. (1969) and with the prediction of diffractive production. Hilpert et al. (1969) were able to separate the elastic cross section on deuterium, that is, $\gamma + d \to d + \rho^0$, from the inelastic cross section. It is possible in this case to use the impulse approximation to calculate this cross section from the proton and the (assumed) neutron cross sections. Their data are in three energy bins, $1.8 < E_\gamma < 2.5$ GeV, $2.5 < E_\gamma < 3.5$ GeV and $3.5 < E_\gamma < 5.3$ GeV and there is some evidence for a deviation from pure diffraction in their preliminary results, but the uncertainties are such that it cannot be taken as significant, particularly in the higher energy bins. Specifically, writing the nucleonic amplitudes as a sum of terms for which $I = 0$ $[A^{(0)}]$ or $I = 1$ $[A^{(1)}]$ is exchanged,

$$A^{(p)} = A^{(0)} + A^{(1)}, \qquad A^{(n)} = A^{(0)} - A^{(1)}, \tag{III-72}$$

it is possible to estimate how much of the $A^{(1)}$ amplitude is present. The preliminary results are

$$\begin{aligned}
|A^{(1)}/A^{(0)}| &\simeq 0.32 \quad \text{at} \quad 2.15 \quad \text{GeV}, \\
|A^{(1)}/A^{(0)}| &\simeq 0.17 \quad \text{at} \quad 4.4 \quad \text{GeV},
\end{aligned} \tag{III-73}$$

the errors in the latter case being of the same order as the number itself. Some degree of nondiffractive production in the lowest energy bin would not be surprising, since it overlaps considerably with the $F_{37}(1940)$ resonance, which we have already noticed in the ρ-photoproduction total cross section.

Regge-pole models of vector-meson photoproduction excluding π exchange have been proposed by Daboul (1968) and by Maheshwari (1968). Those including π exchange have been proposed by Ader and Capdeville (1970). Ader and Capdeville (1970) considered both evasive and conspiring mechanisms, held strictly to $SU(3)$. They assumed not only that the contribution of the pomeron was diagonal with respect to the helicities, but that the two helicity amplitudes remaining were identical, and required that at $t = m_{\pi^2}$ the usual perturbation calculation for the pion exchange term should be recovered, i.e., they imposed a very restrictive parametrization. Within this restrictive framework, the conspiring solution of Ader and Capdeville (1970) for ω photoproduction was preferred. A typical comparison with the data is shown in Fig. 88. However, it is clear that this is too restrictive, since for ρ photoproduction it yields, for example, for the density matrix elements $X_{00} = \mathrm{Re}\, X_{10} = X_{1\,-1} \simeq 0$ everywhere, which is not borne out by the data. However, the data on the whole are not sufficiently precise or extensive to justify a more detailed approach.

2. ρ, ω, ϕ Photoproduction on Complex Nuclei

The general motivation is based on the fact that since photons and vector mesons have the same quantum numbers, then at high energy and small mo-

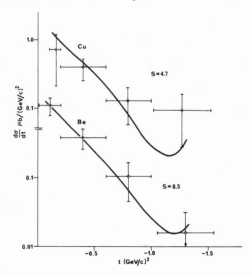

FIG. 88. The differential cross section for ω photoproduction at high energies with the Regge-pole fit of Ader and Capdeville (1970).

mentum transfers the reaction $\gamma A \rightarrow V^0 A$ should be diffraction dominated and should have a similar behavior to that of a high energy elastic scattering cross section, namely

(i) $$|d\sigma/dt|_{t\,\text{small}} \simeq \exp[a(A, t)t], \qquad (\text{III-74})$$

where $a(A, t)$ will be some measure of the nuclear density, its value depending on the t range used and on A;

(ii)   the total cross section should either decrease slowly with increasing energy or be constant;

(iii)   the vector mesons produced should carry the same polarization as the initial photons.

This is given substance by the vector dominance model which says that the electromagnetic current density can be written as a linear combination of the neutral vector-meson fields even for off-mass shell processes.

Specifically for ρ mesons, the relation

$$j_\mu^3 = (m_\rho^2/\gamma_\rho)\rho_\mu, \qquad (\text{III-75})$$

which is a trivial identity on the ρ-meson mass shell, is transcribed into

$$\langle f|j_\mu^3|i\rangle = m_\rho^2/\gamma_\rho(k^2 + m_\rho^2)^{-1}\langle f|j_\mu^{(\rho)}|i\rangle, \qquad (\text{III-76})$$

where $j_\mu^{(\rho)}$ stands for the ρ source, and the equation is to be applied to off-mass-shell processes ($k^2 = m_\rho^2$).

In ρ photoproduction, this allows the initial photon to be considered to be propagating as a ρ meson, and the photoproduction cross section can be related to the cross section for elastic ρ nuclear scattering. To study the detailed mechanism a diffraction model is assumed (Drell and Trefil, 1966; Ross and Stodolsky, 1966; Margolis, 1968a, b; Kölbig and Margolis, 1968; von Bochmann et al., 1969).

The observed differential cross section is composed of several factors and is of the general form

$$(d\sigma/d\Omega\ dm_{\pi\pi})(m_{\pi\pi}, p_{\pi\pi}, t_\perp)$$
$$= \pi^{-1}p_{\pi\pi}^2\, 2m_{\pi\pi}\, R_n(m_{\pi\pi})[f_c + f_{\text{inc}}] + B(m_{\pi\pi}, p_{\pi\pi}, t_\perp), \qquad (\text{III-77})$$

where f_c is the forward coherent (diffractive) production cross section, f_{inc} the forward incoherent production cross section (this is most important for light nuclei when it can contribute up to 10% of the cross section, but decreases with increasing nuclear mass and is negligible for $A > 100$), and B is an arbitrary background term which can be represented as a general polynomial in

$(m_{\pi\pi}, p_{\pi\pi}, t_{\perp})$ space. Here $m_{\pi\pi}$ is the invariant dipion mass, $p_{\pi\pi}$ the dipion momentum, and $t_{\perp}$ the square of the momentum transfer to the nucleus.

The quantity $R_n(m_{\pi\pi})$ has the general form

$$R_n(m_{\pi\pi}) = r(m_{\pi\pi})(m_\rho/m_{\pi\pi})^2 + I(m_{\pi\pi}), \qquad \text{(III-78)}$$

with

$$r(m_{\pi\pi}) = \pi^{-1} m_\rho \Gamma_\rho(m_{\pi\pi})/[(m_\rho^2 - m_{\pi\pi}^2) + m_\rho^2 \Gamma_\rho^2(m_{\pi\pi})]. \qquad \text{(III-79)}$$

The factor $(m_\rho/m_{\pi\pi})^2$ multiplying the Breit-Wigner form is due to Ross and Stodolsky (1966), and the term $I(m_{\pi\pi})$ represents possible interference with the nonresonant background. Either or both of these may be omitted. The important quantity f_c is given by

$$f_c = \left| 2\pi f_0 \int_0^\infty b\,db \int_{-\infty}^\infty dz \rho(z, b) J_0(b\sqrt{|t_\perp|}) \exp[iz\sqrt{|t_\parallel|}] \right.$$
$$\left. \times \exp\left[-\tfrac{1}{2}\sigma_{\rho N}(1 - \alpha_\rho) \int_z^\infty dz' \rho(z', b) \right] \right|^2. \qquad \text{(III-80)}$$

The meaning of this expression is that the ρ meson is produced with an effective forward production cross section $|f_0|^2$ on a single nucleon in a nucleus whose structure is taken into account by the factor $\rho(z, b) J_0[b(|t_\perp|)^{1/2}]$. The factor $\exp[iz(|t_\parallel|)^{1/2}]$ arises from the difference in mass between the initial and final states, and finally the ρ-meson is attenuated by

$$\exp\left\{ -\tfrac{1}{2}\sigma_{\rho N} \int_z^\infty dz' \rho(z'b) \right\}$$

in nuclear matter as it leaves the nucleus.

Gottfried (1968) has shown that the corrections arising from ρ instability within the nucleus are negligible. The quantity α_ρ is the ratio of the real to the imaginary part of the scattering amplitude on a single nucleon and ρ is the nuclear density function. In terms of a Woods-Saxon potential, for example,

$$\rho = \rho^0/\{1 + \exp[(r - R)/s]\}, \qquad \text{(III-81)}$$

where R is the nuclear radius. There is an extension of this formalism by von Bochmann et al. (1969), who take correlations into account. The inclusion of these effects produces a correction of a few percent to the final results.

Measurements of coherent ρ photoproduction on complex nuclei have been obtained by Astbury et al. (1967a) for Be, C, Al, Cu, Ag, and Pb at 2.7, 3.5, and 4.5 GeV, by McClellan et al. (1969a, b) for Be, C, Mg, Cu, Ag, Au, and Pb at 6.2 and 6.5 GeV, by Bulos et al. (1969a) for Be, C, Al, Cu, Ag, and Pb at 9.0 GeV, and by Alvensleben et al. (1970) for Be, C, Al, Ti, Cu, Ag, Cd, In, Ta, W, Au, Pb, and U at a number of energies between 2.6 and 6.8 GeV, this latter experiment giving by far the most comprehensive accumulation of data on this reaction.

One interesting feature noted by Alvensleben *et al.* (1970) is that for the heavier nuclei (that is, excluding Be and C) the nuclear radii determined in fitting to the data are insensitive to the other parameters and to possible changes of the functional form in the fits. In other words, the determination of the radii of heavy nuclei is independent of m_ρ and Γ_ρ, the Breit-Wigner mass distribution formula, the normalization, and the vector-dominance model. The results of Alvensleben *et al.* (1970) are shown in Fig. 89 and yield a value of $R(A) = (1.12 \pm 0.02)A^{1/3}$ F, in excellent agreement with that obtained from electron scattering, namely, $R(A) = 1.18A^{1/3}$ F (Hofstadter, 1957).

The results obtained for $\gamma_\rho^2/4\pi$ and $\sigma_{\rho N}$ in fitting to the data depend very critically on the value assumed for the ratio α_ρ of the real to the imaginary part of the forward scattering amplitude. For example, if the data of McClellan *et al.* (1969) are analyzed assuming $\alpha_\rho = 0$, then it is found that $\alpha_\rho^2/4\pi = 4.4$ and $\sigma_{\rho N} = 37$ mb, but for $\alpha_\rho = -0.45$, then $\gamma_\rho^2/4\pi = 2.6$ and $\sigma_{\rho N} = 27$ mb (Swartz and Talman, 1969). Both are equally good fits to the data, as can be seen in Fig. 90, which is a comparison of the two fits to the data of McClellan *et al.* (1969b) at $0°$.

It is possible to estimate the magnitude and sign of α_ρ by using vector dominance to relate it to the real part of the forward Compton scattering amplitude $\alpha_{\gamma\gamma}$ ($\alpha_\rho \equiv \alpha_{\gamma\gamma}$) and calculating the latter from forward dispersion relations using as input the total proton-photoabsorption cross sections $\sigma_{tot}(\gamma p)$, which are known over the range 1.5–16.0 GeV. With this procedure, Allcock (1969) has estimated $\alpha_\rho \simeq -0.25$, which would imply a result intermediate to the two quoted above from Swartz and Talman (1969).

In their fit to the data, Alvensleben *et al.* (1970) assumed $\alpha_\rho = -0.2$, and obtained

$$\sigma_{\rho N} = 26.7 \pm 2.0 \quad \text{mb}, \qquad \gamma_\rho^2/4\pi = 2.28 \pm 0.40. \tag{III-82}$$

Their data at $0°$ and the fit thereto are shown in Fig. 91.

Incoherent ρ photoproduction at 4.0 and 8.0 GeV on C, Cu, and Ag has been measured by McClellan *et al.* (1969c). As for incoherent charged-pion photoproduction, it is convenient to plot the quantity A_{eff}, defined in this case by

$$A_{eff} = \sigma(\gamma A \rightarrow \rho A')/\sigma(\gamma p \rightarrow \rho p) \tag{III-83}$$

and which is shown in Fig. 92. The principal feature of these data is that there is no evidence of any energy dependence, a feature already observed in incoherent charged-pion photoproduction and in disagreement with theoretical calculations. The results of the calculation of Gottfried and Yennie (1969) are also shown, and the discrepancy at 8.0 GeV is very marked. It should be noted,

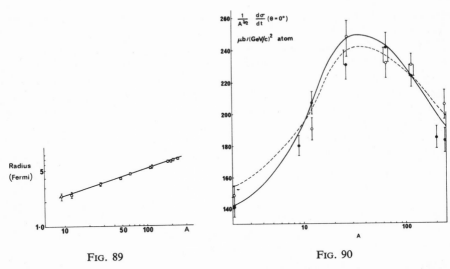

FIG. 89

FIG. 90

FIG. 89. Nuclear radii obtained from ρ photoproduction from complex nuclei at high energies.

FIG. 90. Forward differential cross section for ρ photoproduction from complex nuclei at 6.2 GeV. The solid line is a fit obtained with $\alpha_\rho = 0$, $\sigma_{\rho N} = 37$ mb, and $\gamma_\rho^2/4\pi = 1.10$; the broken line is a fit obtained with $\alpha_\rho = -0.45$, $\sigma_{\rho N} = 27$ mb, and $\gamma_\rho^2/4\pi = 0.65$.

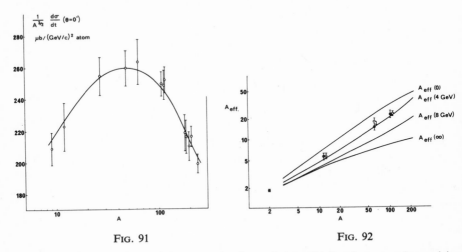

FIG. 91

FIG. 92

FIG. 91. Forward differential cross section for ρ photoproduction from complex nuclei at 7.4 GeV. The fit assumes $\alpha_\rho = -0.2$ and yields $\sigma_{\rho N} = 26.7 \pm 2.0$ mb, $\gamma_\rho^2/4\pi = 0.57 \pm 0.10$.

FIG. 92. Incoherent ρ photoproduction at 4.0 GeV (●) and 8.0 GeV (○). A_{eff} is defined by $\sigma_{\gamma A \rightarrow \rho A} = A_{eff}\, \sigma_{\gamma p \rightarrow \rho p}$. The curves are from the calculation of Gottfried and Yennie (1969).

however, that this calculation involves use of the vector-dominance model, and the curves shown assume a value of the ρ-nucleon total cross section of 38 mb. As we have indicated above and shall explore in more detail in Section III,F, this is probably too large and any reduction in it will reduce the discrepancy but not remove it entirely, since the model will always contain some energy dependence on A_{eff}, in opposition to what is observed.

Coherent ω photoproduction in the near-forward direction (essentially 0°) has been obtained by Bulos *et al.* (1969b) for Be, C, Al, Cu, and Pb, and by Behrend *et al.* (1970) for Be, C, Al, Cu, and Pb, the latter going to sufficiently wide angles to observe the incoherent ω photoproduction as well. These data for Be and Cu at 7.5 GeV are shown in Fig. 93. The data on Cu show the characteristic diffractive forward peak, but this peak is much more difficult to separate out in Be. The ratio of the incoherent to the coherent cross section in ω photoproduction is 3–4 times larger than that in ρ photoproduction, implying that at this energy (7.5 GeV) ω photoproduction on the nucleon has a large nondiffractive amplitude.

The A dependence of ω photoproduction is consistent with that of ρ photoproduction within the rather large errors arising from the data on the former reaction. These uncertainties are such that at the moment it is possible to conclude only that $|\sigma_{\omega N} - \sigma_{\rho N}| > 10$ mb.

ϕ Production has been reported by Asbury *et al.* (1967) and by McClellan *et al.* (1969e). The data analysis is precisely the same as that for ρ photoproduction, and involves the same uncertainty on α_ρ. Assuming $\alpha_\phi = 0$, Asbury *et al.* (1967) obtained $\sigma_{\phi N} = (12.0 \pm 3.9)$ mb, but with the same assumption McClellan *et al.* (1969b) found a much larger value, and obtained $\gamma_\phi^2/4\pi = 37.5 \pm 6$. With $\alpha_\phi = -0.35$, McClellan *et al.* (1969) found $\gamma_\phi^2/4\pi = 14.4 \pm 3.20$ and $\sigma_{\phi N} = (12 \pm 4)$ mb.

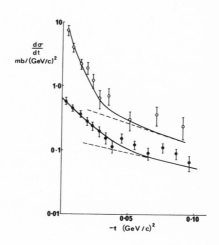

FIG. 93. ω Photoproduction on copper and beryllium at 7.5 GeV. An estimate of the incoherent production is indicated by the broken line.

3. *Other Vector Mesons*

The principal interest here is the search for higher mass vector mesons, and in particular ρ', the daughter of ρ, the existence of which is required in a number of versions of Regge theory. Such a vector meson would be expected to lie in the mass range 1.0–2.0 GeV. Since it is a vector meson, it may be expected to be photoproduced diffractively off complex nuclei and consequently to have a large cross section, comparable with ρ photoproduction.

a. *2π mode.* This mode has been searched for by Hicks *et al.* (1969), with a mass range 1.36–1.78 GeV, by McClellan *et al.* (1969d), with a mass range 1.0–1.8 GeV, by Bulos *et al.* (1969b), with a mass range 1.0–2.0 GeV, and by Ballam *et al.* (1969) with a mass range 1.0–2.0 GeV, all without success, very low cross sections being found. Any conclusions to be drawn from this depend on the unknown branching ratio of ρ' into two pions as well as the ρ'-γ coupling and the ρ' total width. The results have been conveniently summarized by Silverman (1969):

$$\gamma_{\rho'}^{2}/\gamma_{\rho}^{2} > 10^{3}(\Gamma_{\rho'}(2\pi)\Gamma_{\rho})/\Gamma_{\rho'}^{2}\,. \tag{III-84}$$

b. *2μ Mode.* This mode has been investigated by Earles *et al.* (1969), with a mass range 0.9–1.8 GeV, again inconclusive results being obtained. These data have also been summarized by Silverman (1969):

$$\gamma_{\rho'}^{2}/\gamma_{\rho^{2}} > (5 \to 20)(\Gamma_{\rho}/\Gamma_{\rho'}) \tag{III-85}$$

the limit varying smoothly with increasing mass of the ρ'.

c. *All Modes (missing mass).* This has been performed by Anderson *et al.* (1969), with a mass range 1.0–2.0 GeV, no positive result being obtained. Again following Silverman (1969)

$$\frac{(d\sigma/dt)(\rho')}{(d\sigma/dt)(\rho)} < 0.025\Gamma_{\rho'}/\Gamma_{\rho}\,. \tag{III-86}$$

It is clear that if there is a 1^{-} meson in this mass range and if it is not too broad, then its coupling to the photon is at least of one order, and probably two orders of magnitude, smaller than the ρ meson. However, if it is sufficiently broad, say $\Gamma_{\rho'} > 500$ MeV, it is unlikely that it would have been observed in any of these experiments.

F. Vector Meson Dominance in Photoproduction

The starting point for vector dominance is to express the electromagnetic current density as

$$J_{\mu} = (m_{\rho}^{2}/\gamma_{\rho})\rho_{\mu}^{0} + (m_{\omega}^{2}/\gamma_{\omega})\omega_{\mu} + (m_{\phi}^{2}/\gamma_{\phi})\phi_{\mu}, \tag{III-87}$$

where $\rho_\mu{}^0$, ω_μ, and ϕ_μ are the neutral ρ, ω, and ϕ fields. By putting in a smooth-ness assumption, namely that the hadronic matrix elements do not vary very much as we go away from $q^2 = -m_V{}^2$ then, by time reversal invariance (III-87) relates the production of transversely polarized vector mesons in meson-nuclear collisions to meson photoproduction.

The ratios of the coupling constants γ_ρ, γ_ω, and γ_ϕ are given in strict $SU(3)$ with ideal ω-ϕ mixing by

$$\gamma_\rho : \gamma_\omega : \gamma_\phi = 1 : 3 : -\tfrac{3}{2}\sqrt{2}. \tag{III-88}$$

They can be obtained experimentally by studying the reactions $e^+e^- \to 2\pi$ (γ_ρ), $e^+e^- \to 3\pi$ (γ_ω), and $e^+e^- \to K^+K^-$, $K_s{}^0K_L{}^0(\gamma_\phi)$. The current values are (Perez-y-Jorba, 1969)

$$\gamma_\rho{}^2/4\pi = 1.99 \pm 0.11, \qquad \gamma_\omega{}^2/4\pi = 14.9 \pm 2.8, \qquad \gamma_\phi{}^2/4\pi = 11.5 \pm 0.9. \tag{III-89}$$

The ratio $\gamma_\rho : \gamma_\omega$ is, within the error, consistent with the $SU(3)$ values, although results on ω-ρ interference appear to require that the relative sign be negative.

1. *Pion Photoproduction*

The first case we shall consider is the relation between vector meson production by pions and pion photoproduction. The formalism for unpolarized photons has been discussed by Beder (1966) and by Dar *et al.* (1968) and for polarized photons by Krammer and Schildknecht (1968). For unpolarized photons the general form of the relation is

$$(d\sigma/dt)(\gamma N_\alpha \to \pi_\beta N_{\alpha'})$$
$$= \tfrac{1}{2} \sum_{\rho,\,\omega,\,\phi} e^2/2 \cdot (\mathbf{p}_\pi{}^2/\mathbf{p}_V{}^2)(1 - X_{00}^V)(d\sigma/dt)(\pi_\beta N_{\alpha'} \to VN_\alpha)$$
$$\pm \text{ interference terms } ((\rho, \omega), (\rho, \phi), (\omega, \phi)), \tag{III-90}$$

and for polarized photons

$$(d\sigma/dt)(\gamma N_\alpha \to \pi_\beta N_{\alpha'})$$
$$= \sum_{\rho,\,\omega,\,\phi} e^2/2 \cdot (\mathbf{p}_\pi{}^2/\mathbf{p}_V{}^2)(X_{11}^V + \xi X_{1-1}^V)(d\sigma/dt)(\pi_\beta N_{\alpha'} \to VN_\alpha)$$
$$\pm \text{ interference terms } ((\rho, \omega), (\rho, \phi), (\omega, \phi)). \tag{III-91}$$

Here X_{00}^V, X_{11}^V and X_{1-1}^V are the density matrix elements of the outgoing vector meson in the helicity system, ξ is the polarization of the photon ($\xi = +1$ if the photon is polarized perpendicular to the production plane and $\xi = -1$ if the photon is polarized parallel to the production plane) and α, α', and β label the initial and final nucleons and the pion, respectively.

The interference terms cannot be obtained directly from experiment, although they can be estimated using a quark model and assuming additivity for all values of t. A simplification of (III-90) and (III-91) can be made by noting that the ϕ-production cross section is considerably smaller than those for ρ and ω, and it can be omitted.

The various photoproduction reactions and the associated vector-meson production cross sections required are

$$
\begin{aligned}
\text{(i)} \quad & \gamma p \to \pi^0 p: \pi^0 p \to \rho^0(\omega)p; \\
\text{(ii)} \quad & \gamma n \to \pi^0 n: \pi^0 n \to \rho^0(\omega)n; \\
\text{(iii)} \quad & \gamma p \to \pi^+ n: \pi^+ n \to \rho^0(\omega)p; \\
\text{(iv)} \quad & \gamma n \to \pi^- p: \pi^- p \to \rho^0(\omega)n.
\end{aligned}
\tag{III-92}
$$

The vector-meson reaction (iii) is precisely the same as (iv), and (ii) is precisely the same as (i) by charge symmetry. The latter may be determined using the isospin relations

$$
\begin{aligned}
(d\sigma/dt)(\pi^0 p \to \rho^0 p) &= \tfrac{1}{2}\{(d\sigma/dt)(\pi^- p \to \rho^- p) + (d\sigma/dt)(\pi^+ p \to \rho^+ p) \\
&\quad - (d\sigma/dt)(\pi^- p \to \rho^0 n)\}, \\
(d\sigma/dt)(\pi^- p \to \omega p) &= \tfrac{1}{2}(d\sigma/dt)(\pi^- p \to \omega n).
\end{aligned}
\tag{III-93}
$$

Without the ϕ contribution, the interference term is such that it cancels in the linear combinations

$$
\begin{aligned}
(d\sigma/dt)(\gamma p \to p\pi^0) + (d\sigma/dt)(\gamma n \to n\pi^0), \\
(d\sigma/dt)(\gamma p \to n\pi^+) + (d\sigma/dt)(\gamma n \to p\pi^-),
\end{aligned}
\tag{III-94}
$$

and consequently these provide a direct test of the vector dominance model in photoproduction. The comparison for π^0 photoproduction is shown in Fig. 94 and for charged pion photoproduction in Fig. 95.

Krammer and Schildknecht (1968) have calculated the interference term for charged pion production using the quark model, and found the result to be too small to account for the experimental value of the difference

$$
(d\sigma/dt)(\gamma p \to \pi^+ n) - (d\sigma/dt)(\gamma n \to \pi^- p).
\tag{III-95}
$$

They also showed that the upper bound on the interference, term, obtained by assuming maximal interference between the ρ and ω contributions either coincides with or lies above the experimental value of the difference, so there is certainly no immediate inconsistency with the vector dominance model since the validity of the quark model calculation can certainly be questioned.

When dealing with polarized photons, the uncertainties in the density matrix elements are sufficiently large to allow the further simplification of dropping

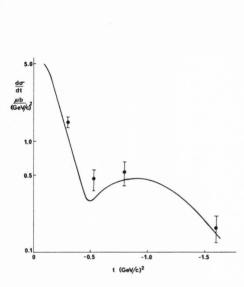

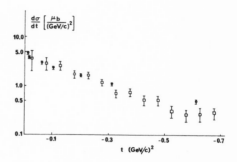

FIG. 94. Comparison of π^0 photoproduction (●) with vector dominance model
predictions (—).

FIG. 95. Comparison of $\pi^\pm$ photoproduction (●) with vector dominance
model predictions (□).

the ω-meson contribution. When this is done, a very simple expression results for the combined π^+ and π^- asymmetry

$$A(\pi^+ + \pi^-) = \frac{\sigma_\perp(\pi^+ + \pi^-) - \sigma_\parallel(\pi^+ + \pi^-)}{\sigma_\perp(\pi^+ + \pi^-) + \sigma_\parallel(\pi^+ + \pi^-)}$$

$$= X^{\rho^0}_{1\,-1}/X^{\rho^0}_{1\,1}. \qquad \text{(III-96)}$$

A series discrepancy for this vector dominance model relation was found by
Gewiniger et al. (1968) when the ρ^0 density matrix elements used were those
evaluated in the helicity frame, which is the natural choice. However, Fraas
and Schildknecht (1969) have shown that this choice of reference frame is not
necessarily unique and Bialas and Zalewski (1969) obtained much better agreement by using density matrix elements evaluated in the frame proposed by
Donohue and Högassen (1967) which is reached from the helicity frame by a
rotation about the normal to the ρ^0 production plane. This has been further
improved by Guiragossian and Levy (1969) who used density matrix elements
obtained by fitting the ρ^0 data of Johnson et al. (1968) directly in the Donohue–
Högassen frame. This result is shown in Fig. 96.

It should be noted that if the Donohue-Högassen frame is used to predict the sum of the unpolarized cross sections then the result is rather poor. To obtain a successful fit in this case it is necessary to use the helicity frame. Now it is not very satisfactory to have to use different frames for different comparisons, and one possible explanation is that it is not vector dominance which is in difficulties, but that the density element $X^{\rho^0}_{1-1}$ is poorly determined.

Cho and Sakurai (1970) suggested reversing the procedure, and gave predictions for ρ^0 production using the photoproduction model of Jackson and Quigg (1969). This idea has been extended by Kellett (1971a, b) and using his own detailed fit to photoproduction (Kellett, 1970), made predictions for vector-meson photoproduction that are in excellent agreement with the full range of high energy data.

To be able to predict vector-meson production from photoproduction, it is necessary to define an extrapolation in the vector mass in such a way that the longitudinal polarization states of the vector particles are obtained from the purely transverse photon. The separation of the polarization states of a vector particle into longitudinal and transverse components is in general a Lorentz-frame dependent procedure, and it is precisely this which has given rise to the ambiguities in the application of the vector dominance model discussed above. It was suggested by Kellett (1971a) that this ambiguity of frame can be reduced to the well-defined kinematic question of the relation between the invariant and the helicity amplitudes in any given frame if the model is formulated in terms of the explicitly Lorentz covariant invariant amplitudes.

Having chosen invariant amplitudes that are free of kinematical singularities, the assumption is made that they do not depend kinematically on the extrapolation in the external mass. This freedom from kinematical dependence on K^2 does not exclude dynamical dependence on K^2 through form factors associated with particular exchange mechanisms. This dynamical mass dependence is not connected with the frame ambiguity, but depends on the dynamical model employed and can be treated in an explicitly covariant way.

Like pion electroproduction, vector-meson production by pions can be described in terms of the six invariant amplitudes B_i of (I-26). These are related to the amplitudes A_i of (I-28) by the inverse of (I-29), which provide at $K^2 = 0$ (recall that then $A_5 = A_6 = 0$) six equations relating the six B_i to the four photoproduction amplitudes $A_1, \ldots, A_4$. These relations are strictly valid only in the zero mass limit, but by the assumption of the model that these amplitudes do not depend kinematically on the extrapolation in K^2 they can be used at the vector-meson mass.

With this approach Kellett (1971a, b) succeeded in obtaining a unified description of all pion photoproduction and all vector meson production by pions (i.e., ρ^0, ρ^-, and ω^0). The model is essentially unique in that the kine-

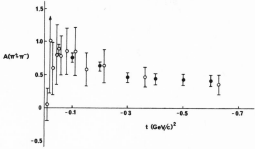

FIG. 96. Comparison of the sum of the polarized photon asymmetry ratios for π^+ and π^- photoproduction ($\bullet$) with vector dominance model predictions ($\circ$).

matics are well defined and there is no frame ambiguity associated with the polarization states of the vector mesons. The only freedom allowed was in the coupling $g_{\rho\omega\pi}$, which is very sensitive to the extrapolation from the photon to the vector meson mass, and affects the results for ρ^- and ω production by pions. This effect is primarily one of normalization, and does not alter the shape of cross sections or of density matrix elements. The predictions for ρ^0 production by pions are absolute.

A final check of vector dominance is made in Fig. 97, which shows of the sum of the transverse cross sections for π^+ and π^-

$$\tfrac{1}{2}[(d\sigma_\perp{}^+/dt) + (d\sigma_\perp{}^-/dt)]$$
$$= \alpha \cdot (4\pi/\gamma_\rho{}^2)(\mathbf{p}_\pi{}^2/\mathbf{p}_\rho{}^2)(X^{\rho^0}_{1\,1} + X^{\rho^0}_{1-1})(d\sigma/dt)(\pi^-p \to \rho^0 n). \qquad \text{(III-97)}$$

A less direct test can be applied to the photoproduction of $(\pi\Delta)$ in particular the reactions $\gamma p \to \pi^- \Delta^{++}$, $\gamma n \to \pi^+ \Delta^-$. The straightforward application

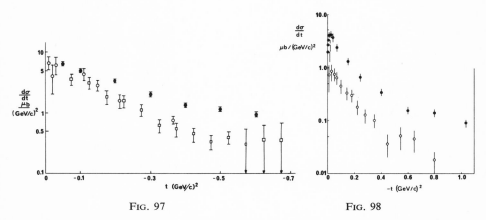

FIG. 97   FIG. 98

FIG. 97. Comparison of $\tfrac{1}{2}(d\sigma_\perp{}^+/d\Omega + d\sigma_\perp{}^-/d\Omega)$ ($\bullet$) with vector dominance model predictions ($\circ$).

FIG. 98. Comparison of $\pi\Delta$ photoproduction ($\bullet$) with vector dominance model predictions ($\circ$).

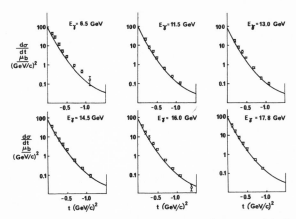

FIG. 99. Comparison of the data with the vector dominance model plus quark model predictions for ρ photoproduction.

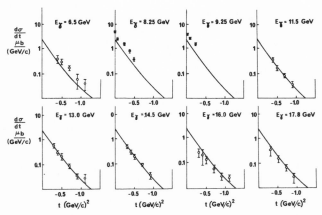

FIG. 100. Comparison of the data with the vector dominance model plus quark model predictions for ϕ photoproduction.

of vector dominance relates these reactions to the reaction $\pi\Delta \to \rho^0 N$, which is not physically accessible. It is necessary to make an analytic continuation from the s-channel of this reaction to the u-channel, which puts the nucleon into the initial state and the Δ into the final state. When this is done, the result is

$$\frac{1}{2}[(d\sigma/dt)(\gamma p \to \pi^-\Delta^{++}) + (d\sigma/dt)(\gamma n \to \pi^+\Delta^-)]$$
$$\simeq 4\pi/\gamma_\rho^2 \cdot \alpha(\mathbf{p}_\pi^2/\mathbf{p}_\rho^2)(1 - X_{0\,0}^{\rho\,0})(d\sigma/dt)(\gamma^+p \to \rho^0\Delta^{++}). \qquad \text{(III-98)}$$

A comparison is given in Fig. 98 and shows a large discrepancy. However, it should be emphasized that because of the analytic continuation necessary this is not a direct test of vector dominance.

2. Vector Meson Photoproduction

The immediate application of vector dominance to photoproduction of vector mesons on hydrogen is obvious:

$$(d\sigma/dt)(\gamma p \to Vp) = \alpha \cdot (4\pi/\gamma_\rho{}^2)(d\sigma/dt)(Vp \to Vp). \tag{III-99}$$

The reaction on the right-hand side of this equation can be related to physically observable reactions by the simple quark model. The results for ρ and ϕ scattering are

$$(d\sigma/dt)(\rho^0 p \to \rho^0 p) = \{\tfrac{1}{2}[(d\sigma/dt)(\pi^+ p \to \pi^+ p)]^{1/2} + \tfrac{1}{2}[(d\sigma/dt)(\pi^- p \to \pi^- p)]^{1/2}\}^2 \tag{III-100}$$

and

$$(d\sigma/dt)(\phi p \to \phi p) = \{[(d\sigma/dt)(K^+ p \to K^+ P)]^{1/2} + [(d\sigma/dt)(K^- p \to K^- p)]^{1/2} - [(d\sigma/dt)(\pi^- p \to \pi^- p)]^{1/2}\}. \tag{III-101}$$

A comparison is made with the data on ρ and ϕ photoproduction in Figs. 99 and 100. The results are clearly good.

The vector dominance model and its relation to ρ and ω photoproduction on complex nuclei has already been discussed at length in Section III,E. It will be recalled that the information which can be extracted from these data are the total vector-nucleon cross section σ_{VN} and the photon vector coupling constant $\gamma_V{}^2/4\pi$ the actual values obtained depending very critically on the assumption made about the ratio α_V of the real to the imaginary part of the forward scattering amplitude. Using values of $\gamma_V{}^2/4\pi$ consistent with those obtained in the storage ring experiments, the overall conclusion reached can be summed up by

$$\begin{aligned}
\sigma_{\rho N} &\simeq 27 \pm 3 \quad \text{mb}, \qquad \alpha_\rho \simeq -0.35 \pm 0.1, \\
\sigma_{\phi N} &\simeq 12 \pm 4 \quad \text{mb}, \qquad \alpha_\phi \simeq -0.2 \pm 0.2.
\end{aligned} \tag{III-102}$$

For comparison the quark model results are $\sigma_{\rho N} \simeq 25$ mb and $\sigma_{\phi N} \simeq 12$ mb, and the dispersion relation calculation of Allcock (1969) gives $\alpha_\rho \simeq -0.25$.

These results can be extracted also from the data on vector-meson photoproduction on hydrogen by extrapolating to $t = 0$. Using the optical theorem, at $t = 0$ (III-99) becomes

$$(d\sigma/dt)(\gamma p \to Vp)_{t=0} = \alpha/16\pi \cdot 4\pi/\gamma_V{}^2 \cdot \sigma_{VN}^2(1 + \alpha_V{}^2). \tag{III-103}$$

The results of the extrapolation for ρ photoproduction are shown in Fig. 101. The term $d\sigma/dt$ $(t = 0)$ falls slowly from 135 ± 15 $\mu b/(GeV/c)^2$ at 5 GeV to 107 ± 7 $\mu b/(GeV/c)^2$ at 17 GeV. This fall is in itself indicative of there being some real part in the forward amplitude. These numbers translate into

$$\sigma_{\rho N}^2(1 + \alpha_\rho{}^2) \simeq 27 \pm 2 \quad \text{mb} \qquad \text{at} \quad 5 \quad \text{GeV}, \tag{III-104a}$$

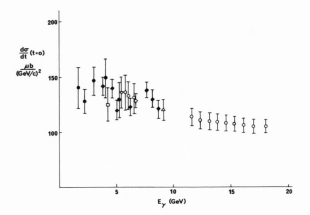

FIG. 101. The differential cross section for ρ photoproduction at zero momentum transfer.

reducing to

$$\sigma_{\rho N}^2(1 + \alpha_\rho^2) \simeq 24 \pm 1 \quad \text{mb} \qquad \text{at} \quad 17 \quad \text{GeV,} \qquad \text{(III-104b)}$$

which are consistent with the previous conclusion.

The extrapolation for ϕ photoproduction is not as precise as that for ρ photoproduction, and it has been made assuming the shape (but not the normalization) implied by (III-101), using the data of Anderson *et al.* (1969a). The results translate into

$$\sigma_{\phi N}^2(1 + \alpha_\phi^2) \simeq 11 \pm 3 \quad \text{mb} \qquad \text{at} \quad 6 \quad \text{GeV,} \qquad \text{(III-105a)}$$

reducing to

$$\sigma_{\phi N}^2(1 + \alpha_\rho^2) \simeq 8 \pm 3 \quad \text{mb} \qquad \text{at} \quad 18 \quad \text{GeV,} \qquad \text{(III-105b)}$$

which is again consistent. The data of McClellan *et al.* (1969b) require either that $\gamma_\phi^2/4\pi \simeq 16$ or that $\sigma_{\phi N}^2(1 + \alpha_\phi^2)$ be about 35% larger than the value given in (III-105).

G. HIGH ENERGY ELECTROPRODUCTION

1. Single-Arm Experiments

It will be recalled that the differential cross section for the inelastic scattering of electrons on protons can be written as

$$(d^2\sigma/d\Omega_e \, dE') = (\alpha/4\pi^2)(k/K^2)(E'/E)\,[2/(1 - \varepsilon)]\{\sigma_T(v, K^2) + \varepsilon\sigma_S(v, K^2)\}$$

$$\text{(III-106a)}$$

or

$$(d^2\sigma/d\Omega_e \, dE') = (4\alpha^2 E'^2/K^4)\{2W_1(v, K^2) \sin^2(\tfrac{1}{2}\theta) + W_2(v, K^2) \cos^2(\tfrac{1}{2}\theta)\},$$

$$\text{(III-106b)}$$

with the connection

$$W_1(v, K^2) = (k/4\pi^2\alpha)\sigma_T(v, K^2),$$

$$W_2(v, K^2) = k/4\pi^2\alpha \cdot K^2/(K^2 + v^2) \cdot \{\sigma_T(v, K^2) + \sigma_S(v, K^2)\} \qquad \text{(III-107)}$$

and

$$v = E - E' = (W^2 - M^2 + K^2)/2M = k + K^2/2M. \qquad \text{(III-108)}$$

The general features of the data are:

1. For small K^2 the cross section is dominated by the resonance peaks which we have already discussed.

2. As K^2 increases, these peaks disappear rapidly, decreasing approximately as K^{-8} (that is, roughly as the square of the nucleon form factors), leaving a smooth spectrum which is large (by large, it is meant that the integrated cross section at fixed K^2 is of the same order of magnitude as the Mott cross section for the scattering of electrons by a point proton).

3. For fixed W (fixed hadronic missing mass), this continuum decreases slowly with increasing K, approximately as K^{-2}.

Prior to there being any data available, Bjorken (1969) related W_1 and W_2 to matrix elements of equal time-current commutators by considering asymptotic sum rules at infinite momentum. He then argued that the infinite momentum limit of these commutators is not divergent, and consequently, if they are nonvanishing, he predicted that

$$W_1(v, K^2) \rightarrow F_1(v/K^2), \qquad vW_2(v, K^2) \rightarrow F_2(v/K^2) \qquad (v, K^2 \rightarrow \infty), \quad \text{(III-109)}$$

that is, these quantities are functions of just one variable v/K^2 for both v and K^2 large and the experimental results should "scale," that is, lie on a universal curve.

For definiteness, we shall use the asymptotic variable

$$\omega = 2Mv/K^2 = x^{-1}. \qquad \text{(III-110)}$$

To separate W_1 and W_2 (or equivalently σ_T and σ_S) at given v, K^2 it is necessary to have data at widely separated angles, This does not apply to most of the published data, which are at 6 and 10° (Bloom et al., 1969). However, some additional data do exist at 48° and by incorporating this Albrecht et al. (1969) have succeeded in making this separation. The ratio $R = \sigma_S/\sigma_T$ which they obtained at $K^2 = 2.0$ (GeV/c)^2 is shown in Fig. 102 as a function of ω, and is clearly consistent with $R \simeq 0$ (or $\sigma_S \ll \sigma_T$). Preliminary results from the

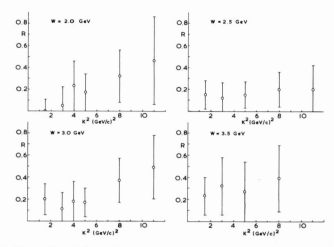

FIG. 102. The ratio of the scalar to the transverse cross section in inelastic electron scattering.

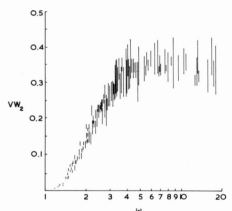

FIG. 103. νW_2 as a function of ω for $\omega > 2.0$ GeV and $K^2 > 1.0$ (GeV/c)^2, assuming $R = 0.18$.

same quantity using data at 18, 26, and 34°, together with the published data at 6 and 10°, have been quoted by Taylor (1969) and agree with this conclusion. Hence, for the remainder of this discussion we shall assume that $\sigma_s \equiv 0$. Note that with this assumption W_1 will scale if W_2 scales for large values of ν such that $\omega/2M \gg 1/\nu$, since

$$W_1 = (1 + R)^{-1}\{\omega/2M + \nu^{-1}\}\nu W_2. \qquad \text{(III-111)}$$

The values of νW_2 obtained from the 6 and 10° data of Bloom *et al.* (1969) assuming $R = 0$ are shown in Fig. 103, and clearly imply the existence of a "universal curve" for $\omega \gtrsim 4$ and $K^2 \gtrsim 0.5$ (GeV/c)^2. What is so surprising about these results is not so much the existence of a universal curve (we have already noted that this was anticipated on theoretical grounds) but that the data should be "scale invariant," even for very nonasymptotic data points.

An intuitive but very illuminating model to apply in this situation is the parton model in which it is assumed that the nucleon is composed of pointlike constituents (partons). This model has been developed in the present context by Bjorken and Paschos (1969), and is based on ideas initially promulgated by Feynman. It is their approach which we outline here.

The trick is to go into the infinite-momentum frame in which the motion of the constituents of the proton is reduced to zero by time dilation, and the proton itself is Lorentz contracted into a flat disk. In this frame, with K^2 large compared to M^2 or to the square of any transverse momentum, the electron can be considered to scatter instantaneously and coherently off the individual pointlike constituents (n in number), leaving their mass and charge unaltered. By ignoring transverse momenta, the ith parton can be considered to have a fraction x_i of the total momentum of the proton

$$p_\mu^{(i)} = x_i P_\mu \tag{III-112}$$

from which it is possible to show that the contribution of a single parton, charge Q_i, to $W_2(v, K^2)$ is

$$W_2^{(i)}(v, K^2) = Q_i^2 \delta(v - K^2/2Mx_i). \tag{III-113}$$

Thus, for a general distribution of partons in the proton

$$W_2(v, K^2) = \sum_n P(n)\left(\sum_i Q_i^2\right)_n \int_0^1 dx f_n(x)\delta(v - K^2/2M_x), \tag{III-114}$$

where $P(n)$ is the probability of finding n partons in the proton, $\left(\sum_i Q_i^2\right)_n$ is the average value of $\sum_i Q_i^2$ in any such configuration and $f_n(x)$ is the probability of finding a parton of momentum xP_μ in this configuration.

Integrating over x yields

$$vW_2(v, K^2) = \sum_n P(n)\left(\sum_i Q_i^2\right)_n f_n(x)$$

$$\equiv F(x) \qquad (x = K^2/2M_v), \tag{III-115}$$

producing the result that vW_2 is a function of the single variable v/K^2.

Normalization requires

$$\int_0^1 dx f_n(x) = 1, \tag{III-116}$$

and if a symmetric distribution of momenta among the partons is assumed, then

$$\int_0^1 dx x f_n(x) = n^{-1}, \tag{III-117}$$

yielding the sum rules

$$\int_0^\infty dv W_2(v, K^2) = \int_0^1 dx \, F(x)/x = \sum_n P(n)\left(\sum_i Q_i^2\right)_n,$$

$$\frac{K^2}{2M} \int_0^\infty dv \, W_2(v, K^2)/v = \int_0^1 dx F(x) = \sum_n P(n)n^{-1}\left(\sum_i Q_i^2\right)_n,$$
(III-118)

which relate moments of the data to the sum of squares of the charges of the partons and the average squared of the partons. Bjorken and Paschos (1969) have evaluated the second of these, giving

$$(K^2/2M) \int dv \, W_2(v, K^2)/v = \int dx F(x) \simeq 0.16.$$
(III-119)

which implies a rather small mean-square charge per parton. An additional feature of this model is that if all the partons have spin $\frac{1}{2}$, then $R \equiv \sigma_S/\sigma_T \to 0$.

Any qualitative predictions are clearly very model dependent and should not be taken too seriously. Bjorken and Paschos (1969) illustrated the possibilities by two examples, for both of which one-dimensional phase space was chosen for the distribution function

$$f_n(x) = (n - 1)(1 - x)^{n-2}.$$
(III-120)

A consequence of this is that if the number of partons is kept finite, then $v W_2(x) \to 0$ as $x \to 0$, in apparent contradiction to the data. This is seen explicitly for a three-quark model with the usual charges, for which

$$v W_2(x) = 2x(1 - x).$$
(III-121)

The other model chosen by Bjorken and Paschos (1969) was the addition of a distribution of quark-antiquark pairs. With this it was possible to obtain the shape of $v W_2(x)$ in fair agreement with experiment, but the overall normalization was off since the model predicts

$$\int_0^1 dx F(x) > 0.22$$
(III-122)

compared to the experimental value of $\simeq 0.16$ quoted above.

However, it is reasonable to consider that there might be neutral particles ("gluons") present to bind the quarks. For example, if it is assumed that three of the n partons are "valence quarks" and, on average, $(1 - \varepsilon)(n - 3)$ are gluons with the remainder forming a sea of quark-antiquark pairs, then the data would lead to

$$\langle 1/n \rangle < 0.18, \qquad \varepsilon < 0.72$$
(III-123)

so that many partons are required.

A parton model has been derived from a canonical field theory of pions and nucleons by Drell *et al.* (1969a, b; 1970a, b). The partons are considered to be bare pions and nucleons, and a cutoff in transverse momenta is imposed *ab initio*. The model is technically very complicated, but it has two very obvious merits: (1) it shows the consistency of the intuitive approach discussed above and (2) it provides a direct connection to processes related by crossing, for example, to $e^+e^- \to p^+$ X (unobserved).

An alternative approach is to treat the reaction like a hadronic reaction, that is, to think of the virtual photon as an incident particle of mass $\{-K^2\}^{1/2}$ and consider the Regge limit $v \to \infty$ with K^2 fixed. If $\alpha(t)$ is the leading trajectory function, then the expected behavior for W_1 and W_2 is

$$W_1(v, K^2) \xrightarrow[v \to \infty]{} \beta_1(K^2)v^{\alpha(0)}, \qquad W_2(v, K^2) \xrightarrow[v \to \infty]{} \beta_2(K^2)v^{\alpha(0)-2},$$

$$\text{(III-124)}$$

with β_1 and β_2 arbitrary functions of K^2. We would expect Pomeron exchange (that is, diffraction processes) to dominate, with $\alpha(0) = 1$, and experimentally we have seen that the absorption cross section for real photons is very nearly constant above 2.0 GeV.

Abarbanel *et al.* (1969) and Harari (1969, 1970) have conjectured that the leading Regge pole at finite K^2 continues to dominate as $K^2 \to \infty$, with ω fixed. The existence of scaling then implies that $\beta_1(K^2) \sim \{K^2\}^{-\alpha(0)}$, $\beta_2(K^2) \sim \{K^2\}^{1-\alpha(0)}$ so that with $\alpha(0) = 1$ we require $W_1(\omega) \sim \omega$ and $vW_2 \sim$ constant. In a simple theory with a scalar photon and a scalar nucleon, Abarbanel *et al.* (1969) have shown that the sum of ladder graphs gives $\beta(K^2) \sim \{K^2\}^{\alpha(0)-1}$. Although not the required result it is sufficiently close to provide encouragement.

Harari (1969, 1970) has taken the analogy with the hadron reactions further by using a finite energy sum rule for vW_2. This is of the usual form

$$\int_0^N vW_2(v, K^2) \, dv = \sum_i \beta_i(K^2)N^{\alpha_i(0)}, \qquad \text{(III-125)}$$

where the sum is over all the Regge trajectories. Since the resonance terms on the left-hand side decrease rapidly with K^2, it is necessary to keep the non-resonant background and associate it with the Pomeron contribution on the right-hand side. If it is assumed that the background contribution does not vary with K^2, at least at large K^2, it follows that the Pomeron contribution is scale invariant, while the remainder, being associated with the resonance terms, will decrease like the square of the resonance form factors and be negligible at large K^2.

The most serious criticism of this approach is that Regge behavior is usually assumed to apply when $v \gg |K^2|$, in which case N, the energy above which the amplitude is assumed to be Regge behaved, must be an increasing function of $|K^2|$. If the level density of resonances increases sufficiently rapidly, as it does, for example, with a Veneziano-type amplitude which allows the non-Pomeron trajectories to give a scale-invariant contribution, then the above reasoning fails.

The idea that the data could be well explained by summing an infinite number of resonances was first explored by Landshoff and Polkinghorne (1970), who summed the resonances by constructing a Veneziano-like amplitude and found that when this was combined with requirements of current algebra, scaling behavior was a natural result. This initial model incorporated many desirable features, but made various assumptions, such as the neglect of the Δ trajectory, and has been superseded by later work.

A more phenomenological approach was developed by Bloom and Gilman (1970). They noted that when using the variable ω, the position of any given resonance moves towards $\omega = 1$ as $|K^2|$ increases, but the nucleon pole, corresponding to elastic scattering, always occurs at a fixed value of $\omega = 1$. They suggested that a more useful variable was $\omega' = (2Mv + M^2)/K^2$, for which the elastic peak occurs for $\omega' > 1$ and moves to smaller values of ω' as $|K^2|$ increases, just as the other resonances do.

The result of plotting vW_2 versus ω' is to exhibit more clearly the resonance region, and it was apparent that the scaling curve obtained from data with large $|K^2|$ fitted the resonances "on average," that is, the prominent resonances do not disappear at large $|K^2|$ relative to a background under them, but fall at roughly the same rate and, resonance by resonance, follow the scaling limit curve. It was therefore proposed that the resonances are not a separate entity, but are an intrinsic part of the scaling behavior of vW_2, and that a substantial part of the observed scaling behavior is nondiffractive in nature. Appropriately averaged, the nucleon and the resonance at low energy build the relevant non-Pomeron exchanges at high energy, in the usual duality sense.

We have already seen that all the resonance excitation form factors have approximately the same behavior as the elastic form factor at large $|K^2|$ and decrease as some power of $|K^2|$, say $|K^2|^{-n/2}$. As $|K^2|$ increases, the resonances are pushed down towards $\omega' = 1$, where vW_2 can be parametrized by

$$vW_2 \xrightarrow[\omega' \to 1]{} c(\omega' - 1)^p. \tag{III-126}$$

An interesting feature of the model is that to obtain consistency it is necessary that

$$n = p + 1. \tag{III-127}$$

Equation (III-127) was first derived by Drell and Yan (1970) in the parton model for the elastic form factor.

Although the resonance contributions may be considerable, it is unlikely that they form the only mechanism, and some Pomeron contribution should be present, each amplitude consisting of a sum of these two components. The Pomeron contribution cannot be calculated, but a model calculation of the resonance part made. can be A particularly successful one is that of Landshoff and Polkinghorne (1971a, b), as an extension of their earlier work (Landshoff and Polkinghorne, 1970).

In duality language, if we imagine that the nucleon consists of three valence quarks plus a background sea of quark-antiquark pairs, it is only the valence quarks which are responsible for generating the resonance terms and only they need be considered. Charge symmetry then can be imposed, and with the additional assumption that the momentum distributions of the p and n quarks within a nucleon have the same shape, it is possible to obtain the resonance parts of the structure function in terms of one function $R(\omega)$.

To calculate $R(\omega)$, Landshoff and Polkinghorne (1971a, b) constructed a Veneziano-like amplitude, putting in the current algebra fixed poles of Fubini, Gell-Mann, and Dashen, which play a crucial role in obtaining scaling when taken in conjunction with the analyticity properties associated with duality. An important feature of the model is that it achieves factorization on the leading trajectory, although other obstacles which arise in constructing the amplitudes are not overcome. However, these do not appear to affect the properties of the amplitude in the scaling limit.

The required function $R(\omega)$ depends to some extent on the Dirac isovector elastic form factor $F_D(t)$, and Landshoff and Polkinghorne (1971a, b)

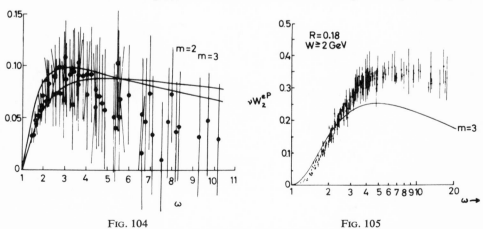

FIG. 104.

FIG. 105.

FIG. 104. Comparison of the theoretical curve for $(\nu W_2{}^p - \nu W_2{}^n)$ with the data.
FIG. 105. The theoretically calculated nondiffractive part of $\nu W_2{}^p$ compared with the data. The difference is to be identified with the diffractive part.

chose a parametrization for it such as that for large t, $F_D(t) \sim t^{-m}$. The integer m is the only parameter in their model, and their result for $(\nu W_2^p - \nu W_2^n)$ is plotted against the data in Fig. 104. Clearly both are compatible with the data. The Pomeron contribution can be estimated from the difference between the calculated value of νW_2^p and the data. The relevant comparison is made in Fig. 105.

2. Coincidence Experiments

To date, the only specific reactions which have been considered in any detail in high energy electroproduction are $e^- + p \to e^- + n + \pi^+$ and $e^- + p \to e^- + \pi^+ + \Delta^0$. The first of these is particularly notable, since in the near-forward direction in the $\pi^+ n$ system the cross section is completely dominated by the pion pole term, and this allows a very good determination of the pion form factor. However, with the electron energies presently available, this can be studied only in a limited region of momentum transfer, since for a fixed value of W, the $\pi^+ n$ center-of-mass energy, the minimum value of t increases with increasing $|K^2|$, and the pion pole thus becomes increasingly farther away from the physical region.

The relevant experiments have been performed by Driver *et al.* (1971a), Brown *et al.* (1971), and Kummer *et al.* (1971). The results of the three experiments agree well within the statistical errors. To display the results explicitly, it is convenient to use the form of the differential cross section

$$d^4\sigma/dW^2 \, dK^2 \, dt \, d\phi$$
$$= \Gamma_t \{ \sigma_U(W, K^2, t) + \varepsilon \sigma_L(W, K^2, t)$$
$$+ \varepsilon \sigma_P(W, K^2, t) \cos 2\phi + [2\varepsilon(\varepsilon + 1)]^{1/2} \sigma_I(W, K^2, t) \cos \phi \}, \quad \text{(III-128)}$$

where, as usual, σ_U is the differential cross section for production by unpolarized transverse virtual photons, σ_L is the cross section for production by longitudinal polarized photons, σ_P is the cross section for production by virtual photons with transverse linear polarization, and σ_I takes account of the interference between the transverse and the longitudinal components of the virtual photon polarization.

The separation of $\sigma_U + \varepsilon \sigma_L$, σ_P, and σ_I was made by measuring the azimuthal dependence of the cross section for fixed W, K^2, and t. In the kinematical regions considered, the variation of ε was not sufficient to allow a separation of σ_U and σ_L. A typical t dependence is shown in Fig. 106, from which it is obvious that $\sigma_U + \varepsilon \sigma_L$ is the dominant part of the cross section. Indeed $\sigma_U + \varepsilon \sigma_L$ exceeds the photoproduction cross section up to $|t| \simeq 0.10$ $(\text{GeV}/c)^2$, beyond which it decreases appreciably faster than the photoproduction cross section. This effect is very apparent in the K^2 dependence, a typical example of which is given in Fig. 107.

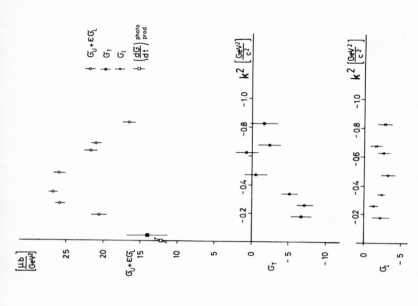

Fig. 107. K^2 dependence of the differential cross sections for $e^+ p \to e^- \pi^+ n$ at $W = 2.2$ GeV, $t = -0.037$ (GeV/c)^2 (data from Driver et al., 1971a).

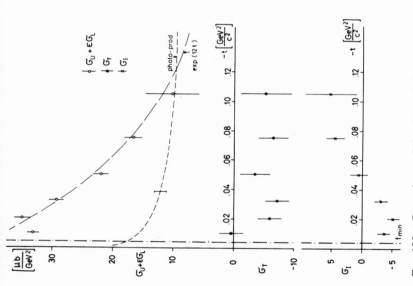

Fig. 106. t Dependence of the differential cross sections for $e^- p \to e^- \pi^+ n$ at $W = 2.2$ GeV, $K^2 = -0.26$ (GeV/c)^2 (data from Driver et al., 1971a).

The obvious model to use in interpreting this data is the electric Born term model, which, as we have seen, can successfully explain the near-forward cross section for real photons, $\gamma + p \rightarrow \pi^+ + n$. This model essentially contains only the electric form factor of the pion as a free parameter, which can then be obtained by a fit to the experimental data. This has been done by Schmidt (1971), Berends (1970), and Devenish and Lyth (1971), although with somewhat different simplifying assumptions concerning the imaginary and isoscalar parts of the amplitude, the contribution of higher resonances to the dispersion integral, and the *ansatz* for the pion form factor. All the fits show qualitative agreement with the experimental data, and it would appear from these results that the pion form factor is not appreciably different from the nucleon isovector form factor up to a momentum transfer of $|K^2| \simeq 1.2$ $(\text{GeV}/c)^2$.

These measurements also provide a useful test for the vector meson dominance (VMD) hypothesis, and predictions have been made by Fraas and Schildknecht (1971), Berends and Gastmans (1971), and Kellett (1971). Fraas and Schildknecht (1971) and Berends and Gastmans (1971) make the conventional assumption that the K^2 continuation is smooth in the helicity frame, so that the major K^2 dependence in the spacelike region comes from the vector meson propagator. Then in the usual way the components of the cross section can be related to the elements of the vector meson-decay density matrix:

$$d\sigma/dt\, d\phi = (e/f_\rho)[m_\rho^2/(m_\rho^2 - K^2)]$$

$$\times \{\rho_{11} + \varepsilon\rho_{00} C^2 - \varepsilon \cos 2\phi\rho_{1\,-1} - 2[\varepsilon(\varepsilon + 1)]^{1/2} \cos \phi \,\text{Re}\, \rho_{1\,0}C\}$$

$$\times (K^2/k)[2W/(s - M^2)](d\sigma/dt)(\pi^- p \rightarrow \rho^0 n). \qquad \text{(III-129)}$$

The factor C which has been introduced is arbitrary in form, but is required to ensure that the longitudinal amplitudes vanish at $K^2 = 0$. Fraas and Schildknecht (1971) took the simplest possible form, $C = (-K^2/m_\rho^2)^{1/2}$, while Berends and Gastmans (1971) considered various possible modifications of it. These do not appear to have too significant an effect on the result, and the model predicts reasonably well the normalization of the electroproduction data, as well as the t, K^2, and energy dependence. Its most notable failure, as stressed by Berends and Gastmans (1971), is that it predicts the wrong sign for the longitudinal-transverse interference term σ_I, although the predicted magnitude is reasonable. Kellett (1971c) took a different viewpoint, using the usual Lorentz invariant amplitudes of Ball (1961) [i.e., the amplitudes defined by (I-20)] to provide a covariant framework.

Kellett (1971c) developed his own model of vector meson dominance in photoproduction (Kellett, 1971a, b), as discussed in Section (III,F). In

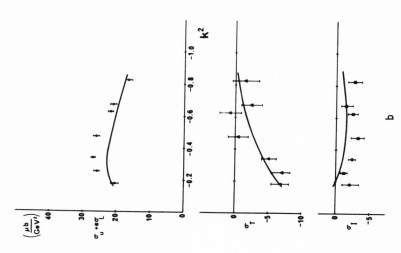

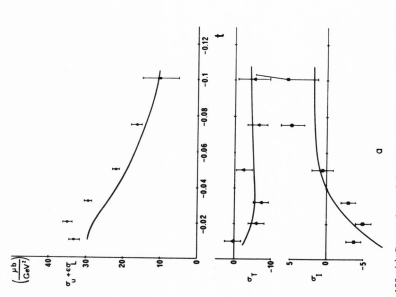

FIG. 108. (a) Comparison of the vector-dominance model calculation of Kellett (1971c) with the data for $e^- p \rightarrow e^- \pi^+ n$ at $W = 2.2$ GeV, $K^2 = -0.26$ (GeV/c)^2 (data from Driver et al., 1971a). (b) Comparison of the vector dominance model calculation of Kellett (1971c) with the data for $e^- p \rightarrow e^- \pi^+ n$ at $W = 2.2$ GeV, $t = -0.037$ (GeV/c)^2 (data from Driver et al., 1971a).

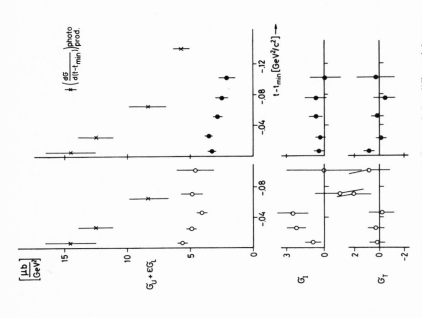

FIG. 110. $(t - t_{min})$ Dependence of the differential cross sections for $e^- p \to e^- \pi^+ \Delta^0$ at $W = 2.35$ GeV: (O) $K^2 = -0.5$ (GeV/c)²; (●) $K^2 = -0.675$ (GeV/c)² (data from Driver et al., 1971b).

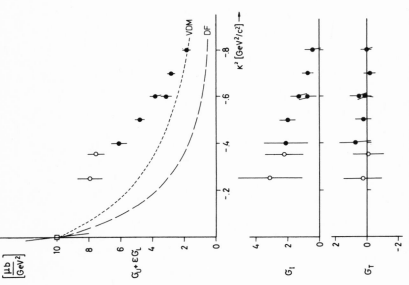

FIG. 109. K^2 dependence of the differential cross sections for $e^- p \to e^- \pi^+ \Delta^0$ at $W = 2.35$ GeV, $(t - t_{min}) = -0.05$ (GeV/c)² (data from Driver et al., 1971b).

this work it is pointed out that some care is required in the kinematic continuation from timelike to spacelike values of K^2. It turns out that the longitudinal amplitudes involve an overall factor of $1/\lambda$, where $\lambda^2 = K^2$, coming from the normalization of the longitudinal polarization vector of the Γ particle. The longitudinal helicity amplitudes are proportional to λ for small K^2, and the important point to remember is that $\lambda = (-K^2)^{1/2}$ for negative K^2. With this, the overall K^2 dependence of the amplitudes given by the explicit kinematic factors solves the problem of the sign of $\sigma_{\rm I}$. An interesting point is that the helicity-frame density matrix element Re $\rho_{1\,0}$ for $\pi^- p \to \rho^0 n$, which is related to $\sigma_{\rm I}$ by vector-meson dominance, has a zero at $t \approx -0.02$ $({\rm GeV}/c)^2$. In going to electroproduction, the position of this zero is changed by the kinematical dependence on K^2, and occurs at larger values of $-t$. This shift of zero with K^2 can be seen in the data of Driver *et al.* (1971a).

Kellett (1971c) took the relationship between the electroproduction invariant amplitudes and those at $K^2 = 0$ to be of the simplest possible vector-meson dominance form

$$A(K^2 < 0) = m_V^{\,2}/(m_V^{\,2} - K^2)A(K^2 = 0), \qquad \text{(III-130)}$$

and achieved an extremely good description of the π^+ electroproduction data. Some typical results are shown in Fig. 108.

The reaction $ep \to e\pi^+ \Delta^0$ has been measured by Driver *et al.* (1971b) in much the same kinematical region as the reaction $ep \to e\pi^+ n$. Measurement of the ϕ dependence again allows separation of the cross section elements $\sigma_{\rm U} + \varepsilon\sigma_{\rm L}$, $\sigma_{\rm T}$, and $\sigma_{\rm P}$, and once again $\sigma_{\rm U} + \varepsilon\sigma_{\rm L}$ is the dominant cross section. Typical K^2 dependence is shown in Fig. 109 and the dependence on $t - t_{\min}$ in Fig. 110. In this case the variable $t - t_{\min}$ is to be preferred to t, since $t_{\min}$ depends strongly on K^2 and on the mass variation across the Δ peak. Unlike single π^+ electroproduction, the cross section decreases rather rapidly with increasing $|K^2|$. The vector dominance model can again be applied here, and has been done so, by Barth and Schildknecht (1972). The agreement with the data is very good.

Although $\sigma_{\rm L}$ and $\sigma_{\rm U}$ cannot be separated experimentally, they can be extracted from the various model-dependent fits to the data. These indicate in both cases that the ratio $\sigma_{\rm L}/\sigma_{\rm U}$ is large, and the resulting longitudinal cross section is sufficient to saturate, within the errors, the total longitudinal cross section obtained from measurements of single-arm scattering.

REFERENCES

Abarbanel, H. D. I., Goldberger, M. L., and Treiman, S. B. (1969). *Phys. Rev. Lett.* **22**, 500.
Ader, J. P., and Capdeville, M. (1968). *Nuovo Cimento* **56A**, 315.
Ader, J. P., and Capdeville, M. (1970). *Nucl. Phys.* **B17**, 127.
Ader, J. P., Capdeville, M., and Navelet, H. (1968). *Nucl. Phys.* **56A**, 315.

Adler, S. L., (1966). *Phys. Rev.* **143**, 1144.
Adler, S. L., (1968). *Ann. Phys. (New York)* **50**, 189.
Akerlof, C. W., Ash, W. W., Berkelman, K., Lichtenstein, C. A., Ramanauskas, A., and Siemann, R. H. (1967). *Phys. Rev.* **163**, 1482.
Albrecht, W., Brasse, F. W., Dorner, H., Flauger, W., Frank, K. H., Gayler, J., Hultschig, H., and May, J. (1969). Paper (70) submitted to Int. Symp. Electron Photon Interaction at High Energies, 4th, Daresbury, 1969 (Braben, D. W., ed.).
Albrecht, W., Brasse, F. W., Dorner, H., Flauger, W., Frank, K. H., Gayler, J., Korbel, V., May, J., Zimmerman, P. D., Courau, A., Diaczek, A., Dumas, J. C., Tristram, G., Valentin, J., Aubret, C., Chazelas, E., and Ganssauge, E. (1971a). *Nucl. Phys.* **B25**, 1.
Albrecht, W., Brasse, F. W., Dorner, H., Fehrenbach, W., Flauger, W., Frank, K. H., Gayler, J., Korbel, V., May, J., Zimmerman, P. D., Courau, A., Diaczek, A., Dumas, J. C., Tristram, G., Valentin, J., Aubret, C., Chazelas, E., and Ganssauge, E. (1971b). *Nucl. Phys.* **B27**, 615.
Allcock, G. R. (1969). *Proc. Int. Symp. Electron Photon Interactions High Energies, 4th, Daresbury, 1969*, 261.
Alvensleben, H., Becker, V., Bertram, W. K., Chen, M., Cohen, K. J., Knasel, T. M., Marshall, R., Quinn, D. J., Rohde, M., Sanders, G. H., Schubel, H., and Ting, S. C. C. (1970). *Nucl. Phys.* **B18**, 333.
Amati, D., Cohen-Tannoudji, G., Jengo, R., and Salin, P. (1968). *Phys. Lett.* **26B**, 510.
Anderson, R. L., and Prepost, R. (1969). *Phys. Rev. Lett.* **23**, 46.
Anderson, R. L., Gustavson, D., Johnson, J., Ritson, D. M., Wick, B. H., Jones, W. G., Kreinick, D., Murphy, F., and Weinstein, R. (1968a). *Phys. Rev. Lett.* **21**, 384.
Anderson, R. L., Gustavson, D., Johnson, J., Ritson, D. M., Weinstein, R., Jones, W. G., and Kreinick, D. (1968b). *Phys. Rev. Lett.* **21**, 479.
Anderson, R. L., Gustavson, D., Johnson, J., Overman, I., Ritson, D. M., and Wick, B. H. (1969). *Phys. Rev. Lett.* **23**, 721.
Asbury, J. G., Bertram, W. K., Becker, V., Joos, P., Rhode, M., Smith, A. J. S., and Ting, S. C. C. (1967a). *Phys. Rev. Lett.* **19**, 869.
Asbury, J. G., Becker, V., Bertram, W. K., Binkley, M., Coleman, E., Jordan, C. L., Rohde, M., Smith, A. J. S., and Ting, S. C. C. (1967b). *Phys. Rev. Lett.* **20**, 227.
Ash, W. W., Berkelman, K., Lichtenstein, C. A., Ramanauskas, A., and Siemann, R. H. (1967). *Phys. Lett.* **24B**, 165.
Bacci, C., Baldino-Celio, R., Mencuccini, C., Reale, A., Spinetti, M., and Zallo, A. (1969). *Phys. Lett.* **28B**, 687.
Bajpai, R. P., and Donnachie, A. (1969a). *Nucl. Phys.* **B12**, 274.
Bajpai, R. P., and Donnachie, A. (1969b). Daresbury Preprint DNPL/P15.
Bajpai, R. P., and Donnachie, A. (1970). *Nucl. Phys.* **B17**, 453.
Ball, J. S. (1961). *Phys. Rev.* **124**, 2014.
Ball, J. S., Frazer, W. R., and Jacob, M. (1968). *Phys. Rev. Lett.* **20**, 518.
Ballam, J., Chadwick, G. B., Guiragossian, Z. G. T., Levy, A., Menke, M., Seyboth, P., and Wolf, G. (1969). Paper (108) submitted to Int. Symp. Electron Photon Interactions at High Energies, 4th, Darebury, 1969 (Braben, D. W., ed.).
Ballam, J., Chadwick, G. B., Gearhart, R., Guiragossian, Z. G. T., Menke, M., Murray, J. J., Seybooth, P., Shapira, A., Skillicorn, I. O., Wolf, G., Millburn, R. H., Sinclair, C. K., Bingham, H. H., Fretter, W. B., Graves, W. R., Moffeit, K. C., Podolsky, W. J., Rabin, M. S., Rosenfeld, A. H., and Windmolders, R. (1970). *Phys. Rev. Lett.* **24**, 1364.
Barbour, I. M. (1963). *Nuovo Cimento* **27**, 1382.
Barger, V., and Weiler, P. (1969). *Phys. Lett.* **30B**, 105.
Bartel, W., Dudelzak, B., Krehbiel, H., McElroy, J., Meyer-Berkhout, V., Schmidt, W., Walther, V., and Weber, G. (1968). *Phys. Lett.* **28B**, 148.
Barth, A., and Schildknecht, D. (1972). DESY Rep., in preparation.

Bar-Yam, Z., de Pagter, J., Hoenig, M. M., Kern, W., Luckey, P. D., and Osborne, L. S. (1967). *Phys. Rev. Lett.* **19**, 40.

Bar-Yam, Z., de Pagter, J., Dowd, J., and Kern, W. (1970a). *Phys. Rev. Lett.* **24**, 1078.

Bar-Yam, Z., de Pagter, J., Dowd, J., Kern, W., Osborne, L. S., and Luckey, P. D. (1970b). *Phys. Rev. Lett.* **25**, 1053.

Beaupre, J. V., and Paschos, E. A. (1969). SLAC-PUB-632.

Beder, D. S. (1966). *Phys. Rev.* **149**, 1203.

Behrend, H. J., Lobkowicz, F., Thorndike, E. H., Wehmann, A. A., and Norbderg, M. E. (1970). *Phys. Rev. Lett.* **24**, 1246.

Bellenger, D., Deutsch, S., Luckey, P. D., Osborne, L. S., and Schwitters, R. (1968). *Phys. Rev. Lett.* **21**, 1205.

Belletini, G., Bemporad, C., Braccini, P. L., Bradaschia, C., Foa, L., Lübelsmeyer, K., and Schmitz, D. (1969). Int. Symp. Electron Photon Interactions at High Energies, 4th, Daresbury, 1969 (Braben, D. W., ed.).

Berends, F. A. (1970). *Phys. Rev.* **D1**, 2550.

Berends, F. A., and Donnachie, A. (1967). *Phys. Lett.* **25B**, 278.

Berends, F. A., and Donnachie, A. (1969). *Phys. Lett.* **30B**, 555.

Berends, F. A., and Gastmans, R. (1971). CEA Preprint.

Berends, F. A., and Weaver, D. L. (1971a). *Nucl. Phys.* **B30**, 5.

Berends, F. A., and Weaver, D. L. (1971b). CEA Preprint.

Berends, F. A., Donnachie, A., and Weaver, D. L. (1967a). *Nucl. Phys.* **B4**, 1.

Berends, F. A., Donnachie, A., and Weaver, D. L. (1967b). *Nucl. Phys.* **B4**, 54.

Berends, F. A., Donnachie, A., and Weaver, D. L. (1967c). *Nucl. Phys.* **B4**, 103.

Bialas, A., and Zalewski, K. (1969). *Phys. Lett.* **28B**, 436.

Bingham, H. H., Fretter, W. B., Moffeit, K. C., Podolsky, W. J., Rabin, M. S., Rosenfeld, A. H., Windmolders, R., Ballam, J., Chadwick, G. B., Gearhart, R., Guiragossian, Z. G. T., Menke, M., Murray, J. J., Seybooth, P., Shapira, A., Sinclair, C. K., Skillicorn, I. O., Wolf, G., and Millburn, R. H. (1969). *Phys. Rev. Lett.* **24**, 955.

Bjorken, J. D. (1969). *Phys. Rev.* **179**, 1547.

Bjorken, J. D., and Paschos, E. A. (1969). SLAC-PUB-572.

Blackmon, M. L., Kramer, G., and Schilling, K. (1969). ANL 514.

Blankenbecler, R., Gartenhaus, S., Huff, R., and Nambu, Y. (1960). *Nuovo Cimento* **17**, 775.

Bloom, E. D., and Gilman, F. J. (1970). *Phys. Rev. Lett.* **25**, 1140.

Bloom, E. D., Coward, D. H., de Staebler, H., Drees, J., Litt, J., Miller, G., Mo, L. W., Taylor, R. E., Breidenbach, M., Friedman, J. I., Kendall, H. W., and Loken, S. (1968). Paper (563) submitted to Int. Conf. High Energy Phys., 14th, Vienna, CERN, 1968 (Prentki, J., and Steinberger, J., eds.).

Bloom, E. D., Coward, D. H., de Staebler, H., Drees, J., Miller, G., Mo, L. W., Taylor, R. E., Breidenbach, M., Friedman, J. I., Hartmann, H. C., and Kendall, H. W. (1969). *Phys. Rev. Lett.* **23**, 930.

Bolon, G. C., Garelick, D. A., Homma, S., Lewis, R. A., Lobar, W., Luckey, D., Osborne, L. S., Schwitters, R., and Uglum, J. (1967). *Phys. Rev. Lett.* **18**, 926.

Bolon, G. C., Bellenger, D., Lobar, W., Luckey, P. D., Osborne, L. S., and Schwitters, R. (1971). *Phys. Rev. Lett.* **24**, 964.

Booth, P. S. L., Butler, M. F., Carroll, L. J., Holt, J. R., Jackson, J. N., Range, W. H., Tait, N. R. S., Williams, E. G. H., and Wormald, J. R. (1969). *Nuovo Cimento Lett.* **2**, 66.

Booth, P. S. L., Barton, J., Carroll, L. J., Holt, J. R., Jackson, J. D., Range, W. H., Sprakes, K., and Wormall, J. R. (1971). *Nucl. Phys.* **B25**, 510.

Borgese, A., and Colocci, M. (1969). CERN Preprint TH-1024.

Boyarski, A. M., Bulos, F., Busza, W., Diebold, R., Ecklund, S. D., Fischer, G. E., Rees, J. R., and Richter, B. (1968a). *Phys. Rev. Lett.* **20**, 300.

Boyarski, A. M., Diebold, R., Ecklund, S. D., Fischer, G. E., Murata, Y., Richter, B., and Williams, W. S. C. (1968b). *Phys. Rev. Lett.* **21**, 1767.

Boyarski, A. M., Diebold, R., Ecklund, S. D., Fischer, G. E., Murata, Y., Richter, B., and Williams, W. S. C. (1969a). *Phys. Rev. Lett.* **22**, 148.

Boyarksi, A. M., Bulos, F., Busza, W., Diebold, R., Ecklund, S. D., Fischer, G. E., Murata, Y., Rees, J. R., Richter, B., and Williams, W. S. C. (1969b). *Phys. Rev. Lett.* **22**, 1131.

Boyarski, A. M., Diebold, R., Ecklund, S. D., Fischer, G. E., Murata, Y., Richter, B., and Sands, M. (1969c). Paper (90) submitted to Int. Symp. Electron and Photon Interactions at High Energies, 4th, Daresbury, 1969 (Braben, D. W., ed.).

Boyarski, A. M., Diebold, R., Ecklund, S. D., Fischer, G. E., Murata, Y., Richter, B., and Sands, M. (1969d). *Phys. Rev. Lett.* **23**, 1343.

Boyarski, A. M., Diebold, R., Ecklund, S. D., Fischer, G. E., Murata, Y., Richter, B., and Sands, M. (1970). *Phys. Rev. Lett.* **25**, 695.

Braunschweig, M., Braunschweig, W., Husmann, D., Lübelsmeyer, K., and Schmitz, D. (1966). *Phys. Lett.* **22**, 705.

Braunschweig, M., Braunschweig, W., Husmann, D., Lübelsmeyer, K., and Schmitz, D. (1968). *Phys. Lett.* **26B**, 405.

Braunschweig, W., Erlewein, W., Freese, H., Lübelsmeyer, K., Schmitz, D., Schultz von Dratzig, A., and Wessels, G. (1969). Preliminary results.

Breidenbach, M., Freidman, J. I., Kendall, H. W., Bloom, E. D., Coward, D. H., de Staebler, H., Drees, J., Mo, L. W., and Taylor, R. E. (1969). Paper (101) submitted to Int. Symp. Electron and Photon Interactions at High Energies, 4th, Daresbury, 1969 (Braben, D. W., ed.).

Brouwer, R., and Dash, J. M. (1968). UCRL 18199.

Brown, C. N., Canizares, C. R., Cooper, W. E., Eisner, A. M., Feldman, G. J., Lichtenstein, C. A., Litt, L., Lockeretz, W., Montana, V. B., and Pipkin, F. M. (1971), *Phys. Rev. Lett.* **26**, 987.

Bulos, F., Busza, W., Giese, R. F., Larsen, R. R., Leith, D. W. G. S., Richter, B., Perez-Mendez, V., Stetz, A., Williams, S. H., Beniston, M., and Rettberg, J. (1969a). *Phys. Rev. Lett.* **22**, 490.

Bulos, F., Busza, W., Giese, R. F., Kluge, E., Leith, D. W. G. S., Larsen, R. R., Richter, B., Stetz, A., Williams, S. H., and Beniston, M. (1969b). Paper submitted to Conf. High Energy Phys., Boulder, Colorado, 1969.

Burfeindt, H., Buschorn, G., Geweniger, C., Heide, P., Kotthaus, R., Wahl, H., and Wegener, K. (1969). Paper (87) submitted to Int. Symp. Electron and Photon Interactions at High Energies, 4th, Daresbury, 1969 (Braben, D. W., ed.).

Burfeindt, H., Buschorn, G., Geweniger, C., Kotthaus, R., Wahl, H., and Wegener, K. (1970). *Phys. Lett.* **33B**, 509.

Buschorn, G., Carroll, L. J., Eandi, R. D., Heide, P., Hübner, R., Kern, W., Kötz, U., Schmüser, P., and Skronn, H. J. (1966). *Phys. Rev. Lett.* **17**, 1027.

Buschorn, G., Carroll, L. J., Eandi, R. D., Heide, P., Hübner, R., Kern, W., Kötz, U., Schmüser, P., and Skronn, H. J. (1967a). *Phys. Rev. Lett.* **18**, 571.

Buschorn, G., Heide, P., Kötz, U., Lewis, R. A., Schmüser, P., and Skronn, H. J. (1967b). *Phys. Lett.* **25B**, 201.

Capella, A., and Tran Thanh Van, J. (1969). *Lett. Nuovo Cimento* **1**, 321.

Carreras, B., and Donnachie, A. (1970). *Nucl. Phys.* **B16**, 35.

Chau, A. Y. C., Dombey, N., and Moorhouse, R. G. (1967). *Phys. Rev.* **163**, 1632.

Chew, C. F., Goldberger, M. L., Low, F. E., and Nambu, Y. (1956). *Phys. Rev.* **106**, 1345.

Cho, C. F., and Sakurai, J. J. (1970). *Phys. Rev.* **D2**, 517.

Clegg, A. B. (1969). *Proc. Int. Symp. Electron and Photon Interactions at High Energies, 4th, Daresbury, 1969*, 123.

Cohen-Tannoudji, G., Morel, A., and Navelet, H. (1968). *Ann. Phys. (New York)* **46**, 239.

Collins, P. D. B., and Squires, E. J. (1968). "Regge Poles in Particle Physics; Springer Tracts in Modern Physics" Vol. 45. Springer-Verlag, Berlin.

Cone, A. A., Chen, K. W., Dunning, J. R., Hartwig, G., Ramsey, N. F., Walker, J. K., and Wilson, R. (1967a). *Phys. Rev.* **156**, 1490; **163**, 1854(E).

Copley, L. A., Karl, G., and Obryk, E. (1969). *Phys. Lett.* **29B**, 117.

Daboul, J. (1968). *Nucl. Phys.* **B7**, 651.

Dalitz, R. H. (1963). *Ann. Rev. Nucl. Sci.*, **13**, 346.

Dalitz, R. H., and Sutherland, D. G. (1966). *Phys. Rev.* **146**, 1180.

Dar, A., Weisskopf, V., Levinson, C. A., and Lipkin, H. J. (1968). *Phys. Rev. Lett.* **20**, 1261.

Davies, A. T., and Moorhouse, R. G. (1967). *Nuovo Cimento* **52A**, 1112.

Deans, S. R. (1969). *Phys. Rev.* **177**, 2623.

Deans, S. R., and Holladay, W. G. (1967). *Phys. Rev.* **161**, 1466.

Deans, S. R., and Holladay, W. G. (1968). *Phys. Rev.* **165**, 1886.

Dennery, P. (1961). *Phys. Rev.* **124**, 2000.

Devenish, R. C. E., and Lyth, D. H. (1971). Lancaster University Preprint.

Devenish, R. C. E., Lyth, D. H., and Rankin, W. A. (1972). Daresbury Preprint, in preparation.

de Vries, C., Hofstadter, R., and Johnson, A. (1964). *Phys. Rev.* **134**, B848.

Diambrini-Palazzi, G., McClellan, G., Mistry, N., Mostek, P., Ogren, H., Swartz, J., and Talman, R. (1969). Paper (96) submitted to Int. Symp. Electron and Photon Interactions at High Energies, 4th, Daresbury, 1969 (Braben, D. W., ed.).

Dietz, K., and Korth, W. (1968). *Phys. Lett.* **26B**, 394.

Dombey, N. *Proc. Int. Symp. Electron and Photon Interactions at High Energies, 4th, Daresbury, 1969* (Proc. Abstr. no. 26).

Donnachie, A. (1968). *Proc Int. Conf. High Energy Phys., 14th, Vienna, CERN, 1968*.

Donnachie, A. (1970). Internal Daresbury Rep. (unpublished).

Donnachie, A., and Shaw, G. (1966). *Ann. Phys. (New York)*, **37**, 333.

Donnachie, A., and Shaw, G. (1967). *Nucl. Phys.* **87**, 556.

Donohue, J. T., and Högassen, H. (1969). *Phys. Lett.* **28B**, 436.

Drell, S. D., and Trefil, J. S. (1966). *Phys. Rev. Lett.* **16**, 552, 832(E).

Drell, S. D., and Walecka, J. D. (1964). *Ann. Phys. (New York)* **28**, 18.

Drell, S. D., and Yan, T. M. (1970). *Phys. Rev. Lett.* **24**, 181.

Drell, S. D., Levy, D. J., and Yan, T. M. (1969a). *Phys. Rev. Lett.* **22**, 744.

Drell, S. D., Levy, D. J., and Yan, T. M. (1969b). *Phys. Rev.* **187**, 2159.

Drell, S. D., Levy, D. J., and Yan, T. M. (1970a). *Phys. Rev.* **D1**, 1035.

Drell, S. D., Levy, D. J., and Yan, T. M. (1970b). *Phys. Rev.* **D1**, 1617.

Driver, C., Heinloth, K., Höhne, K., Hofmann, G., Karow, P., Schmidt, D., Specht, G., and Rathje, J. (1971a). *Phys. Lett.* **35B**, 77.

Driver, C., Heinloth, K., Höhne, H., Hofmann, G., Karow, P., Schmidt, D., and Specht, G. (1971b). DESY Rep. 71/25.

Earles, D., Faissler, W., Lutz, G., Ken Min Moy, Tang, Y. W., von Briesen, H., von Goeler, E., and Weinstein, R. (1969). Paper (64) submitted to Int. Symp. Electron and Photon Interactions at High Energies, 4th, Daresbury, 1969 (Braben, D. W., ed.).

Eisenberg, Y., Ronat, E. E., Brandstetter, A., Levy, A., and Gotsman, E. (1966a). *Phys. Lett.* **22**, 217.

Eisenberg, Y., Ronat, E. E., Brandstetter, A., Levy, A., and Gotsman, E. (1966b). *Phys. Lett.* **22**, 223.

Elings, V. B., Cohen, K. J., Garelick, D. A., Homma, S., Lewis, R. A., Luckey, P. D., and Osborne, L. S. (1966). *Phys. Rev. Lett.* **16**, 474.

Engels, J., and Schmidt, W. (1968). *Phys. Rev.* **169**, 1296.

Engels, J., Schwiderski, G., and Schmidt, W. (1968). *Phys. Rev.* **166**, 1343.

Faiman, D., and Hendry, A. W. (1968). *Phys. Rev.* **173**, 1720.

Faiman, D., and Hendry, A. W. (1969). *Phys. Rev.* **180**, 1572.

Fidecaro, G., Fidecaro, M., Poirier, J. A., and Schiavon, P. (1966). *Phys. Lett.* **23**, 163.

Finkler, P. (1964). UCRL Rep. 7953T.

Fischer, G. E., Fischer, H., von Holtey, G., Kämpgen, H., Knop, G., Schultz, P., Wessels, H., Braunschweig, W., Genzel, H., and Wedemeyer, R. (1968). Univ. of Bonn preprint.

Fischer, G. E., Fischer, H., von Holtey, G., Kämpgen, H., Knop, G., Schulz, P., Wessels, H., Braunschweig, W., Genzel, H., and Wedemeyer, R. (1970). *Nucl. Phys.* **B16**, 93.

Fraas, H., and Schildknecht, D. (1969). DESY Rep. 69/18.

Fraas, H., and Schildknecht, D. (1971). *Phys. Lett.* **35B**, 72.

Frautschi, S., and Jones, L. (1967). *Phys. Rev.* **163**, 1820.

Frøyland, J. (1969). *Nucl. Phys.* **B11**, 204.

Frøyland, J., and Gordon, D. (1969). *Phys. Rev.* **177**, 2500.

Fubini, S., Nambu, Y., and Wataghin, V. (1958). *Phys. Rev.* **111**, 329.

Fujii, T., Okuno, H., Sasaki, H., Nozaki, T., Takasaki, F., Takikawa, K., Amako, K., Endo, I., Yoshida, K., Higuchi, M., Sato, M., and Sumi, Y. (1971). *Phys. Rev. Lett.* **26**, 1672.

Geweniger, C., Heide, P., Kötz, U., Lewis, R. A., Schmüser, P., Skronn, H. J., Wahl, H., and Wegener, K. (1968). *Phys. Lett.* **28B**, 155.

Geweniger, C., Heide, P., Kötz, U., Lewis, R. A., Schmüser, P., Skronn, H. J., and Wahl, H. (1969). *Phys. Lett.* **29B**, 41.

Gilman, F. J. (1968). *Phys. Rev.* **167**, 1365.

Goitein, M., Dunning. J. R., and Wilson, R. (1967). R., *Phys. Rev. Lett.* **18**,1018.

Gorczyka, B., and Hayashi, M. (1968). Univ. of Krakow Preprint, TJPU 14/68.

Gottfried, K. (1968). *Bull. Amer. Phys. Soc.* **13**, 175.

Gottfried, K., and Jackson, J. D. (1964). *Nuovo Cimento* **33**, 309.

Gottfried, K., and Yennie, D. R. (1969). *Phys. Rev.* **182**, 1595.

Gourdin, M., and Salin, P. (1963). *Nuovo Cimento* **27**, 193, 309.

Guiragossian, Z. G. T., and Levy, A. (1969). SLAC-PUB-581.

Gutbrod, F. (1969a). DESY Rep. 69/22.

Gutbrod, F. (1969b). DESY Rep. 69/33.

Gutbrod, F., and Simon, D. (1967). *Nuovo Cimento* **51A**, 602.

Harari, H. (1969). *Phys. Rev. Lett.* **22**, 1078.

Harari, H. (1970). *Phys. Rev. Lett.* **24**, 286.

Heide, P., Kötz, U., Lewis, R. A., Schmüser, P., Skronn, H. J., and Wahl, H. (1968). *Phys. Rev. Lett.* **21**, 248.

Hellings, R. D., Allison, J., Clegg, A. B., Foster, F., Hughes, G., Kummer, P. S., Siddle, R., Dickinson, B., Ibbotson, M., Lawson, R., Montgomery, H. E., Shuttleworth, W. J., Sofair, A., and Fannon, J. (1971). *Nucl. Phys.* **B32**, 179.

Henyey, F. (1968). *Phys. Rev.* **170**, 1619.

Heusch, C. A., Prescott, C. Y., Rochester, L. S., and Winstein, B. D. (1969). Paper (58) submitted to Int. Symp. Electron and Photon Interactions at High Energies, 4th, Daresbury, 1969 (Braben, D. W., ed.).

Hicks, N., Feldman, G. J., Litt, L., Lockeretz, W., Pipkin, F. M., and Randolph, J. K. (1968). *Proc. Int. Conf. CERN, High Energy Phys., Vienna.*

Hicks, N., Feldman, G., Litt, L., Lockeretz, W., Pipkin, F. M., Randolph, J. K., and Stanfield, K. C. (1969). Paper (146) submitted to Int. Symp. Electron and Photon Interactions at High Energies, 4th, Daresbury, 1969 (Braben, D. W., ed.).

Hilger, E., Roegler, H. J., Simons, L., and Tonutti, M. (1972). Bonn Preprint, in preparation.

Hilpert, H. G., Lauscher, P., Matziolis, M., Idschok, V., Muller, K., Knies, G., Kolb, A., Raulefs, P., Spitzer, H., Braun, O., Stieve, J., Schlamp, P., and Weigl, J. (1969). Paper (94) submitted to Int. Symp. Electron and Photon Interactions at High Energies, 4th, Daresbury, 1969 (Braben, D. W., ed.).

Hofstadter, R. (1957). *Ann. Rev. Nucl. Sci.* **4**, 231.

Höhler, G., and Dietz, K. (1960). *Z. Physik* **160**, 453.

Höhler, G., and Müllensiefen, A. (1959). *Z. Physik* **157**, 30.

Höhler, G., and Schmidt, W. (1964). *Ann. Phys. (New York)* **28**, 34.

Höhler, G., Dietz, K., and Müllensiefen, A. (1960). *Z. Physik* **159**, 77.

Imrie, D., Mistretta, C., and Wilson, R. (1968). *Phys. Rev. Lett.* **20**, 1074.

Jackson, J. D., and Quigg, C. (1969). *Phys. Lett.* **29B**, 236.

Jacob, M., and Wick, G. C. (1959). *Ann. Phys. (New York)* **7**, 404.

Johnson, P. B., Poirier, J. A., Biswas, N. N., Cason, N. M., Groves, T. H., Kenney, V. P., McGahan, J. T., Shephard, W. D., Gutay, L. J., Campbell, J H., Eisner, R. L., Loeffler, F. J., Peters, R. E., Sahni, R. J., Yen, W. L., Derado, I., and Guiragossian, Z. G. T. (1968). *Phys. Rev.* **176**, 1651.

Joseph, P. M., Hicks, H., Litt, L., Pipkin, F. M., and Russell, J. J. (1967). *Phys. Rev. Lett.* **19**, 1206.

Kane, G. (1969). SLAC topical Conf. Backward Processes.

Kellett, B. H. (1970). *Nucl. Phys.* **B25**, 205.

Kellett, B. H. (1971a). Daresbury Preprint DNPL/P68.

Kellett, B. H. (1971b). Daresbury Preprint DNPL/P76.

Kellett, B. H. (1971c). Daresbury Preprint DNPL/P81.

Kim, Y. T. (1968). Ph.D. Thesis, Univ. of Bonn.

Kölbig, K. S., and Margolis, B. (1968). *Nucl. Phys.* **B6**, 85.

Korth, W. To be published.

Krammer, M. (1969). Paper (125) submitted to Int. Symp. Electron and Photon Interactions at High Energies, 4th, Daresbury, 1969 (Braben, D. W., ed.).

Krammer, M., and Schildknecht, D. (1968). *Nucl. Phys.* **B7**, 583.

Kummer, P. S., Clegg, A. B., Foster, F., Hughes, G., Siddle, R., Allison, J., Dickinson, B., Evangelides, E., Ibbotson, M., Lawson, R., Meaburn, M. S., Montgomery, H. E., Shuttleworth, W. J., and Sofair, A. (1971). Daresbury preprint DNPL/P67.

Landshoff, P. V., and Polkinghorne, J. C. (1970). *Nucl. Phys.* **B19**, 432.

Landshoff, P. V., and Polkinghorne, J. C. (1971a). *Nucl. Phys.* **B28**, 225.

Landshoff, P. V., and Polkinghorne, J. C. (1971b). *Phys. Lett.* **34B**, 621.

Le Bellac, M. (1967). *Phys. Lett.* **25B**, 254.

Levinson, C. A., Lipkin, H. J., and Meshkov, S. (1963). *Phys. Lett.* **7**, 81.

Lohrmann, E. (1969). *Proc. Lund Int. Conf. Elementary Particles.*

Lovelace, C. (1969), "Pion-Nucleon Scattering" (G. L. Shaw and D. Y. Wong, eds.), p. 27. Wiley, New York.

Lübelsmeyer, K. (1969). *Proc. Int. Symp. Electron and Photon Interactions at High Energies, 4th, Daresbury, 1969*, 45.

Luming, M. (1964). *Phys. Rev.* **136B**, 1120.

Lynch, H. L., Allaby, J. V., and Ritson, D. M. (1967). *Phys. Rev.* **164**, 1635.

McClellan, G., Mistry, N., Mostek, P., Ogren, H., Silverman, A., Swartz, J., Talman, R., Gottfried, K., and Lebedev, A. I. (1969a). *Phys. Rev. Lett.* **22**, 374.

McClellan, G., Mistry, N., Mostek, P., Ogren, H., Silverman, A., Swartz, J., and Talman, R. (1969b). *Phys. Rev. Lett.* **22**, 377.

McClellan, G., Mistry, N., Mostek, P., Ogren, H., Osborne, A., Silverman, A., Swartz, J., Talman, R., and Diambrini-Palazzi, G. (1969c). *Phys. Rev. Lett.* **23**, 554.

McClellan, G., Mistry, N., Mostek, P., Ogren, H., Osborne, A., Silverman, A., Swartz, J., Talman, R., and Diambrini-Palazzi, G. (1969d). *Phys. Rev. Lett.* **23**, 718.

McClellan, G., Mistry, N., Mostek, R., Ogren, H., Osborne, A., Swartz, J., Talman, R., and Diambrini-Palazzi, G. (1969e). Cornell Univ. Preprint CLNS-70.

McKinley, J. (1962). Tech. Rep. No. 38, Univ. of Illinois, Urbana, Illinois.

McNeely, W. A., Heusch, C. A., and Yellin, S. J. (1969). Paper (59) submitted to Int. Symp. Electron and Photon Interactions at High Energies, 4th, Daresbury, 1969 (Braben, D. W., ed.).

Margolis, B. (1968a). *Nucl. Phys.* **B4**, 433.

Margolis, B. (1968b). *Phys. Lett.* **26B**, 524.

Maheshwari, A. N. (1968). *Phys. Rev*, **170**, 1533.

Meyer, R. F. (1969). *Nuovo Cimento Lett.* **2**, 76.

Mistretta, C., Imrie, D., Appel, J., Budnitz, R., Carroll, L. J., Chen, J., Dunning, J. R., Goitein, M., Hanson, K., Litke, A., and Wilson, R. (1968a). *Phys. Rev. Lett.* **20**, 1070.

Mistretta, C., Imrie, D., Appel, J., Budnitz, R., Carroll, L., Goitein, M., Hanson, K., and Wilson, R. (1968b). *Phys. Rev. Lett.* **20**, 1523.

Moorhouse, R. G. (1966). *Phys. Rev. Lett.* **16**, 772.

Moorhouse, R. G., and Rankin, W. A. (1970). *Nucl. Phys.* **B23**, 181.

Morand, R., Erikson, E. F., Pahin, J. P., and Croissiaux, K. C. (1969). *Phys. Rev.* **180**, 1299.

Mushkelishvili, N. I. (1953). *In* "Singular Integral Equations." Groningen, Noordhoff.

Noelle, P. (1971). Bonn Preprint P1, 2–92.

Noelle, P., and Pfeil, W. (1970). Bonn Preprint P1, 2–87.

Noelle, P., Pfeil, W., and Schwela, D. (1971). *Nucl. Phys.* **B26**, 461.

Omnès, R. (1958). *Nuovo Cimento* **8**, 316.

Orito, S. (1969). Paper (25) submitted to *Int. Symp. Electron and Photon Interactions at High Energies, 4th, Daresbury, 1969* (Braben, D. W., ed).

Panofsky, W. K. H. (1968). *Proc. Int. Conf. High Energy Physics, Vienna, CERN, 1968*.

Paschos, E. A. (1968). *Phys. Rev. Lett.* **21**, 1855.

Perez-y-Jorba, J. (1969). *Proc. Int. Symp. Electron and Photon Interactions at High Energies, 4th, Daresbury, 1969*, 213.

Pfeil, W. (1968). Ph.D. Thesis, Univ. of Bonn.

Pritchett, P. L., Walecka, J. D., and Zucher, P. A. (1969). Stanford University Preprint ITP-328.

Proia, A., and Sebastiani, F. (1970). *Nuovo Cimento Lett.* **3**, 483.

Proia, A., and Sebastiani, F. (1971). *Nuovo Cimento. Lett.* **11**, 560.

Rankin, W. A. (1970). Ph.D. Thesis, Univ. of Glasgow.

Rankin, W. A. (1972). Daresbury Preprint, in preparation.

Renard, F. M., and Renard, Y. (1971). *Nucl. Phys.* **B25**, 490.

Rollnik, H. (1967). *Proc. Int. Conf. Elementary Particles, Heidelberg, 400*.

Ross, M., and Stodolsky, L. (1966). *Phys. Rev.* **149**, 1172.

Salin, P. (1963). *Nuovo Cimento* **28**, 1294.

Schilling, K., and Storim, F. (1968). *Nucl. Phys.* **B7**, 559.

Schmidt, W. (1971). DESY Rep. 71/22.

Schorsch, W., Tietge, J., and Weilnbock, W. (1971). *Nucl. Phys.* **B25**, 179.

Schwela, D. (1968). Ph.D. Thesis, Univ. of Bonn.

Schwela, D. (1969). *Z. Physik* **221**, 158.

Schwela, D. (1971). *Nucl. Phys.* **B26**, 525.

Schwela, D., and Weizel, R. (1969). *Z. Physik* **221**, 71.

Schwela, D., Rollnik, H., Weizel, R., and Korth, W. (1967). *Z. Physik* **202**, 452.

Shaw, G. (1966). Ph.D. Thesis, Univ. of London (unpublished).

Silverman, A. (1969). *Proc. Int. Symp. Electron and Photon Interactions at High Energies, 4th, Daresbury, 1969*, 71.

Stichel, P. (1964). *Z. Physik* **180**, 170.

Swartz, J., and Talman, R. (1969). Cornell University Preprint CLNS-79.

Taylor, R. E. (1969). *Proc. Int. Symp. Electron and Photon Interactions at High Energies, 4th, Daresbury, 1969*, 251.

Thom, H. (1966). *Phys. Rev.* **151**, 1322.

Thornber, N. S. (1968a). *Phys. Rev.* **169**, 1096.

Thornber, N. S. (1968b). *Phys. Rev.* **173**, 1414.

Tompkins, D., Anderson, R. L., Gittelman, B., Litt, J., Wick, B. H., and Yount, D. (1969). *Phys. Rev. Lett.* **23**, 725.

von Bochmann, G., Margolis, B., and Tang, L. C. (1969). *Phys. Lett.* **30B**, 254.

von Gehlen, G. (1969). *Nucl. Phys.* **B9**, 17.

von Gehlen, G. (1970). *Nucl. Phys.* **B20**, 102.

Wagner, F., and Lovelace, C. (1971). *Nucl. Phys.* **B25**, 411.

Walecka, J. D., and Zucher, P. A. (1968). *Phys. Rev.* **167**, 1479.

Walker, R. L. (1969a). *Phys. Rev.* **182** (1729).

Walker, R. L. (1969b). *Proc. Int. Symp. Electron and Photon Interactions at High Energies, 4th, Daresbury, 1969*, 23.

Watson, K. M. (1954). *Phys. Rev.* **95**, 228.

Zagury, N. (1966). *Phys. Rev.* **145**, 112; **150**, 1406(E).

Zagury, N. (1968). *Phys. Rev.* **165**, 1934(E).

REGGE PHENOMENOLOGY

R. J. N. Phillips and G. A. Ringland

I. Introduction

This chapter deals with high energy phenomenology, using Regge poles and related ideas. The subject was introduced by Bertocchi and Ferrari earlier (1967, p. 71), but there have been several important developments since, notably those pertaining to finite energy sum rules (FESR), duality, the Veneziano model, and Regge cuts. These topics are our main concern, and are discussed in Sections II–V. For completeness, however, some other developments and general questions will be mentioned here, which include a discussion of motivation and a description of the ideas of exchange degeneracy, wrong-signature zeros, and conspiracy.

This work is not addressed to theoretical experts, but rather to those with a general interest in the subject, especially experimenters. Our aim is to concentrate on the physical ideas, giving just a few illustrations, rather than enumerating the myriad applications that have been made. (We apologize to the authors whose work is omitted, and refer to the proverb about the wood and the trees.) The reader is referred to some recent reviews, books, and summer school lectures (Frazer, 1967; Collins and Squires, 1968; Barger and Cline, 1968; Jacob, 1969a; Harari, 1969b; Kugler, 1970a), for further information. Conference summary talks are also very helpful. Those by Drell (1962), Van Hove (1966), Bertocchi (1967), Chan (1968), Jacob (1969b), Lovelace (1970), and Jackson (1970) chart the progress of the subject. The discussion here is limited to two-body and quasi-two-body reactions. For an introduction to multiparticle Regge phenomenology, see the review by Chan (1970). On units, $\hbar = c = \text{GeV} = 1$, unless otherwise stated.

A. WHY USE REGGE POLES?

The first applications of Regge phenomenology were very simple and appealing. Lately the art has become rather complicated in trying to explain increasingly more diverse data, and one may ask if it is still worth following.

The justification begins with the idea of particle exchange. Whenever the selection rules allow a known particle to be exchanged, high energy cross sections show a corresponding forward (or backward) peak; when exchange is forbidden, the peak is absent or strongly suppressed. The conclusion that these peaks are related to particle exchange is very attractive.

On the other hand, the scattering amplitudes for single-particle exchange calculated by the Feynman rules in field theory prove quite unsuitable. They are essentially real, and for high-spin exchanges (spin > 1) they increase rapidly with energy, contradicting both experiment and theoretical bounds based on analyticity.

The only known amplitudes that combine reasonable behavior in the scattering region with particle poles in the crossed channels are those based on Regge poles or something very similar. This was one of the original reasons for trying Regge poles, and it still seems valid. Whether a single Regge pole gives a good approximation to the complete amplitude in any particular case is quite a different question.

In the following sections the terminology appropriate to the Regge-pole model is not defined. For the background to the model and definitions of the terms introduced, the reader is referred to Bertocchi and Ferrari (1967, p. 160).

B. Exchange Degeneracy

Regge poles with opposite signatures are not generally related because "exchange forces," contributing with opposite signs to even-J and odd-J scattering, are generally present. If such effects are absent, however, as in a potential model with "direct" forces only, the trajectory and residue functions are the same for both signatures: the Regge poles are then "exchange degenerate."

Arnold (1965) argued that this simple picture should apply to meson Regge poles if they are generated mainly by baryon-antibaryon interactions. In this case, meson exchanges give "direct" forces and dibaryon exchanges give "exchange" forces, but the latter are presumably weak since no dibaryon resonances are known. More recently, the duality principle has provided new arguments for exchange degeneracy in many circumstances (see Section III).

If two Regge poles are exchange degenerate in one particular reaction, the degeneracy of their trajectories carries over to all reactions; but the degeneracy of residues does not carry over in general. There is evidence for exchange degeneracy of trajectories in the boson mass spectrum. The mesons $\rho(765)$, $A_2(1300)$, and $g(1660)$ have $J^P = 1^-$, 2^+, and 3^-. If values 4^+, 5^-, and 6^+ are assigned to the mesons S(1930), T(2200), and U(2375), respectively, all six are compatible with a single almost linear exchange-degenerate trajectory. The Y^* spectra for isospin $I = 0$ and 1 also show exchange degeneracy (see Figs. 1 and 2).

Suppose there are two t-channel Regge poles, with opposite signatures, giving high energy amplitudes in the s and u channels of the forms

$$T^s = \sum_{i=1,\,2} \gamma_i [1 + \tau_i \exp(-i\pi\alpha_i)](s - u)^{\alpha_i}/(\sin \pi\alpha_i),$$

$$T^u = \sum_{i=1,\,2} \tau_i \gamma_i [1 + \tau_i \exp(-i\pi\alpha_i)](u - s)^{\alpha_i}/(\sin \pi\alpha_i), \tag{I-1}$$

where $\alpha_i(t)$, $\gamma_i(t)$, and $\tau_i = \pm 1$ are, respectively, the trajectory, residue, and signature of Regge pole i. Exchange degeneracy implies $\alpha_1 = \alpha_2$, $\gamma_1 = \gamma_2$. (An alternative for the residues $\tau_1 \gamma_1 = \tau_2 \gamma_2$ is equally possible in principle; in practice the choice depends on experiment, or some further theoretical argument, as with duality: see Section III.)

Let us examine some consequences:

1. Imposing $\alpha_1 = \alpha_2 = \alpha$ (exchange degeneracy of the trajectories), but leaving the residues free, results in the phases of the two Regge terms differing by $\pi/2$. Hence, the cross terms in $|T|^2$ vanish, and $|T^s|^2 = |T^u|^2$ at corresponding energies. If the s and u channels differ simply by crossing spinless mesons, then

$$d\sigma^s/dt = d\sigma^u/dt. \tag{I-2}$$

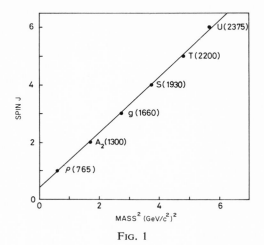

FIG. 1

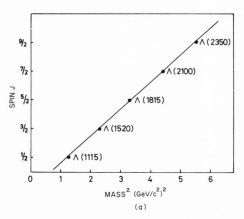

(a)

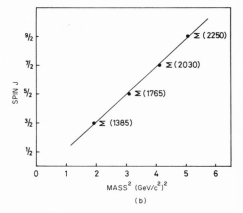

(b)

FIG. 2

FIG. 1. A plot of spin J versus mass squared for natural-parity mesons with isospin $I = 1$. (Data from Roos *et al.*, 1970).

FIG. 2. Plots of spin J versus mass squared for strange baryons: (a) isospin $I = 0$; (b) isospin $I = 1$. (Data from Roos *et al.*, 1970.)

When particles with spin are crossed, further spin-weight factors may appear in Eq. (1-2).

2. Imposing further $\gamma_1 = \gamma_2$,

$$T^s = 2\gamma(s - u)^\alpha/(\sin \pi\alpha), \qquad T^u = 2\gamma \exp(-i\pi\alpha)(u - s)^\alpha/(\sin \pi\alpha). \quad (1\text{-}3)$$

Thus, T^s becomes purely real, and T^u has a nontrivial phase. These phases apply equally to all spin-flip and nonflip amplitudes if all the residues are degenerate. Hence, the vector polarization P that depends on interference terms of the form $\mathrm{Im}(T_i\, T_j^*)$ vanishes:

$$P^s = 0, \qquad P^u = 0. \tag{I-4}$$

This result is well known for a single Regge pole: here it follows also for an exchange-degenerate pair of poles.

If there are more than one pair of exchange-degenerate poles, Eqs. (I-2)–(I-4) are not generally true, but the reality of T^s and the vanishing of P^s remain.

The discussion above refers to boson Regge poles. For the fermion case, let us replace α by $(\alpha - \frac{1}{2})$ in Eq. (I-1) and also replace t by u, since it has become conventional to speak of fermion poles in the u channel. Now, however, two new and related complications arise. According to the conventional theory (Gribov, 1963), fermion u-channel Regge poles occur in pairs, with α and γ related by complex conjugation for $u < 0$. Furthermore, α and γ are not expected to be real for $u < 0$, and the phase of each contribution is not given simply by the signature factor $1 + \tau \exp[-i\pi(\alpha - \frac{1}{2})]$. However, it turns out that some results analogous to the boson case still hold. If there is exchange degeneracy between two conjugate pairs of fermion poles, with equality of residues for s-channel scattering, then the usual invariant amplitudes are purely real in the s channel. If the further condition that α is real (in fact is an even function of $\sqrt{u}$) is imposed, then the amplitudes in the crossed t channel are again equal to the s-channel amplitudes at the corresponding energy, apart from an overall phase factor $\exp[-i\pi(\alpha - \frac{1}{2})]$.

C. WRONG-SIGNATURE ZEROS

Integer values of $\alpha(t)$ are special points, since here the Regge pole denominator $1/[\sin \pi\alpha(t)]$ gives poles. "Ghost" poles must be prevented from appearing in the region $t < 0$ (since they would correspond to t-channel particles of imaginary mass, making the physical s-channel scattering amplitude infinite). There are a plethora of mechanisms available to eliminate such ghosts by giving zeros in the Regge residues; these mechanisms differ in their treatment of various spin amplitudes and in the order of zeros produced. Dreschsler (1969) reviews them extensively. When one takes into consideration the signature factor $1 + \tau \exp(-i\pi\alpha)$, that also vanishes at half the integer values of α, the net effect can produce a zero in some cases. There is no general argument for the occurrence of such zeros: some physical assumptions are necessary—of which perhaps the most convincing is exchange degeneracy.

Complete exchange degeneracy leads to zeros in all spin amplitudes at wrong-signature points. Consider a typical Regge pole amplitude

$$T = \gamma(s - u)^\alpha [1 + \tau \exp(-i\pi\alpha)]/(\sin \pi\alpha). \tag{I-5}$$

At ghost values of $\alpha(t)$ with "right signature," where the signature factor $1 + \tau \exp(-i\pi\alpha)$ itself does not vanish, there must be a zero in the residue function $\gamma(t)$. Now if there is an exchange-degenerate pair of Regge poles, one of them always has right signature, and $\gamma(t)$ must vanish at all ghost values of $\alpha(t)$. At such points, each Regge pole then gives a finite contribution for right signature and zero for wrong signature.

Thus, zeros occur at wrong-signature ghost points in processes where both α and γ are exchange degenerate. Through factorization, the zeros then propagate into other processes where the residues are not equal, and one of the pair may even be forbidden. The simplest solution compatible with exchange degeneracy and absence of ghosts is to have a single zero in t for all spin amplitudes at wrong-signature ghost points—this is termed "the Gell-Mann mechanism."

The best-known cases where zeros occur are the following:

1. There is a minimum in $(d\sigma/dt)(\pi^- p \to \pi^0 n)$ near $t = -0.6$, where the dominant ρ trajectory passes through zero. This is a ghost point since t is negative; $\rho - A_2$ exchange degeneracy in NN and KN scattering, plus factorization, require the ρ spin-flip and nonflip amplitudes to vanish at $\alpha = 0$.

2. A very sharp dip is seen in $(d\sigma/du)(\pi^+ p \to p\pi^+)$ backward scattering at $u = -0.15$, where the dominant nucleon trajectory passes near $\alpha = -\frac{1}{2}$. This is a ghost wrong-signature point, and both flip and nonflip amplitudes should vanish here. In fact, since $\alpha(u)$ is complex for real $u < 0$ (a special baryon property: Gribov, 1963), it passes near $\alpha = -\frac{1}{2}$ rather than through it.

D. Conspiracies

It can be shown that t-channel helicity amplitudes must satisfy linear relations (constraints) at certain values of t, corresponding to t-channel thresholds and pseudothresholds, in order to satisfy analyticity and crossing.

When Regge poles are introduced in the t channel, these relations restrict the Regge parameters. There are essentially two ways to satisfy these constraints:

1. *Evasion.* Each Regge pole independently satisfies the constraints.

2. *Conspiracy.* The individual Regge pole contributions do not satisfy the constraints, but their sum does. Each pole is less constrained, but correlations between different poles are required.

A very simple example of this is seen in nucleon-nucleon scattering. There are five independent s-channel helicity amplitudes:

$$\phi_1 = \phi(+ +, + +), \qquad \phi_2 = \phi(+ +, - -), \qquad \phi_3 = \phi(+ -, + -),$$
$$\phi_4 = \phi(+ -, - +), \qquad \phi_5 = \phi(+ +, + -), \tag{1-6}$$

where the symbols $\pm$ denote final and initial helicities $\pm\frac{1}{2}$. In this case the constraint equations at $t = 0$ are simply stated in terms of the above amplitudes:

$$\phi_4 = \phi_5 = 0 \qquad \text{at} \quad t = 0. \tag{I-7}$$

These are familiar consequences of angular momentum conservation; amplitudes with net helicity flip vanish at $\theta = 0$.

Now consider the π Regge pole, and another Regge pole c with the opposite parity (corresponding to particles with $J^P = 0^+$, 2^+, etc., and $I = 1$) in the t channel. It turns out that π contributes only to ϕ_2 and ϕ_4, with

$$\phi_2^{\pi} = \phi_4^{\pi}, \tag{I-8}$$

whereas c contributes to all five amplitudes, with

$$\phi_2^{c} = -\phi_4^{c}. \tag{I-9}$$

If π and c separately satisfy the constraint $\phi_4 = 0$ at $t = 0$, it follows from Eqs. (I-8) and (I-9) that their contributions to ϕ_2 must also vanish at $t = 0$, although ϕ_2 has no net helicity flip and is in general not required to vanish. This is termed *evasion*.

On the other hand, if it can be arranged that $\phi_2^{\pi} = \phi_2^{c}$ at $t = 0$, then the constraint $\phi_4^{\pi} + \phi_4^{c} = 0$ will be satisfied by a cancellation, and there is no need for $\phi_2^{\pi} + \phi_2^{c}$ to vanish. However, to ensure $\phi_2^{\pi} = \phi_2^{c}$ at $t = 0$ for all energies, we must have $\alpha_{\pi}(0) = \alpha_c(0)$ and also an appropriate relation between the π and c residue functions. This is a *conspiracy*.

The difficulties with ϕ_2 come from Eqs. (I-8) and (I-9), which hold because π and c each belong to a pure parity class. When the constraint $\phi_4 = 0$ is satisfied with a pure-parity exchange, $\phi_2 = 0$ also. The $\pi + c$ conspiracy, regarded as a single object, is a mixed-parity exchange, and $\phi_2 \neq 0$.

It was once thought that the sharp forward peaks in $n + p \rightarrow p + n$ and $\gamma + p \rightarrow \pi^+ + n$ might be due to just such a $\pi + c$ conspiracy, since with evasion the pion pole—and all poles with which it can interfere—give zero contribution at $t = 0$. However, because of factorization constraints, it seems impossible to reconcile such a conspiracy with the role of π in other reactions. The presently accepted view is that Regge poles are evasive, but that Regge cuts, which do not have unique parity and can readily conspire, are important (see Section V). (See also Volkov and Gribov, 1963; Leader, 1968; Phillips, 1967; Ball *et al.*, 1968; Le Bellac, 1967; and Aderholz *et al.*, 1968.)

II. Finite Energy Sum Rules (FESR)

A. DERIVATION

Can we learn something about high energy scattering from studying low energy scattering, and vice versa? Analyticity says we can. There must be some relations between the low and high energy regions, since the scattering amplitude is an analytic function of the energy variable, and, in principle, can be continued from one region into the other.

Finite energy sum rules (FESR) are a class of such relations. The special merit of these rules is that the high energy parameters enter in a very simple way. To derive them it is necessary to assume only analyticity and Regge asymptotic behavior. With analyticity, Cauchy's theorem can be used for the integral of a scattering amplitude around a contour in the energy plane; this integral involves both low and high energies. With the Regge behavior, one can evaluate the high energy part of the integral exactly. The result is to equate a low energy integral directly to some high energy parameters.

Suppose the amplitude $f(v)$ is a real analytic function of the variable v throughout the v plane except for cuts from v_0 to ∞ and from $-v_0'$ to $-\infty$, plus perhaps some isolated poles, along the real axis. Typically, if one is concerned with scattering at fixed momentum transfer t, a suitable choice for v is the crossing-antisymmetric energy variable $v = (s - u)/4m$, where m is the target mass. Then the right- and left-hand cuts in the v plane are related to physics in the s and u channels, respectively.

Suppose also that $f(v)$ has a Regge-pole asymptotic expansion

$$f(v) = \sum_k \gamma_k\, v^{\alpha_k}[1 + \tau_k \exp(-i\pi\alpha_k)]/(\sin \pi\alpha_k) \qquad \text{(II-1)}$$

valid for $|v| \geqslant N$. Here k labels the t-channel Regge poles, and $\alpha_k(t)$, $\gamma_k(t)$, and $\tau_k = \pm 1$ are the trajectory, residue, and signature, respectively. Strictly speaking, Eq. (II-1) is taken to be the form for s-channel scattering, that is, on the right-hand cut approached from above.

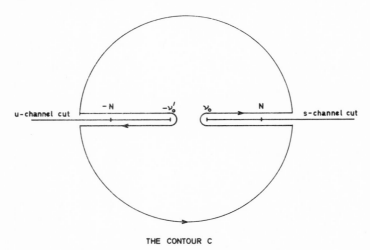

THE CONTOUR C

FIG. 3. The integration contour C.

Now by Cauchy's theorem the integral around the finite closed contour C of Fig. 3 is zero, since C encloses no singularities. This remains true if $f(v)$ is multiplied by any positive integral power v^m:

$$\int_c f(v)v^m \, dv = 0 \qquad (m = 0, 1, 2, \ldots). \qquad \text{(II-2)}$$

These relations involve the scattering amplitude along the physical cuts on the real axis and also along apparently unphysical arcs in the upper and lower half-planes. However, these unphysical integrals can be expressed in terms of physical Regge parameters by analytically continuing Eq. (II-1) into the upper and lower half-planes. If the Regge form is used for all of the high energy region $|v| \geqslant N$, the exact position of the high energy arcs does not matter since the integrand is analytic; in fact, it is convenient to take these arcs to be upper and lower semicircles of radius N. Equation (II-2) now can be rewritten as

$$0 = (1/2i) \int_c f(v)v^m \, dv$$

$$= \int_{-N}^{-v_0'} \text{Im} \, fv^m \, dv + \int_{v_0}^{N} \text{Im} \, fv^m \, dv$$

$$+ \sum_k [N^{\alpha_k + m + 1}/(\alpha_k + m + 1)] \, \gamma_k \, [\tau_k - (-1)^m], \qquad \text{(II-3)}$$

where isolated poles are understood to be included with the continuum integrations. These are FESR, relating low energy integrals ($|v| < N$) to high energy Regge parameters. Because of their symmetry, Regge pole terms with signature even (odd) vanish from Eq. (II-3) when m is even (odd). Indeed, if the whole integrand were symmetric in v, the integral would vanish identically and Eq. (II-2) would hold trivially.

The FESR become simpler when $f(v)$ is exactly symmetric or antisymmetric in v because the left- and right-hand cut integrals are then trivially related. For example, if $f(v)$ is antisymmetric, $f(v) = -f(-v)$, uniquely $\tau_k = -1$ in the Regge expansion Eq. (II-1), and

$$\int_{v_0}^{N} \text{Im} \, f(v)v^m \, dv = \sum_k \gamma_k \, [N^{\alpha_k + m + 1}/(\alpha_k + m + 1)] \qquad (m = 0, 2, 4, \ldots).$$

$$\text{(II-4)}$$

This is the best-known form of FESR, which equates an integral of low energy amplitudes directly to a simple function of high energy parameters. If one can be measured, it gives information about the other.

The FESR integrals above involve only $\text{Im} \, f(v)$. If $\text{Re} \, f(v)$ is to be exploited instead, one can go through analogous steps, with $(-iq)f(v)$ in place of $f(v)$, where $q^2 = v^2 - v_0^2$. This is still a real analytic function with cuts in the same places as before, and instead of Eq. (II-4), one gets the sum rules

$$\int_{v_0}^{N} \text{Re}\, f(v) q v^m \, dv = \sum_k [\gamma_k N^{\alpha_k + m + 2} / (\alpha_k + m + 2)] \tan(\pi \alpha_k / 2), \quad \text{(II-5)}$$

omitting terms of order v_0^2 / N^2 on the right.

More generally, to exploit arbitrary linear combinations of $\text{Re}\, f(v)$ and $\text{Im}\, f(v)$, one can start with the weighted amplitude $(-iq)^\beta f(v)$, where β is a continuous real parameter. Since this class of weight functions essentially includes the ordinary moments v^m, the latter can now be dropped, putting $m = 0$. Then

$$\int_{v_0}^{N} (\cos(\tfrac{1}{2}\pi\beta)\, \text{Im}\, f(v) - \sin(\tfrac{1}{2}\pi\beta)\, \text{Re}\, f(v)] q^\beta \, dv$$

$$= \sum_k [\gamma_k N^{\alpha_k + \beta + 1} / (\alpha_k + \beta + 1)] \cos \tfrac{1}{2}\pi(\alpha_k + \beta) / (\cos \tfrac{1}{2}\pi\alpha_k), \quad \text{(II-6)}$$

omitting terms of order v_0^2 / N^2. This is called a *continuous moment sum rule* (CMSR) and includes both (II-4) and (II-5) as special cases. The term FESR will be used generally to refer to all FESR, CMSR, and other modified sum rules derived in the same spirit.

The reader is reminded that Eqs. (II-4)–(II-6) refer explicitly to an anti-symmetric amplitude $f(v)$, with the particular Regge pole expansion (II-1) and signatures $\tau_k = -1$. Other cases differ in small but significant details.

The question of asymptotic convergence has not arisen because a finite contour has been used. To integrate along the semicircles it was necessary to continue the Regge pole form to complex v, but this is no worse than having to extrapolate a Regge form to infinity along the real axis, as one does in evaluating an ordinary dispersion relation.

The idea of using dispersion sum rules to learn about high energy parameters began with Igi (1962). The present forms of FESR were developed by Logunov *et al.* (1967), Igi and Matsuda (1967), Dolen *et al.* (1967, 1968) and Liu and Okubo (1967); Horn (1969) has given a recent review. Meshcheriakov *et al.* (1967) emphasized that only a finite contour is needed for the derivation. Other authors have constructed superconvergent amplitudes by formally subtracting the leading Regge pole terms, and then expanded the semicircular arcs to infinity; the result is the same.

The above derivations of FESR fail when the integrand is symmetric in v; that is, Eq. (II-4) is not proved for m odd. However, Schwarz (1967) showed how some of these "missing" sum rules could be derived when the amplitude satisfies a Mandelstam representation with no third double-spectral function. There has been some discussion by Dolen *et al.* (1967, 1968) and Michael (1968). This matter will not be pursued, except to remark that the Regge cuts derived by Mandelstam (1963) are associated with a nonvanishing third double-spectral function, and should invalidate the Schwarz sum rules.

B. FESR Bootstraps

The Regge pole parameters appearing on the right-hand side of the FESR are related to bound-state and resonance properties in the t-channel. If the low energy integrals on the left are assumed to be dominated by s-channel bound states and resonances, the FESR look like "bootstrap" or self-consistency relations, especially in simple cases where the s and t channels are the same.

For an example of how the s- and t-channel properties fit together, consider $\pi^+\pi^0 \to \pi^0\pi^+$ charge exchange. The t-channel amplitude is given by ρ exchange, and has a zero at the wrong-signature point $\alpha = 0$ (at $t = m_\rho^2 - 1/\alpha'$, where α' is the slope); this zero comes from $\rho - f^0$ exchange degeneracy, as explained in Section III. Suppose that the imaginary part of the s-channel amplitude is dominated by the ρ resonance, giving a P-wave amplitude with a zero at $\cos\theta_s = 0$ (at $t = -2k_s^2 = 2m_\pi^2 - \frac{1}{2}m_\rho^2$, where k_s is the center-of-mass momentum at ρ). The FESR bootstrap, in this approximation, requires these two zeros to coincide and thus determines either m_ρ^2 or α' in terms of the other:

$$(m_\rho^2 - 1/\alpha') = (2m_\pi^2 - \tfrac{1}{2}m_\rho^2). \tag{II-7}$$

This does not agree closely with experiment; with a linear trajectory the left-hand side is near -0.5, whereas the right-hand side is -0.25. Nevertheless, as a relation between m_ρ^2 and α', this is not a bad first approximation.

Alternatively, the argument can be reversed, saying that FESR with ρ saturation require a zero in the ρ residue near $t = -0.25$.

We shall not pursue the subject here, but simply point out some ways in which this approach differs from the conventional bootstrap, using the N/D equation or the strip approximation

1. Analyticity and crossing is emphasized here; unitarity is neglected.

2. No scale is established; the equations are homogeneous.

3. Coupled processes, such as $\overline{K}N \to \overline{K}N$ and $\overline{K}N \to \pi Y$, are treated separately. However, in each process, only the s- and t-channel particles that couple strongly play an essential role.

(See also Dolen et al., 1967; Ademollo et al., 1967; Freund, 1968; Schmid, 1968a; Rubinstein et al., 1968; Mandelstam, 1969.)

Further development of this approach leads to the concept of duality and to the Veneziano model (Section IV).

C. Determining Regge Parameters

When the low energy amplitudes are known, it is tempting to use a set of FESR or CMSR to determine the Regge parameters, and thus to predict high energy scattering. This approach is quite successful, within certain limits, as our illustrations show; but first, some practical remarks:

1. Sum rules have the advantage of directly giving amplitudes. Scattering experiments usually give nonlinear functions of several amplitudes, and some unscrambling is needed. Sum rules exploit the unscrambling previously done at low energy.

2. Although an infinite number of independent sum rules can be written in principle, they contain only a few independent pieces of information in practice; the finer details are lost in experimental error. Hence, the Regge pole series in Eq. (II-1) must be truncated, and, at best, only a few terms can be determined. This imposes an inevitable coarseness on any solution.

3. For very high moments ($\beta \to \infty$), the low energy integral comes overwhelmingly from near the upper limit $v = N$, and the sum rules reduce simply to Eq. (II-1) evaluated at $v = N$. This means that Regge parameters are fitted directly to data at $v = N$. Since non-Regge fluctuations may still be present at $v = N$ in practice, very high moments are unreliable. With lower moments, such fluctuations have a chance to average out.

4. In the FESR integral there is often an unphysical region at lower energies where the amplitude is not directly measurable and has to be found by extrapolation. The lower-moment sum rules are more affected by such regions than the higher-moment sum rules.

5. Regge cuts can be included in the sum rules as continua of poles. However, since they introduce continuous parameters, they cannot be determined without simplifying assumptions.

6. The moment β in CMSR can be chosen to enhance or suppress a particular pole, if its trajectory α is known, through the factor $\cos \frac{1}{2}\pi(\alpha + \beta)$ in Eq. (II-6).

7. The β dependence of a single Regge pole contribution is essentially like $N^x \sin(\frac{1}{2}\pi x)/x$, with $x = \alpha + \beta + 1$, in Eq. (II-6). Hence, in principle, a single Regge pole can be recognized from the β dependence of the CMSR integral— and can be distinguished from, say, two poles, or a branch cut. However, a pair of adjacent poles or a cut with a localized weight function may be indistinguishable from a single pole in practice. (For an example, see Jackson and Quigg, 1969.)

8. Although a continuous set of CMSR do not provide infinitely more information than a discrete set of FESR, Remarks 6 and 7 above show that they may present the information in a more convenient way.

9. Consider the relative importance of different Regge terms on the right-hand side of the CMSR Eq. (II-6). Apart from the fluctuating factor $\cos \frac{1}{2}\pi(\alpha + \beta)/(\alpha + \beta + 1)$, the relative importance is the same for all moments β, and is the same as that in the Regge expansion of $f(v)$ at $v = N$ in Eq. (II-2). Thus, a Regge pole that is important (negligible) in $f(N)$ will be equally important (negligible) in the sum rules, taken as a whole.

10. If the leading Regge pole has a very small coefficient, and is therefore negligible at laboratory energies, it cannot be found from FESR—although it dominates asymptotically. This example shows that FESR cannot predict all the way to infinity, in practice.

11. Our apparent ability to predict up to very high energies depends on the validity of the Regge pole expansion Eq. (II-2). If the Regge poles are actually parametrizing Regge cuts that contain logarithmic factors as well as simple powers v^α (see Section V), this approximation can only be valid in a limited range of v. For the FESR to be valid, it is enough that the Regge pole approximation be good on the finite contour, with $|v| = N$. There is no guarantee that one can safely extrapolate to $|v| \gg N$.

12. In effect one is trying to continue analytically from the approximately known low energy amplitude. It is intuitively clear that the results will become less reliable as v increases.

13. Many variations on the sum rules can be made, such as: (a) other weight functions in place of v^m or q^β; (b) other " Regge " forms, depending perhaps on s^α or u^α instead of v^α—the corresponding expansions are presumably equivalent, but some converge better than others.

14. The left-hand sides of sum rules can be regarded as another kind of data, supplementing high energy scattering data. Both are expressible in terms of Regge parameters. When both are available, they can be analyzed together to get the Regge parameters.

D. Examples

Below are just a few examples, out of very many in the literature, to show the sort of results achieved. Most of the examples are found in πN-scattering, where low energy amplitudes are now known quite well, up to $N \approx 2$ GeV. Here one uses the invariant amplitudes B and A', related to the conventional amplitudes by $A' = A + vB/(1 - t/4m_N^2)$ (Chew et al., 1957; Singh, 1963). A' appears in the s- and u-channel optical theorems: $\sigma_T = \text{Im } A'(t = 0)/q$. Because both the s and u channels represent πN scattering, it is convenient to form symmetrized combinations:

$$A'^{\pm} = \frac{1}{2}[A'(\pi^- p) \pm A'(\pi^+ p)], \tag{II-8}$$

and similarly for B. Through isospin invariance, the $\pi^- p \to \pi^0 n$ amplitudes are simply $-\sqrt{2}\,A'^-$, $-\sqrt{2}\,B^-$. It turns out that A'^+, B^- are symmetric in v, while A'^-, B^+ are antisymmetric. Thus, FESR can be constructed for the antisymmetric functions $f(v) = vA'^+$, A'^-, B^+, and vB^-.

Bose statistics in the $\pi\pi$ system relate t-channel isospins $I_t = 0$, 1 uniquely to signature $\tau = +1$, -1, respectively. The amplitudes A'^+, B^+ thus are often parametrized at high energy by even-signature t-channel vacuum Regge poles P and P', with $\alpha_P(0) = 1$ and $\alpha_{P'}(0) \approx 0.5$. A'^- and B^- are supposed to be dominated by the odd-signature ρ Regge pole, with $\alpha_\rho(0) \approx 0.5$, but the observed polarization in πN charge exchange indicates interference with other terms, sometimes parametrized as a second Regge pole ρ' or as a cut.

The easiest sum rule applications are found in elastic scattering at $t = 0$, using the spin-average amplitude. Here the imaginary part can be measured directly from the total cross section by the optical theorem, and the real part can often be measured through Coulomb interference. Thus, there is no need for a complete low energy analysis, and the upper limit of the sum rule integral is not particularly restricted.

For example, several researchers studied CMSR for $A'^+(\pi N)$ at $t = 0$, with various cutoffs up to $N = 6$ GeV (Della Selva et al., 1968; Olsson, 1968; Olsson and Yodh, 1968). They found that the usual vacuum poles P and P' could not accurately fit the sum rules, and deduced that some other term P'' is present. However, the CMSR constraints were not enough to determine all six parameters (treating P'' as another pole). The real parts of A'^+ used by these authors in fact were not measured directly, but were derived from imaginary parts through dispersion relations. This procedure is a bit deceptive, since the CMSR in this case can be reexpressed in terms of FESR involving Im A'^+ alone (Ferrari and Violini, 1969).

Elastic scattering applications at $t = 0$, although firmly based, are rather dull. It is much more exciting to investigate the t-dependence of Regge parameters, to discover trajectory slopes, and to see if the residues have any wrong-signature zeros or other dynamical zeros. For this one needs a set of FESR, each at a different t-value. Low energy amplitudes for $t < 0$ can usually be found only through a detailed phase-shift (or partial-wave) analysis. At present such analyses exist for only a few reactions; therefore, the scope is limited. Further, since these analyses currently stop at or below 2 GeV, the FESR can be used only if the Regge parametrization Eq. (II-2) is extrapolated right down to $N = 2$ GeV, well below the usually accepted Regge region. This bold step was first taken by Dolen et al. (1967, 1968) and proved remarkably successful. Many similar applications followed.

1. Dolen et al. (1967, 1968) investigated the πN charge-exchange amplitudes A'^- and B^-, using phase-shift analyses up to $N = 1.1$ GeV as input. Em-

pirical high energy analyses (Phillips and Rarita, 1965; Hohler *et al.*, 1966; Arbab and Chiu, 1966) had indicated that the ρ trajectory is $\alpha(t) \approx 0.5 + t$, that vB^- is an order of magnitude larger than A'^- at small t, and that A'^- and B^- have zeros near $t = -0.2$ and -0.5 (GeV/$c)^2$, respectively.

Assuming a single ρ pole dominates in the Regge region, Dolen, Horn, and Schmid used FESR to predict all the above empirical features for A'^- and B^- (see Fig. 4). This was remarkable because the cutoff $N = 1.1$ GeV was so low. There are substantial resonance terms in the πN amplitudes above this cutoff; apparently their effects—smoothed out by the sum rule integrals— are approximated successfully by the Regge pole terms. This is where the idea of duality began (see Section III).

2. The amplitude $B^+(\pi N)$ has a chequered history. Pre-1969 high energy data constrained B^+ rather little, but various arguments and least-square data fitting led to values that were large, negative, and rapidly t dependent compared to A'^+/v (Phillips and Rarita, 1965). The use of FESR showed, however, that all these conclusions were wrong, and indicated instead the relation $B^+ \approx A'^+/v$ (Barger and Phillips, 1968). Later measurements of πN spin rotations sensitive to B^+ confirmed the sum rule predictions (Amblard *et al.*, 1969; see Fig. 5).

3. Ferro-Fontan *et al.* (1968) set out to predict all high energy πN scattering from CMSR alone, with $N < 2$ GeV. They achieved quite good agreement

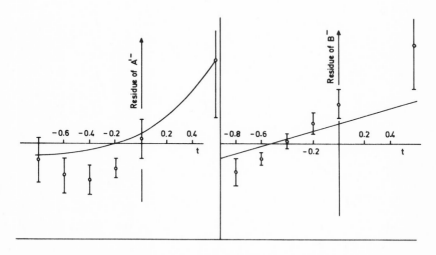

FIG. 4. FESR predictions compared with high energy data fitting for the ρ residues in the πN charge-exchange amplitudes A'^- and B^-. The points with errors are FESR predictions of Dolen *et al.* (1967). The curves are from the high energy model of Hohler *et al.* (1966). (After Dolen *et al.*, 1967, Fig. 2.)

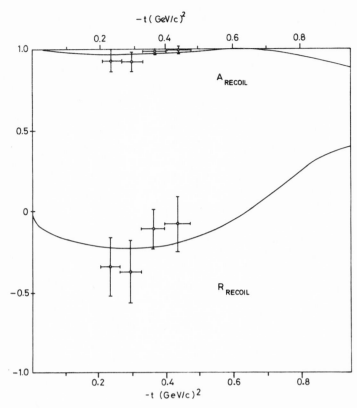

FIG. 5. Measurements of the $\pi^- p$ rotation parameters R_{recoil} and A_{recoil} at 6 GeV/c (Amblard *et al.*, 1969), compared with predictions of a Regge pole model satisfying FESR (Barger and Phillips, 1969a).

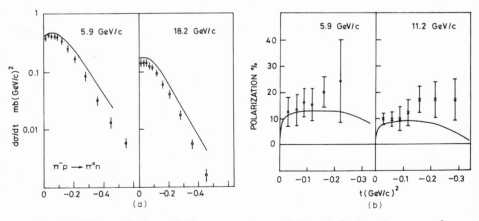

FIG. 6. FESR predictions of (a) cross sections and (b) polarizations in $\pi^- p \to \pi^0 n$, compared to experiment. (From Ferro-Fontan *et al.*, 1968.)

with data up to 20 GeV. Figure 6 illustrates some of their results, compared with data, and shows that sum rules have real predictive value.

A logical next step was to combine CMSR and high energy data in the same analysis. On this basis Barger and Phillips (1969a) fitted πN scattering up to 25 GeV plus CMSR, with $N = 2$ GeV, using P, P', P'', ρ, and ρ' Regge poles.

A rude shock came with the first results in the 25–65-GeV range from the Serpukhov accelerator (Allaby et al., 1969). The $\sigma_T(\pi N)$ data stayed constant, whereas Regge pole models—satisfying FESR constraints—predicted a steady fall with increasing energy, well outside the quoted errors. One explanation for this is that the Regge poles simply approximate more complicated things, such as Regge cuts, and that the approximation breaks down at Serpukhov energies. Recall that, to satisfy FESR, the approximation need only be good near $|v| = N$.

4. Pion photoproduction $\gamma N \to \pi N$ is especially interesting. Many Regge poles (π, ω, ρ, A_1, A_2, B) can be exchanged in the t channel. There are many independent amplitudes for different spin and isospin configurations that can be exploited to separate the poles. A low energy analysis also exists, up to $N = 1.2$ GeV (Walker, 1969).

Di Vecchia et al. (1968a) evaluated FESR for ω exchange. A simple theory of the crossover of $d\sigma/dt$ for pp and $\bar{p}p$ scattering, assuming one dominant factorizable Regge pole with the ω quantum numbers, predicts a universal zero at $t = -0.15$ in all ω-exchange amplitudes (Phillips and Rarita, 1965; Barger and Durand, 1967). In contradiction, Di Vecchia and his co-workers found no such zero in the ω exchange in photoproduction, suggesting the crossover mechanism is not simply an ω zero.

Another interesting contributor is π exchange. Ball et al. (1968) invoked a conspiracy between π and an opposite-parity Regge pole π_c to explain the sharp forward peak in $\pi^{\pm}$ photoproduction (it defies explanation by non-conspiring poles alone). Di Vecchia et al. (1968b) found that CMSR for the π and π_c contributions were well fitted by a pair of Regge poles, supporting the conspiracy hypothesis, and seeming to exclude cuts. However, Jackson and Quigg (1969) made a model with conspiring cuts and nonconspiring poles that fitted both high energy $\pi^{\pm}$ production data and also CMSR at small t, showing that CMSR cannot discriminate between poles and cuts in practice.

5. Backward elastic scattering $\pi N \to N\pi$, $KN \to NK$ can be studied by FESR at fixed u. There are important differences from the previous examples.

a. The exchanged Reggeons are now baryons, with strangeness $S = 0$, -1, $+1$ for πN, KN, and $\overline{K}N$ scattering, respectively. As long as no $S = 1$ particles are confirmed, one expects to approximate the right-hand side of the $\overline{K}N$ FESR by zero or by Regge cuts.

b. The appropriate variable is $v = (s - t)/4m_N$, or something similar. The right-hand cut still represents s-channel elastic scattering, but the left-hand cut now represents the t channel, $\overline{N}N \to \pi\pi$ or $\overline{N}N \to \overline{K}K$.

c. The t cuts begin at $t = 4m_\pi^2$ or $4m_K^2$, far below the physical threshold $t = 4m_N^2$, so that there is a big unphysical region in the FESR integrals.

Though baryon exchanges raise important questions, these cannot be answered reliably by FESR because there are no reliable low energy t-channel analyses, either in the unphysical region or above it. Resonance approximations have been tried, but contain many unknown elements. Perhaps the FESR for backward scattering should be regarded as putting constraints on low energy resonance parameters in terms of high energy exchanges, rather than vice versa.

Several groups have analysed these FESR. Their results are broadly consistent with high-energy analyses of u-channel exchanges (see, for example, Chiu and Der Sarkissian, 1968; Barger *et al.*, 1969).

III. Duality

A. THE IDEA

The principle of duality is that direct-channel resonances and crossed-channel Regge pole contributions are equivalent—in a sense defined more carefully below. They are alternative approximations to the same thing. Therefore, properties of resonances state something about Regge poles, and vice versa. Furthermore, it is wrong to add both resonance and Regge contributions in a scattering amplitude; this is double counting.

The duality principle evolved from FESR. Resonances are assumed to dominate at least the imaginary part of the low energy amplitude, and hence the left-hand side of FESR. Regge terms are introduced as an analytic continuation of the low energy amplitude, so the idea that they are the same thing is implicit. Explicitly, the right-hand side of FESR can be rewritten as a *low-energy integration* over Regge terms—at least, for moments high enough to make it converge at $v = 0$. For example, with resonance approximation on the left, Eq. (II-4) can be written

$$\int_0^N \operatorname{Im} f(\text{res}) v^m \, dv = \int_0^N \sum_i \gamma_i v^{\alpha_i + m} = \int_0^N \operatorname{Im} f(\text{Regge}) v^m \, dv, \qquad \text{(III-1)}$$

provided $\alpha_i + m + 1 > 0$ for all Regge poles in the sum. Bound-state poles are included with the resonance terms. Thus, in the average sense defined by the integrals,

$$\langle \operatorname{Im} f(\text{res}) \rangle = \langle \operatorname{Im} f(\text{Regge}) \rangle. \qquad \text{(III-2)}$$

The idea came from Dolen, Horn, and Schmid (1967, 1968) who first noticed

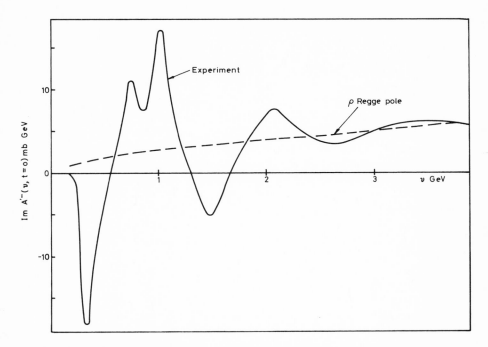

FIG. 7. An illustration of duality, for the πN amplitude A'^-. Experimentally, Im A'^- fluctuates at low energy but is given in the mean by the smooth ρ Regge pole term. (From Dolen *et al.*, 1968.)

that Regge terms approximated the complete amplitude right down into the resonance region. The name *duality* came soon after.

Figure 7 illustrates how duality works for Im A'^- in πN scattering at $t = 0$. This quantity is given by the $\pi^{\pm}p$ cross-section difference Im $A'^- = \frac{1}{2}p_{\text{lab}}\,\Delta\sigma_T$. At low energies it fluctuates, but is described in the mean by a Regge pole term. Jacob (1969a), Kugler (1970a), and Schmid (1970) have recently reviewed this subject.

B. Various Definitions

The general idea that resonances equal Regge poles has been sharpened and stated in various ways—some stronger than others:

1. *Global duality.* This means equality of resonance and Regge integrals in FESR. It rests on the usual FESR assumptions, plus resonance dominance of Imf(v) for $|v| < N$:

$$\int \text{Im } f(\text{res})v^m \, dv = \int \text{Im } f(\text{Regge})v^m \, dv. \tag{III-3}$$

The integrals are taken in general from $v = -N$ to $+N$ [Eqs. (II-4) and (III-1) are for the simple case of an antisymmetric amplitude].

2. *Semiglobal duality.*   This means equality of resonance and Regge integrals in FESR, for the left- and right-hand cuts separately.   Thus, relations of the form (III-3) are assumed to hold, with the integrals taken from $v = -N$ to 0, and also with the integrals taken from 0 to N. This separation is clearly valid for very high moments m, since then it simply states the equality of resonance and Regge parametrizations at $v = -N$ and $v = +N$, separately.

3. *Semilocal duality.*   Here we assume that resonance fluctuations in $\mathrm{Im} f(v)$ average out over relatively small energy intervals $N_1 \leqslant v \leqslant N_2$, so that Eq. (III-3) is supposed to hold approximately when the integrals run from N_1 to N_2.

4. *Local duality: Schmid loops.*   One cannot have local equality between fluctuating resonance terms and smooth Regge terms. However, Schmid (1968b) noticed that the partial-wave amplitudes $f^J(v)$ resulting from a Regge-pole exchange trace out loops in the Argand diagram, as energy varies. These loops are essentially generated by the second term of the signature factor $1 + \tau \exp(-i\pi\alpha)$ (Chiu and Kotanski, 1968), and somewhat resemble the loops generated by resonances (Adair, 1959). Although Regge pole amplitudes have no second-sheet poles and therefore cannot match resonance behavior in detail, Schmid (1968b, 1969) conjectured that the Regge-induced loops do approximate the true resonance loops:

$$f^J(\text{Regge}) \approx f^J(\text{res}). \qquad\qquad (\text{III-4})$$

This is local duality, which is very controversial. Schmid suggested that it would apply to peripheral partial waves, not to central partial waves (small J), so one cannot sum to get the full amplitude in Eq. (III-4). Note that both real and imaginary parts of f^J are involved.

One objection to Eq. (III-4) is that t-channel Regge poles give no bumps in total cross sections, whereas real resonances do. Further, the velocity of $f^J(\text{Regge})$ in the Argand diagram is smooth, whereas a narrow resonance gives a velocity maximum; this is because $f^J(\text{Regge})$ has no second-sheet poles, as already mentioned. A possible solution to these difficulties is to add multiple Regge exchanges (Regge cuts) to the initial Regge-pole term, according to the eikonal or K-matrix prescriptions (see Section V). The resulting pole-plus-cut amplitudes do give bumps in cross sections; also, in the K-matrix case, they have second-sheet poles (Drago, 1970).

5. *Role of the pomeron.*   The Pomeranchuk pole P (" pomeron ") seems to be in a class of its own: indeed it may not be a simple Regge pole at all. Freund (1968) and Harari (1968) conjectured that it plays a separate role in duality; ordinary Regge poles are dual to resonances, but the pomeron is dual

to nonresonant background terms. This postulate leads to a more orderly scheme, and has been widely accepted. The definitions of duality (1–4 above) are modified accordingly.

The Freund-Harari hypothesis has been tested using known πN amplitudes. Qualitatively, if we plot the experimental partial-wave amplitudes corresponding to t-channel isospin $I_t = 0$ and 1, the former show resonance loops superposed on smooth backgrounds, whereas the latter have resonance loops with no background. Thus, all the nonresonant background seems to have $I_t = 0$, as required. Quantitatively, a prescription is found for separating the resonance and background parts in the low energy $I_t = 0$ amplitude; through FESR, the former then generates the P' Regge pole and the latter generates the expected pomeron term (Gilman et al., 1968; Harari and Zarmi, 1969; Harari, 1970). There is room for disagreement about the prescription for separating backgrounds (Dance and Shaw, 1968).

6. *Regge cuts.* The role of Regge cuts in duality is not yet clear. In the first formulations of duality, as in the first applications of FESR, Regge cuts have been ignored. We shall return to this point later.

C. Applications

1. *Exchange Degeneracy and Exotic Particles*

If there are no particles (resonances or bound states) in a particular channel, semiglobal duality says the crossed-channel Regge pole terms have no net imaginary part. For this to happen systematically, over a range of t values, Regge poles can only contribute in exchange-degenerate pairs (Schmid, 1968b; Harari, 1968). Note that this fixes the relative sign of the residues.

We can readily classify the "exotic" channels, with no known particles, using the quark model. Since all the established mesons and baryons can be regarded as the $\bar{q}q$ and qqq configurations in this model, all other configurations are exotic. Examples of exotic channels are $\pi^+\pi^+$, π^+K^+, and K^+N.

The exchange degeneracy required by duality is more general than that originally proposed, for dynamics without "exchange" forces (see Section I,B). The latter relates poles of the same isospin, and the same spin-parity sequence. Duality can relate poles with different isospins.

Consider, for example, $\pi^+\pi^+$ and π^+K^+ scattering that are exotic in the s-channel. The only t-channel Regge poles (excluding P) are P' and ρ; therefore, these must be exchange degenerate in both cases, and, by factorization, in K^+K^+ scattering as well. Now think about K^+K^+ and K^+K^0 scattering: both are exotic in the s-channel, and the t-channel Regge poles must be exchange-degenerate for each isospin separately, that is, in pairs (P', ω) and (A_2, ρ). Putting all these conditions together, all four trajectories must be the same, and two independent residues fix all their couplings in $\pi\pi$, πK, and KK scattering. Duality has given very strong constraints! In fact, the constraints are almost too strong. With just the four Regge poles P', ω, A_2, and ρ, it turns out that their contributions to high energy $K^+\overline{K}^0$ scattering cancel to zero. This is undesirable, since low energy resonances all contribute with the same sign at $t = 0$ and cannot cancel. The only way out is to introduce one more exchange-degenerate pair of poles that couple to $\overline{K}K$ but not to $\pi\pi$; these can be identified with the ϕ and f' mesons, known experimentally to decay into $\overline{K}K$, but essentially not to $\pi\pi$. In SU_3 language, singlets are added to the vector and tensor meson octets: the requirement of pure $\overline{K}K$ decay for one of the isoscalar mesons defines the singlet-octet mixing angle to have $\tan^2 \theta = \frac{1}{2}$. Thus duality leads to nonets of mesons, with the standard mixing angle of the quark model.

In some cases, the constraints from duality can be too strong—that is, logically inconsistent and permitting only null solutions. This is one of the problems mentioned in Section III,D. However, no such troubles of principle appear in meson-meson and meson-baryon scattering.

Consider also K^+p and K^+n scattering, where the dominant Regge poles are P', ω, ρ, and A_2 in the t channel and Y_0^*, Y_1^* in the u channel. Since the s channel is exotic, for each KN isospin state, duality requires exchange degeneracy for each isospin in both the t and u channels: (ρ, A_2), (P', ω), $(Y_1^*$ pair$)$, $(Y_0^*$ pair$)$ degeneracy. The mass spectra confirm degeneracy of the trajectories (see Figs. 1 and 2). There are in fact some Y^* resonances without exchange-degenerate partners, but these are either weakly coupled to the $\overline{K}N$ system or have very low spin compared to their mass, and play only a weak role in FESR and duality constraints. This raises a point already made in Section II,B, that duality constraints are only meaningful for particles that play a major role in FESR for the reaction being considered.

Here, briefly, are more consequences (Harari, 1968; Schmid, 1968b, 1969):

a. Total cross sections are given by forward elastic amplitudes through the optical theorem: $\sigma_T = (4\pi/k)\,\mathrm{Im}\,f(0)$. For an exotic channel at high energy, only the pomeron contributes, and one expects σ_T to be flat. This is approximately true for K^+p, K^+n, pp, and pn data up to 20 GeV/c (Galbraith et al., 1965).

b. Since P has isospin $I = 0$, σ_T is the same for exotic channels that differ only by isospin rotations. Thus, $\sigma_T(K^+p) = \sigma_T(K^+n)$; $\sigma_T(pp) = \sigma_T(pn)$. The data confirm this to be a good approximation.

c. When resonances are present, one expects them to contribute positively to σ_T, so that the corresponding Regge poles contribute positively at high energy and σ_T approaches its asymptotic limit from above. This rule is obeyed by the σ_T data up to 25 GeV.

d. In exotic channels, pairs of exchange-degenerate Regge poles contribute real amplitudes, with no net wrong-signature zeros [see Eq. (1-3)]. If P

exchange is allowed (for example, K^+p, pp elastic; $pp \rightarrow pN^*$), these Regge terms are out of phase with the mainly imaginary P term. Hence, P dominates $d\sigma/dt$, and the secondary poles give negligible t-dependent structure; any shrinking of $d\sigma/dt$ is due to P. If there is no P exchange (for example, $K^+n \rightarrow K^0p$, $K^+p \rightarrow K^0\Delta^{++}$), the whole amplitude is real, and has no wrong-signature dips. The phase of an inelastic amplitude is hard to measure directly, but uniformly real phase will suppress some polarization effects.

e. Schmid loops are consistent with this picture if P gives no loops (a flat P trajectory would give no loops). In an exotic channel (for example, K^+p), the exchange-degenerate pairs of Regge poles give real amplitudes; therefore, their partial wave projections are also real and generate no loops in the Argand diagram. In the crossed channel (K^-p) there are generally resonances, but here the same Regge poles combine with opposite relative signs giving rotating phase factors $\exp(-i\pi\alpha)$, and readily generate loops.

f. Patterns of exchange degeneracy (see especially the extensive work by Mandula *et al.*, 1970). If there were no exchange forces in any process, all Regge poles would be exchange degenerate simply in pairs with the same internal-symmetry quantum numbers. However, the absence of exotics corresponds only to the absence of exchanges in selected representations; this leads through crossing to exchange degeneracies between Regge poles in different representations of internal symmetry. Examples, such as P' and ρ in $\pi\pi$ scattering, have already been given where Regge poles with different isospins are related. There are also correlations between Regge poles—and hence particles—in different SU_3 multiplets. These relations resemble the predictions of higher symmetry schemes.

Finally, a caution—duality is an approximation, and its constraints only apply to Regge poles and resonances which figure prominently in the processes considered. Weakly coupled reggeons and resonances are not effectively constrained. Resonances on daughter trajectories several units below the leading trajectory have relatively low spin and therefore low weighting compared to the leading resonances in the same mass region, and play unimportant roles.

2. Duality Diagrams

A simple graphical way to represent duality constraints, and to find solutions to them, was proposed by Harari (1969a) and Rosner (1969). For a given process, the incoming and outgoing particles are represented as composites of quarks, following the usual $\bar{q}q$ and qqq assignments, drawing a directed line for each quark. The path of each incoming quark is traced through the diagram to link up with a corresponding outgoing quark or incoming antiquark. A quark line must not begin and end on the same external particle. Thus, the scattering is represented graphically in terms of quark exchanges. This generally can be done in many ways; some examples for meson-baryon scattering are shown in Fig. 8.

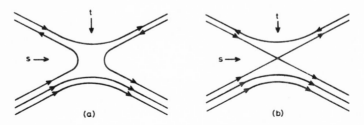

FIG. 8. Two duality diagrams for meson-baryon scattering.

The rule for interpretation is this. If one can draw a "legal" (s, t) graph, meaning a graph that has no exotic states in the s and t channels simultaneously, then there is a nontrivial duality relation between the imaginary parts of the s-channel resonances and t-channel Regge poles. If no such legal graph can be drawn, however, there is deemed to be only a trivial relationship: the imaginary parts are equal but zero. Thus, the t-channel Regge poles must be exchange degenerate, such as to give no imaginary part in high energy s-channel scattering (this fixes relative signs). Similarly, the s-channel Regge poles are also exchange degenerate, giving no imaginary part in t-channel scattering.

If there are no legal diagrams for *two pairs* of channels, say (s, t) and (s, u), one gets stronger results. There is exchange degeneracy in all three channels, but since the s-channel Regge poles must combine to be purely real for high energy scattering in both the t and u channels, they in fact must vanish. If there are no legal diagrams for *any pairs* of channels, the rules allow only a null solution: purely real amplitudes, with no resonances in any channel. Note, by the way, that if two quark lines cross (forming a "nonplanar" graph, in the jargon), the intermediate state in at least one channel is exotic and the graph is illegal. In Fig. 8b, for example, the s channel has a $qqqq\bar{q}$ state.

Duality diagrams were introduced more or less intuitively, and their use is justified by the results. They give all the usual predictions, based on the absence of exotics: if the s-channel is exotic, there cannot be any legal (s, t) or (s, u) diagrams and the argument above gives exchange degeneracy in both t and u channels, with nothing at all in the s channel. Duality diagrams also give many new predictions. The latter can alternatively be derived by adding factorization, SU_3 symmetry, and various quark model assumptions. Thus, it seems that duality diagrams are equivalent to sets of normally acceptable assumptions, but it is not clear exactly what is the minimum set of assumptions.

Below are some examples of the new predictions from duality diagrams.

a. $K^-p \to \pi^- \Sigma^+$, $K^-p \to \pi^0 \Sigma^0$. For these reactions, and all others of the general form $\overline{K}N \to \pi Y$, there are no legal s-t duality diagrams. This is because the λ quark in the incident $\overline{K}$ must emerge on the outgoing hyperon, at the opposite corner of the diagram, necessarily crossing another quark line (as in Fig. 8b). Hence, t-channel Regge poles are exchange degenerate, and the amplitude is purely real at large s and small t.

b. $\pi^- p \to \phi^0 n$, $\pi^+ p \to \phi^0 \Delta^{++}$. No legal duality diagrams can be drawn at all. The difficulty is that ϕ^0 is a $\lambda\bar{\lambda}$ quark pair, and none of the other external particles contain λ or $\bar{\lambda}$. Hence, there are no resonances or Regge poles in any channel. The cross sections are zero (or given by duality-violating background terms, presumably suppressed).

c. *Baryon-antibaryon scattering*: $B\overline{B} \to B\overline{B}$. Here there are so many duality constraints that they cannot be satisfied with normal nonexotic particles. This difficulty of principle was known previously, but duality diagrams reveal the point very quickly and simply. They also suggest a possible resolution of the dilemma.

We expect particles only in the s and t channels since the u channel is exotic, with baryon number $= 2$, but no legal (s, t) diagrams can be drawn; the best that can be done is Fig. 9. Hence, we are left with the null solution,

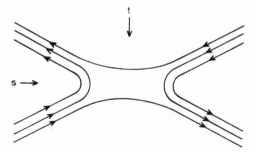

FIG. 9. A duality diagram for baryon-antibaryon scattering, with exotic meson states in the t channel.

real amplitudes, and no particles in any channel. This contradicts experiment, since resonances are observed in $\overline{N}N$ scattering. Also factorization requires P', ω, ρ, and A_2 to be exchanged in $NN \to NN$ and $\overline{N}N \to \overline{N}N$, since they are exchanged in $KN \to KN$.

One way out, suggested, in fact, by the duality diagram Fig. 9, is to have exotics. Rosner (1968) suggested that there could be meson resonances belonging to SU_3 multiplets $\{10\}$, $\{\overline{10}\}$, and $\{27\}$, coupling almost exclusively to the $\overline{B}B$ channels and therefore not seen elsewhere.

Kugler (1970b) suggested a more drastic solution. Arguing that duality applies to "effective" Regge poles (that is, poles plus cut corrections) rather than to pure poles, he proposed to abandon factorization. This allows a solution with nonexotic resonances in $\bar{B}B$ channels, when both baryons belong to an octet, but without any resonances in channels involving a decuplet baryon. For the latter channels, Kugler proposed to accept the duality constraint, and pointed out that high energy amplitudes for $NN \rightarrow N\Delta$, $N\bar{N} \rightarrow N\bar{\Delta}$ and $NN \rightarrow \Delta\Delta$ seem to be dominated by π exchange and hence are essentially real, as predicted. The remarkable absence of ρ or A_2 exchange is then to be seen as a consequence of duality. (In this argument, π is considered to be negligible except for its contribution to the real parts, anomalously enhanced by the nearness of the pole.) Kugler's solution does not correspond to any duality diagrams.

3. *Interference Model*

Regge exchanges are used in the interference model as an additive background to resonances terms at low and intermediate energies (Carroll *et al.*, 1966; Barger and Cline, 1966, 1967; Barger and Olsson, 1966; Baacke and Yvert, 1967). In this model it is postulated that

$$f = f(\text{res}) + f(\text{Regge}).\qquad\qquad (\text{III-5})$$

This contradicts duality. If duality is good, Eq. (III-5) is double-counting. The Freund-Harari conjecture (above) might be regarded as a modified interference model, compatible with duality: $f = f_P + f(\text{res})$. However, the emphasis there is on the separate, aristocratic nature of the pomeron, which is not what is normally meant by the interference model.

Many arguments have been raised against the interference model (Dolen *et al.*, 1968; Chiu and Stirling, 1968). For example, let $f(v)$ be a forward elastic scattering amplitude that has a positive imaginary part from unitarity and satisfies an FESR. Substituting $f = f(\text{res}) + f(\text{Regge})$ on the left-hand side of the FESR, the arguments that previously gave Eq. (III-1) now give

$$\int_0^N \text{Im } f(\text{res})v^m \, dv = 0.\qquad\qquad (\text{III-6})$$

However, with a normal Breit-Wigner form, each resonance contribution to elastic scattering is positive definite and the integral cannot be zero. For sufficiently high moments m, only the top of the integration region counts, so it does not matter if the interference-model prescription is taken down to threshold or not.

Strictly speaking, when there are strong backgrounds, each resonance term does not need to be unitary by itself, and thus its contribution does not need to be positive. However, the idea of a resonance contributing negatively to Im f—and hence giving a dip rather than a bump in σ_T—is unattractive to many people.

The interference model also enters into the controversy about Schmid loops. In this model, Regge poles are the nonresonant background, and hence the Schmid loops are also background effects—and are especially dangerous since they may be mistaken for true resonance loops. This philosophy requires that the Schmid loops be subtracted before looking for genuine resonance loops. The consequences can be drastic. If experimental results initially resemble a Schmid loop, the loop is destroyed and any corresponding resonance is discredited. Conversely, the subtraction can generate new loops, suggesting unsuspected new resonances. (For further discussion, see Alessandrini et al., 1968; Collins et al., 1968a, b; Schmid, 1969; Donnachie and Kirsopp, 1969; Phillips and Ringland, 1969.)

Finally, the interference model is unattractive to many people because it is not predictive, suggesting no fruitful relation between particles in different channels. Duality is destroyed and all its successes are regarded as fortuitous.

Various modified interference models have been suggested. For example, Coulter et al. (1969) consider the origins of a typical t-channel Regge pole signature factor $1 + \tau \exp[-i\pi\alpha(t)]$, in the Veneziano model (see Section IV). The first element, 1, comes from a term that also gives u-channel resonances. The second element, $\tau \exp(-i\pi\alpha)$, comes from a term that gives s-channel resonances; furthermore, it is this element that generates the Schmid loops (Chiu and Kotanski, 1968). Hence, Coulter et al. (1969) argue that the danger of double-counting is restricted to the second element, which therefore should be omitted, taking

$$f = f(s\text{-res}) + f(\text{unsignatured } t\text{-Regge}) \qquad \text{(III-7)}$$

at small t. The effect for exchange-degenerate pairs of Regge poles is interesting. When the pair combines to be real, as in K^+p scattering, the second elements of the signature factors cancel, and the same result is obtained as in the original interference model. In the crossed channels, such as K^-p scattering, the first elements cancel instead; the prescription is now to omit the Regge poles completely.

4. The Deck Effect

The $A_1(1300)$ $\pi\rho$-enhancement, observed in $\pi N \to \pi\rho N$, has been controversial. One view is that this is a resonance, produced by P exchange (Fig. 10a). Another view is that it is wholly or partly a nonresonant enhancement, caused by virtual $\pi \to \rho\pi$ dissociation, π then scattering elastically from N (Fig. 10b): the "Deck effect" (Deck, 1964; Berger, 1968). It was argued that the background from the Deck effect should be calculated and subtracted before looking for any genuine A_1 resonance.

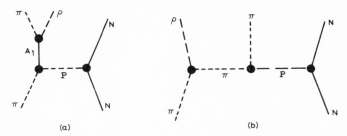

FIG. 10. Duality in $\pi N \to \pi \rho N$: (a) production of the $A_1(1300)$ resonance; (b) $\pi \rho$ production by the Deck mechanism.

If we set aside the nucleon line, however, Figs. 10a and b simply show the virtual scattering $\pi P \to \rho \pi$, in one case via the A_1 resonance and in the other case via π exchange. Chew and Pignotti (1968) argued that these are not distinct contributions, but the same thing (semilocal duality). It therefore would be wrong to subtract the Deck term before looking for A_1, because (Deck) $\approx (A_1)$. This is the interference-model controversy again, in a new context.

It may be a little dubious to invoke semilocal duality here, since the A_1 is close to the $\pi \rho$ threshold. One would hesitate to suggest a semilocal duality between Regge ρ exchange and the lowest resonance $\Delta(1236)$, in the case $\pi^- p \to \pi^0 n$ (compare Fig. 7).

D. TROUBLES

1. Some Exchange Degeneracy Predictions Fail

The transition $K^- n \to \pi^- \Lambda^0$ has no legal s-t duality diagram, and the high energy small-t amplitudes should be purely real; consequently, there should be no polarization. On the contrary, a large polarization is observed at 4.5 GeV/c (Yen *et al.*, 1969). Similarly $n + p \to p + n$ should have a purely real amplitude and no polarization, but polarization on the order of 30% is measured at up to 5.5 GeV/c (Robrisch *et al.*, 1970).

Exchange degeneracy for the K*, K** trajectories, without any condition but reality on the residues, demands that the cross sections for $K^- p \to \pi^- \Sigma^+$ and $\pi^+ p \to K^+ \Sigma^+$ be equal. In fact, the differential cross sections differ by a factor of 2, $K^- p \to \pi^- \Sigma^+$ being the larger (Kirz, 1969). A similar discrepancy occurs in the reactions $K^- n \to \Lambda \pi^-$ and $\pi^- p \to K^0 \Lambda^0$ (Lai and Louie, 1970).

KN and $\overline{K}$N scattering are supposed to be given by P plus exchange-degenerate (P', ω) and (ρ, A_2) pairs of Regge poles. Furthermore, these trajectories coincide, from duality arguments for $\pi\pi$, πK, and KK scattering (Section III,C). Hence, at $\alpha = 0$, near $t = -0.5$, the odd-signature Regge amplitudes have wrong-signature zeros. Hence, one should have $d\sigma/dt(K^+ p) =$

$d\sigma/dt(K^-p)$ at this point, but the crossover occurs experimentally at $t = -0.15$ instead. (Exactly the same discrepancy exists in pp and $\bar{p}p$ scattering.)

The charge-exchanges $K^-p \to \overline{K}^0 p$ and $K^+n \to K^0 p$ are supposed to be dominated by exchange-degenerate $\rho + A_2$ at small t, and, therefore the cross sections should be equal. The cross sections differ by a factor 2 at about 2–3 GeV/c, but are compatible at 5.5 and 12 GeV/c, so perhaps the prediction is a success after all. Also the high energy forward $K^+n \to K^0 p$ amplitude is measured to be essentially real, as predicted (Cline *et al.*, 1969; Firestone *et al.*, 1970).

2. *Unnatural-Parity Trajectories*

The type of argument that works so well for the natural-parity trajectories (see Section III,C,1) runs into problems with the unnatural-parity trajectories π, η, etc. (Dass *et al.*, 1970).

Consider $\pi^+\rho^-$ scattering. The exotic u channel imposes exchange degeneracy on the s- and t-channel trajectories. This is readily satisfied by the (A_2, ω) and (ρ, P') pairs. However, with π, a partner is required with the same G but opposite C, corresponding to a meson H with $J^P I^G = 1^+ 0^-$ (not established experimentally, but with mass presumably near 1 GeV). Consider also $\rho^+\rho^-$ scattering, again with an exotic u channel. Here η appears in the s-channel, and is exchange-degenerate with B(1235). So far so good, but if we now go to K*K scattering, the absence of exotics requires (π, B) and (η, H) exchange degeneracy. All four trajectories must coincide, but the π-η mass splitting makes this impossible.

3. *Limited Applicability*

As one approach to duality breaking, we can look for reasons why the arguments may fail in some cases.

Mandula *et al.* (1969) emphasized that duality is most credible when there is an intermediate energy region where resonance and Regge-pole pictures overlap. This criterion seems to hold in the πN and $\overline{K}$N channels, for example. However, in $\overline{B}B$ channels all the most prominent meson resonances are actually below threshold; if any overlap region exists, it involves high mass resonances of which we know little. Hence, resonance-Regge duality arguments will be least reliable when the resonances are in a $\overline{B}B$ channel.

As t increases to large values, we again expect duality to become unreliable, for several reasons:

a. The physical thresholds in s and u rise, making an overlap region less plausible.

b. The s- and u-channel cuts begin to overlap, making a separation of their effects (semiglobal duality) less plausible.

c. Regge cuts become more important relative to Regge poles (see Section V).

4. *Regge Cuts*

The original duality arguments ignore Regge cuts, although the latter are now known to be important experimentally (Section V). It is not clear how to include them in any simple way. Cuts are associated with multiple exchanges; therefore, their quantum numbers are not those of single particle states, and can easily be exotic. Thus, cuts provide an immediate excuse for any failures of duality. On the other hand, we now have the problem of understanding why duality has been so successful.

The strongest cuts seem to be those generated by multipomeron exchange. These have vacuum quantum numbers and can be regarded as modifying the P term. Since P stands outside duality anyway, these cuts also may do this.

The next most important cuts are those given by adding multi-P exchange to single Regge-pole exchanges. These have the same quantum numbers (apart from parity) as the input Regge poles: none is exotic. If we assume for simplicity that the P amplitude is purely imaginary, these cuts also preserve exchange degeneracy. If the input Regge pole amplitudes are purely real, as with an exchange-degenerate pair, the resulting cuts are also purely real. In fact, these cuts then can be written as continua of exchange-degenerate poles.

Thus, these leading Regge cuts preserve the important duality relationship between exchange degeneracy and absence of exotics (Barger and Phillips, 1969b). However, the P amplitude is not strictly imaginary for $t \neq 0$, and the above argument is only approximate (Kryzwicki and Tran Thanh Van, 1969). Also, cuts involving more than one exchange particle are not all negligible.

Finally, although multipomeron cuts can perhaps be classified with the pomeron, they still disturb the situation indirectly, altering the strength of the P' Regge pole required to fit data, and thus destroying the empirical exchange degeneracy of (P', ω) couplings in KN scattering (Barger and Phillips, 1970).

IV. Veneziano Model

A. ORIGIN AND MOTIVATION

The Veneziano model is a form of amplitude that combines both low energy resonances and high energy Regge behavior in an analytic and crossing-symmetric way. It therefore satisfies FESR and provides a concrete realization of duality.

This model originated as an educated guess. After some experience in trying to satisfy FESR bootstrap conditions for meson-meson scattering (Ademollo *et al.*, 1967; Rubinstein *et al.*, 1968), Veneziano (1968) noticed that a particular combination of Γ functions had all the required properties. The model was first formulated for the symmetrical process $\pi + \pi \to \pi + \omega$, where the s, t, and u channels are the same. Only the ρ trajectory appears, as a Regge exchange or as a sequence of resonances, according to the channel considered. The T matrix can be written in terms of a single invariant amplitude $A(s, t, u)$:

$$T = \varepsilon_{\mu\nu\rho\sigma} e_{\mu} p_{1\nu} p_{2\rho} p_{3\sigma} A, \qquad (\text{IV-1})$$

where e_{μ} is the ω polarization vector and $p_1 \cdots p_3$ are the pion momenta. For resonance behavior in the s channel, poles are required in s when $\alpha_s \equiv \alpha(s) = 1, 3, 5, \ldots$, where α denotes the ρ trajectory. For Regge behavior at fixed t, $A \sim s^{\alpha_t - 1}$ is required as $s \to \infty$ (note that this is a helicity-flip process). Similarly, $A \sim s^{\alpha_u - 1}$ is required as $s \to \infty$ at fixed u.

Veneziano proposed the closed expression

$$A(s, t, u) = \beta\{B(1 - \alpha_s, 1 - \alpha_t) + B(1 - \alpha_t, 1 - \alpha_u) + B(1 - \alpha_u, 1 - \alpha_s)\}, \qquad (\text{IV-2})$$

where $B(x, y) = \Gamma(x)\,\Gamma(y)/\Gamma(x + y)$ is the Euler B function and β is a constant. This is obviously analytic and crossing symmetric between the three channels. The ρ trajectory is taken to be linear:

$$\alpha_s = \alpha_0 + \alpha's. \qquad (\text{IV-3})$$

In the s channel the first and third terms on the right-hand side of Eq. (IV-2) give resonance poles at $\alpha_s = 1, 2, 3, \ldots$ because $\Gamma(x)$ has simple poles at $x = 0, -1, -2, \ldots$. In fact, we did not ask for poles at $\alpha_s =$ even integers; they can be removed by a subsidiary condition (Veneziano, 1968)

$$\alpha_s + \alpha_t + \alpha_u = 2. \qquad (\text{IV-4})$$

An alternative view is that these extra resonances represent real physics and should be allowed to remain. The residue of each pole is a polynomial in t (or u), and hence is a polynomial in $\cos \theta_s$; when this is decomposed into a sum of appropriate Legendre functions, we find that each resonance is a superposition of resonances (all with odd J, because of the symmetry of A), with $0 \leqslant J \leqslant \alpha_s$.

The leading resonances, with $J = \alpha_s$, clearly represent the ρ trajectory. The other resonances with $J < \alpha_s$ can be classified on parallel linear "daughter" trajectories $\alpha_n(s) = \alpha(s) - n$ ($n = 1, 2, 3 \ldots$). The "even daughters" ($n = 2, 4, \ldots$) give lower-spin resonances degenerate with the Regge recurrences of ρ. The "odd daughters" ($n = 1, 3, \ldots$) give the resonances at $\alpha_s = 2, 4, 6 \ldots$ that were mentioned above.

The existence of sets of degenerate daughter particles is a general prediction of Veneziano models (and with exchange degeneracy, daughters appear with all spins $0 \leqslant J < \alpha$). However, it should be stressed that there is little experimental support for this feature. There is probably a broad $J = 0$, $I = 0$ state degenerate with the ρ, but vigorous searches have found no evidence for the predicted $J = 1$, $I = 1$ daughter of f^0. Similarly, there is no sign of daughters for K*(890) or K*(1400). The absence of such daughters creates a problem for any locally dual model without exotics (Harari, 1969b).

When α_s is real, all the above poles lie on the real s axis. This is an idealization, a narrow-width approximation. Physical resonance poles are displaced below the real axis onto the second sheet, according to their widths Γ. A Breit-Wigner resonance denominator $E_R - E - \frac{1}{2} i \Gamma$ is proportional to $s_R - s - i E_R \Gamma$ (where $E^2 = s$), showing the pole to be at $s = s_R - i E_R \Gamma$.

Now consider high energy behavior at fixed t. On the real s axis the first and third terms of Eq. (IV-2) have no smooth asymptotic dependence, just successive poles at regular intervals. However, if $s \to \infty$ along a ray just above the real axis, smoothness results. Using the relations $\Gamma(\alpha)\Gamma(1 - \alpha) = \pi/(\sin \pi\alpha)$ and $\Gamma(z + a)/\Gamma(z + b) \sim z^{a-b}$ as $|z| \to \infty$, the first and second terms of Eq. (IV-2) reduce to

$$A = \frac{\pi\beta}{\Gamma(\alpha_t) \sin \pi\alpha_t} [(-\alpha_s)^{\alpha_t - 1} + (-\alpha_u)^{\alpha_t - 1}] = \frac{\pi\beta}{\Gamma(\alpha_t)} (\alpha_s)^{\alpha_t - 1} \frac{1 - \exp(-i\pi\alpha_t)}{\sin \pi\alpha_t},$$

$$(\text{IV-5})$$

which is the required form for a t-channel Regge pole. It is crucial, however, that α_s be linear in s, even asymptotically. Similarly, the second and third terms of Eq. (IV-2) behave like a u-channel Regge pole exchange.

Duality is apparent, in that the terms which give the s-channel resonances are also required for building the t-channel and u-channel Regge exchanges. It would be double counting simply to add all three. Alternatively, duality is seen in the fact that each B function in Eq. (IV-2) has particle poles in two channels simultaneously. (Double poles are avoided, at the intersection of two crossed-channel poles, by the Γ function in the denominator of B.)

The Veneziano model above is explicitly for $\pi + \pi \to \pi + \omega$, but the idea can be adapted to other reactions. For example, $\pi^+ \pi^- \to \pi^+ \pi^-$ has a single invariant amplitude, with exchange degenerate ρ and f^0 trajectories in the s and t channels, while the u channel is exotic. Thus, there is only s-t duality, and an appropriate Veneziano amplitude is

$$A(\pi^+ \pi^- \to \pi^+ \pi^-) = \beta \Gamma(1 - \alpha_s)\Gamma(1 - \alpha_t)/\Gamma(1 - \alpha_s - \alpha_t). \qquad (\text{IV-6})$$

This is not exactly a B function any more. The argument of Γ in the denominator has been changed, to get the correct Regge behavior $A \sim s^{\alpha_t}$ for this nonflip amplitude (the previous $\pi\pi \to \pi\omega$ case had helicity flip).

There is also freedom to add "satellite" terms to the amplitude, such as constant multiples of the following:

$$\Gamma(1 - \alpha_s)\Gamma(1 - \alpha_t)/\Gamma(2 - \alpha_s - \alpha_t), \qquad\qquad (IV\text{-}7)$$

$$\Gamma(2 - \alpha_s)\Gamma(2 - \alpha_t)/\Gamma(3 - \alpha_s - \alpha_t). \qquad\qquad (IV\text{-}8)$$

Satellite (IV-7) has all the direct-channel poles, but Regge behavior $s^{\alpha_t - 1}$. Satellite (IV-8) has all the direct-channel poles except those at $\alpha_s = 1$, and Regge behavior s^{α_t}, and so on (Altarelli and Rubinstein, 1969a). If the Pomeranchuk term is dual to nonresonant backgrounds, one would not expect it to appear in a Veneziano amplitude. It must somehow be added afterwards, in reactions where it appears. This is a problem in itself.

Factorization is an important question. It first was observed by Freund (1969) that, for two-body scattering, factorization could only be imposed on the leading Regge pole. The essential reason is that the residue of each pole in s is a polynomial in α_t, and thus a polynomial in $\cos \theta$ (for linear α_t). Only for the leading power of $\cos \theta$ are the external masses absent from the coefficient. For daughter particles the residue is a function of the external masses, which appear in a nonfactorizing form, making factorization impossible here. Fubini and Veneziano (1969) discussed this problem, using the powerful N-point formalism. They found that factorization at the daughter level holds only if there is a degeneracy of states for given mass and spin, the multiplicity increasing exponentially with mass.

A great deal of work has been done on multiparticle generalizations of the Veneziano model, that is, for amplitudes with $N > 4$ external particles. The reviews by Chan (1970) and Lovelace (1970) give an excellent introduction to this development.

B. Unitarization

With α real and all the resonance poles on the real axis, the Veneziano amplitude violates unitarity. Various prescriptions have been used to correct this.

1. Modification of α

A positive imaginary part can be introduced in α, by an empirical formula such as

$$\alpha(s) = \alpha_0 + \alpha's + \lambda(s_0 - s)^{1/2}, \qquad\qquad (IV\text{-}9)$$

where s_0 is the s-channel threshold. This moves the resonance poles off the axis, but also creates new problems.

a. *Ancestors.* The numerator of each direct-channel pole no longer is a polynomial in $\cos \theta$. Thus, each resonance has components with arbitrarily high spin, apparently lying on linear trajectories above the original leading trajectory α, contrary to experiment.

b. *Non-Regge Asymptotics.* Any imaginary part behaving like s^x $(x < 1)$ causes the third term of Eq. (IV-2) to dominate over the first two at fixed t. This can be corrected by arranging Im $\alpha \sim s(\ln s)^{-\nu}$, with a more complicated parametrization (Roskies, 1968).

c. *Crossing.* For s-channel scattering, we can avoid the above difficulties a and b by putting an imaginary part in α_s but not in α_t or α_u. This then violates crossing symmetry. Even so, many authors have chosen to use this prescription.

2. *K-Matrix Method*

Elastic unitarity in one channel can be imposed by regarding the Veneziano amplitude as the K matrix, related to the S and T matrices in a particular partial wave by

$$S = 1 + iT = (1 + \tfrac{1}{2}iK)/(1 - \tfrac{1}{2}iK). \qquad \text{(IV-10)}$$

The term K should be real for elastic unitarity; the Veneziano amplitude is indeed real for real α. This prescription works for elastic scattering (ignoring the pomeron) in one channel, but violates crossing symmetry.

3. *Smoothing with Respect to* α

The poles on the real axis can be removed by a suitable averaging. In $\pi\pi$-scattering, for example, Martin (1969) defines the variables $\sigma = s - 4m_\pi^2/3$, $\tau = t - 4m_\pi^2/3$, $\upsilon = u - 4m_\pi^2/3$, and writes the Veneziano amplitude as

$$A = \beta\Gamma(1 - \alpha_s)\Gamma(1 - \alpha_t)/\Gamma(1 - \alpha_s - \alpha_t) = W(\sigma, \tau, \upsilon). \qquad \text{(IV-11)}$$

He then forms the smoothed amplitude

$$\overline{W} = \int_{\lambda_0}^{1} d\lambda\phi(\lambda)W(\lambda\sigma, \lambda\tau, \lambda\upsilon), \qquad \text{(IV-12)}$$

where $\phi(\lambda)$ is some weight function vanishing at both limits $\lambda = 1$ and $\lambda_0 > 0$. Clearly $W(\lambda\sigma, \lambda\tau, \lambda\upsilon)$ is just the original amplitude, with the trajectory $\alpha_s = \alpha_0 + \alpha's$ replaced by $\alpha_s = \alpha_0 + 4(1 - \lambda)m_\pi^2/3 + \lambda\alpha's$ (and similarly for α_t). We have averaged over the trajectory.

As a result the resonance poles are smeared out into cuts on the real axis. Regge pole exchanges are also smeared into Regge cuts. Martin (1969) showed further that $\overline{W}$ preserves positivity (Im $\overline{W}$ has positive Legendre coefficients in the physical region), and that the resonance poles can be located by analytic continuation through the cuts onto the second sheet, for suitable weight functions ϕ. However, the physical relevance of this mathematical device is not established.

4. *Kikkawa, Sakita, Virasoro Program*

Kikkawa *et al.* (1969) draw an analogy between the Veneziano formulas and the Born approximations in field theory, and suggest that unitarity can be achieved by adding iterations that correspond to diagrams with closed loops. This is a very ambitious program, with great technical difficulties, and is far from finished. However, the theory is claimed to be complete up to the level of diagrams with one closed loop, and has been reviewed by Alessandrini *et al.* (1971).

C. SOME APPLICATIONS

1. *ππ Scattering*

It is enough to define the amplitude $A(s, t)$ for $\pi^+\pi^-$ scattering in the s channel. By symmetry and crossing, the amplitudes A^I for s-channel i-spin I are

$$A^0 = \tfrac{3}{2}\{A(s, t) + A(s, u)\} - \tfrac{1}{2}A(t, u), \qquad A^1 = A(s, t) - A(s, u),$$
$$A^2 = A(t, u). \tag{IV-13}$$

For $A(s, t)$ itself there is a choice of terms, starting with

$$A(s, t) = \beta\Gamma(1 - \alpha_s)\Gamma(1 - \alpha_t)/\Gamma(1 - \alpha_s - \alpha_t)$$
$$+ \gamma\Gamma(1 - \alpha_s)\Gamma(1 - \alpha_t)/\Gamma(2 - \alpha_s - \alpha_t) + \text{(other satellites)}. \tag{IV-14}$$

Lovelace (1968) noted that this amplitude does not depend on the external particle masses, except perhaps through the coupling constants β, γ, etc. Hence, only the latter can change if the mass of an external particle is extrapolated.

Adler's (1965) self-consistency condition requires that $A(s, t)$ vanish at the point $s = t = u = m_\pi^2$ when one of the pions has zero mass. Lovelace (1968) remarked that the first term of Eq. (IV-14) has a zero at $\alpha_s + \alpha_t = 1$, coming from a pole of the denominator, that will coincide with the required Adler zero if $\alpha(m_\pi^2) = \tfrac{1}{2}$. This is rather close to experiment. In fact, if a linear trajectory is taken through the ρ position, and the slope fixed by the Adler condition, we obtain

$$\alpha_s = 0.483 + 0.885s. \tag{IV-15}$$

The second Veneziano term in (IV-14) and other satellites, however, do not share this zero.

Lovelace and his co-workers went on to practical calculations of $\pi\pi$ scattering, and also of coupled $\pi\pi$, πK, and KK scattering (Lovelace, 1969a; Wagner, 1969; Roberts and Wagner, 1969). They took just the first term of Eq. (IV-14), normalized β to the universal ρ coupling constant, ignored the pomeron, unitarized by the K-matrix trick, and thus predicted the amplitudes. Specifically, they predicted phase shifts, scattering lengths, resonance masses, and partial widths with considerable success. Virtual $\pi\pi$ scattering, as in $\pi N \to \pi\pi N$, was also treated with an *ansatz* for the off-mass-shell extrapolation. The K-matrix method was also extended into the inelastic region, by adding an empirical pomeron and absorptive corrections (Roberts, 1970).

2. $\bar{p}n \to \pi^- \pi^- \pi^+$

Since $\bar{p}n$ annihilation at rest has isopin 1 and is observed to occur in S states, G parity restricts $\bar{p}n$ to the $^1S_0$ state—with the same quantum numbers as a heavy pion.

Lovelace (1968) therefore proposed to describe this process with a Veneziano model like that for $\pi\pi$ scattering, since external particle masses enter the model only through the couplings β, γ, etc. Experimentally the ρ-resonance bands are conspicuously absent from the $\pi\pi\pi$ Dalitz plot. Lovelace therefore dropped the first term in Eq. (IV-14), and adopted instead the second (satellite) term to describe $\bar{p}n \to \pi^- \pi^- \pi^+$:

$$A = \gamma\Gamma(1 - \alpha_s)\Gamma(1 - \alpha_t)/\Gamma(2 - \alpha_s - \alpha_t). \tag{IV-16}$$

Thus, the residue of the pole at $\alpha_s = J$ is a polynomial of order $J - 1$ in $\cos\theta$, describing resonances of spin $\leqslant J - 1$. At $\alpha_s = 1$, we have only the spin-0 ε resonance.

With this amplitude, and unitarizing by putting an empirical imaginary part into α_s, Lovelace (1968) secured an approximate fit to data. An important feature of these data is an absence of events in the middle of the Dalitz plot, which is explained in the present model by the zero at $\alpha_s + \alpha_t = 3$, from the denominator $\Gamma(2 - \alpha_s - \alpha_t)$.

Note that in the s-t plane there are lines of poles at fixed s and at fixed t from the Γ functions in the numerator, and diagonal lines of zeros through the intersections of the poles from the Γ function in the denominator. This interlacing network of poles and zeros is an important property of the Veneziano model.

Lovelace's approach is elegant and imaginative, but there are shortcomings in the fit to data. A strong enhancement is seen at the f^0 mass, but there is no f^0 contribution in his model; not just ρ but all the particles on the leading trajectory are absent in the amplitude of Eq. (IV-16). In this model the enhancement comes from a ρ' daughter of f. We have remarked earlier that there is no evidence for such a ρ' state; also the angular distribution of $\pi^+\pi^-$ is too sharply peaked to be given by such a P wave (Berger, 1969). It seems that f^0 should be reintroduced somehow. Altarelli and Rubinstein (1969b) have made a more detailed fit to data, using both the first two terms of Eq. (IV-14) plus other satellites.

3. *High Energy Data Fitting*

The high energy limit of the Veneziano model gives Regge pole exchanges with explicit forms for the residue functions [see Eq. (IV-5)]. This contrasts with the arbitrary forms previously used to fit data, and suggests a more restrictive parametrization. As a result, Γ-function residues have become widely used.

Systematic investigations of Veneziano residues, with all the correlations between different channels that they imply, have been made for πN and KN scattering by Berger and Fox (1969) and by Lovelace (1969b). The former authors represent the invariant amplitudes A and B by sums of Veneziano terms, plus empirical pomeron exchange. They find that sizable satellite terms are needed to approximate the data, and even so the result is not really satisfactory; for example, the relations between s-channel baryon poles and t-channel meson poles do not agree well with experiment. Lovelace (1969b), on the other hand, argues that Regge cuts are essential. Starting from a sum of Veneziano amplitudes plus an empirical pomeron, he iterates this input to generate cuts, by the K matrix or eikonal prescription (see Section V). He claims that this is not the best of all Regge fits, but certainly that it has the fewest parameters.

Further work has been done by Roberts (1970) and by Lovelace and Wagner (1970) in KN scattering, using Veneziano amplitudes with absorptive corrections.

V. Regge Cut Phenomenology

A. Troubles with Pure Regge Pole Exchange

We recall the form of the Regge pole contribution for the simple spinless case

$$A(s, t) = \gamma_{12}(t)[1 \pm \exp -i\pi\alpha(t)]/[\sin \pi\alpha(t)]s^{\alpha(t)}.$$

What are the constraints on this parametrization?

1. $\alpha(t) \leqslant 1$, $t \leqslant 0$ (from the Froissart bound).
2. $\alpha(t)$ real, $t \leqslant 0$, neglecting the possibility of colliding trajectories, and usually assuming $\alpha(t)$ to be linear, $\alpha(t) = \alpha(0) + \alpha' t$.
3. Factorization $\gamma_{12}^2(t) = \gamma_{11}(t)\gamma_{22}(t)$.

For $t \leqslant 0$, γ is real and analytic.

Constraint 3 gives all the trouble, in particular through the propagation of zeros in γ_{ij}. First consider a t-channel process with two identical vertices where the residue $\gamma_{11}(t)$ has a single zero:

$$\gamma_{11}(t) \simeq (t + t_0), \qquad \gamma_{12}^2(t) \simeq (t + t_0)_{11} \gamma_{22}(t).$$

Since $\gamma_{12}(t)$ cannot have a square root branch point (condition 3), and γ_{22} cannot have a pole, then γ_{12} and γ_{22} must have at least single zeros. Thus, the exchange contribution must have a zero in any process to which it contributes.

A double zero in a residue for identical vertices ii need only propagate to a process having one vertex i. For the nonidentical vertex situation a single zero need only propagate to process involving one of the vertices, and one cannot know *a priori* which vertex has the zero. The propagation of zeros via factorization causes difficulties in the understanding of crossover phenomena in $\overline{K}N$, KN, and $\overline{N}N$, NN elastic scattering, as illustrated in Fig. 11.

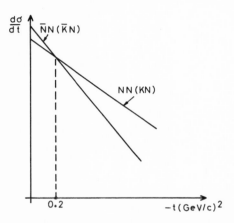

FIG. 11. The crossover phenomenon.

It is known that the amplitudes for these processes are dominantly spin-nonflip and imaginary (diffractive), and $C = -1$, $I = 0$ exchange (ω) is expected to account for the difference between particle-particle and particle-antiparticle cross sections:

$$A_{AB} = A^{pomeron} + A^{\omega}, \qquad A_{\overline{A}B} = A^{pomeron} - A^{\omega},$$

where AB refers to either KN or NN nonflip amplitudes.

Thus, ω exchange for both KN and NN scattering must have single zeros in the residue function at $t \simeq -0.15$ [note that units of t are $(GeV/c)^2$ unless otherwise stated]. It is obvious from the above argument and the single zero in the NN case that this zero must propagate to all processes involving ω exchange. Clearly the KN case is consistent. This can be checked further in the process $\pi N \to \rho N$ (Contogouris *et al.*, 1967).

The linear combination

$$X(s, t) = d\sigma_+/dt + d\sigma_-/dt - d\sigma_0/dt$$

(where the subscripts refer to the charge of ρ) isolates the contribution of

$I = 0$ (and consequently $C = -1$) exchange. The ω is the highest-lying trajectory with these quantum numbers and one would expect that this combination gives the ω exchange contribution in a pure Regge pole model. The data shows a dip at $t = -0.5$, consistent with a wrong-signature zero (WSZ), but at $t = -0.15$, the ω contribution is maximum—in clear contradiction to factorization. As noted in Section II,D,4, the FESR work of Di Vecchia et al. (1968a) also fails to show a zero in the ω-exchange contribution to π photoproduction at $t \simeq -0.15$.

π Exchange presents considerable difficulty for pure Regge pole exchange models. The reactions $np \to pn$ and $\gamma p \to \pi^+ n$ have sharply peaked differential cross sections with a half-width in t of μ^2 (μ = mass of the pion) and have an energy dependence $(d\sigma/dt)F(t)s^{-2}$. Naturally one would like an explanation in terms of π exchange.

However, due essentially to the psuedoscalar nature of the π, the π amplitudes allowed to contribute in the forward direction by angular momentum conservation behave as $A = t/(t - \mu^2)$ for small t, thus giving a sharp dip instead of a sharp peak. Attempts were made to relax the constraints on the π-exchange amplitude at $t = 0$ by adding another trajectory with the same signature and intercept, but opposite parity, and related residue—the "conspiracy" solution. However, extrapolations to the π-pole failed to give the correct π-N coupling constant unless the residue of the π varied very strongly—in fact had a zero at $t \simeq -\mu^2$—a result confirmed by FESR work. This zero should then propagate into other π exchange reactions, at least those with a nucleon-nucleon vertex. One predicts that $\rho_{00}\, d\sigma/dt$ for the reaction $\pi^- p \to \rho^0 n$ should have a minimum at $t \simeq -\mu^2$. The data show a maximum at this point.

Further difficulties with factorization were demonstrated by Le Bellac (1967). He showed that factorization and analyticity at $t = 0$ would force a conspiring π to dip in the forward direction in the reaction $\pi N \to \rho\Delta$. The data peaks with the characteristic sharpness expected from π exchange.

B. General Properties of Regge Cuts

For many years Regge phenomenologists ignored the theoretical arguments which suggested that the existence of only simple J-plane poles was inconsistent with the unitarity condition. Amati et al. (1962) pointed out that if pure Regge pole amplitudes were inserted into the two-particle elastic unitarity integral, a J-plane cut was produced. Mandelstam (1963) later demonstrated that this cut was canceled by higher-order intermediate states in the unitarity integral, but that diagrams which produced Regge cuts on the physical sheet did exist. Only results, not arguments, will be dealt with here [the reader is referred to the paper by Rothe (1968) and a review by Landshoff (1969)].

The properties of Regge cuts based on formal unitarity arguments are the following:

1. Position of the branch point $\alpha_c(t)$: for a cut generated by two trajectories $\alpha_1(t)$, $\alpha_2(t)$, linear in t,

$$\alpha_c(t) = \alpha_1(0) + \alpha_2(0) - 1 + \alpha_1'\alpha_2't/(\alpha_1' + \alpha_2'). \qquad \text{(V-1)}$$

2. Regge cuts do not factorize. Clearly since the factorization condition is nonlinear and a cut may be represented by a sum of poles, the nonlinear condition, in general, cannot be satisfied.

3. The signature of the cut is the product of the signature of the Regge poles producing the cut.

4. Contribution to the amplitude:

$$A \text{ cut} \simeq F(t)s^{\alpha_c(t)}/\ln(s).$$

We know some general properties of cut contributions obtained from unitarity, or from formal representations of unitarity in perturbation theory. However, these arguments give no information on the cut discontinuity—or equivalently the magnitude and the t dependence $F(t)$ of the cut contribution. Indeed, Landshoff (1969) has stated that if the effects of Regge cuts are large, no way is known to calculate them on the above general basis. If something is to be done, a less well-based model of Regge cuts must be accepted, which leads naturally to Section V,C—the absorption model.

C. The Absorption Model of Regge Cuts

The absorption model proposed by Sopkovich (1962) for elementary particle exchange may be extended to the case of Regge-pole exchange, and gives Regge cuts with the required general properties; it also has the fortunate feature that it actually can be calculated.

The basic idea is that the pole exchange is modified by the strong absorptive effect of elastic scattering in the initial and final states, as illustrated in Fig. 12. Although the basic idea of the absorption model has considerable intuitive appeal, we must emphasize that there is no remotely satisfactory derivation of its quantitative formulation. The prescription was motivated by work on the distorted wave Born approximation (see Gottfried and Jackson, 1964; Durand and Chiu, 1965); however, this has no necessary relevance to high energy hadronic scattering. We therefore will state the absorption model prescription and proceed to a discussion of its qualitative features.

Assuming, as one usually must, that elastic scattering is identical in initial and final states, and suppressing spin indices for simplicity, the absorptively corrected partial wave amplitude may be expressed as

$$a_J(s) = a_J^{\text{Regge}}(s)\, S_J^{\text{elastic}}, \qquad \text{(V-2)}$$

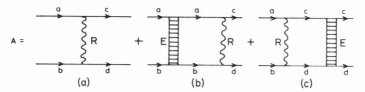

FIG. 12. The intuitive idea of the absorption model: (a) Regge pole exchange; (b) initial elastic scattering, followed by Regge pole exchange; (c) Regge pole exchange, followed by final elastic scattering.

or equivalently in terms of the full amplitudes

$$A(s, t) = A^{\text{Regge}}(s, t) + (2i/\pi s) \iint A^{\text{Regge}}(s, t') A^{\text{elastic}}(s, t'') \, dt' \, dt'' \, \theta(\tau)/\tau^{1/2},$$

$$\tau = -(t^2 + t'^2 + t''^2 - 2tt' - 2t't'' - 2tt'') + 0(1/s). \tag{V-3}$$

Schematically (V-3) is written

$$A = A^{\text{Regge}}(s, t) + iA^{\text{Regge}} \otimes A^{\text{elastic}}. \tag{V-4}$$

Now in order to get some idea of the sort of cut corrections that may be expected, assume pure imaginary elastic scattering of the form

$$A^{\text{elastic}} = iBe^{At/2}, \qquad B = \sigma_T q/4\pi.$$

Then

$$S_J \simeq 1 - C \exp -\gamma J^2, \qquad C = \sigma_T/4\pi A, \qquad \gamma = \tfrac{1}{2}q^2 A. \tag{V-5}$$

The variation of S_J with J is illustrated in Fig. 13.

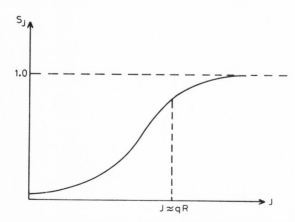

FIG. 13. A typical partial wave S-matrix element for elastic scattering (q is the center-of-mass momentum; R is the characteristic radius of interaction).

Thus we see that low partial waves are very strongly absorbed compared with high partial waves, or in terms of the full amplitude, the cut contribution is comparatively isotropic and interferes destructively with the pole. It is now easy to see that such contributions can help solve both the π-exchange problem and the problem of the crossovers.

We recall that the contribution to $np \to pn$ and $\gamma p \to \pi^+ n$ had the form

$$A = t/(t - \mu^2) = 1 + \mu^2/(t - \mu^2).$$

Now by taking the limiting case of strong absorption and eliminating only the S wave which in the nonflip amplitude is mainly given by the constant, the modified amplitude is then

$$A \simeq \mu^2/(t - \mu^2),$$

which has the desired peak. Figure 14 shows the results of a full calculation of the effects of absorption on an evasive π-exchange amplitude.

Similarly, a fairly isotropic, destructively interfering cut contribution can help in the case of the crossover problem. One way in which this could be done is illustrated in Fig. 15, where a zero in the imaginary part of the pole amplitude, a WSZ, for instance, is moved to a more forward t value. As another possibility, although the input Regge amplitude has no zero in its

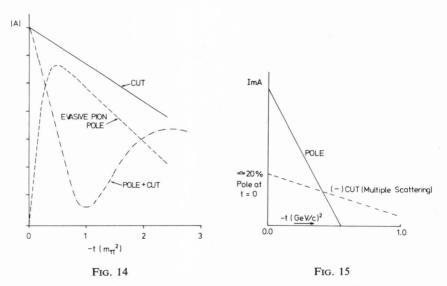

FIG. 14   FIG. 15

FIG. 14. The mechanism for generating a steep peak from an evasive exchange (π exchange) by using an absorptive cut.

FIG. 15. A possible mechanism for moving a WSZ in an imaginary part of nonflip amplitude to a more forward point.

imaginary part, the destructively interfering cut generates this zero. These two possibilities lead naturally to a discussion of the ambiguity of the Regge input in an absorptive model framework. Note however that no matter which of the two possibilities occurs, the cuts do not factorize, and therefore the zero does not propagate to other processes.

The Ambiguity of Regge Pole Input: The Argonne and Michigan Models

There is a wide variety of possible Regge inputs, connected with the presence or absence of residue zeros linked to pole position $\alpha(t)$. Only two are distinguished—the Argonne and Michigan models.

a. *The Argonne Model.* The class of absorptive cut models with a dual Regge pole input will be referred to as the *Argonne model*. This designation arises from the pioneering work of Arnold and his co-workers at the Argonne National Laboratory on models of the above class (see, for instance, Arnold, 1967; Arnold and Blackmon, 1968). The essential feature of the dual input is the existence of wrong-signature zeros (WSZ), and, where relevant, strong exchange degeneracy. The linking of WSZ of the Regge pole input (see Sections I,B and C) with absorptive-cut corrections is an attractive notion; indeed without absorptive cuts, the WSZ Regge model would be in trouble. The point is that with WSZ in all spin amplitudes a pure Regge pole model would be zero in, for instance,

$$(d\sigma/dt)(\pi^- p \to \pi^0 n) \qquad \text{at} \quad t = -0.6,$$

whereas the data show a dip, not a zero.

One role of the cuts in the Argonne model is to transform the zero of the modulus of the input amplitude into a minimum. Therefore, the possibility of an arbitrary normalization constant λ for the cut term should be introduced. Thus, generalizing Eq. (V-4), we have

$$A = A^{\text{Regge}}(s, t) + i\lambda A^{\text{Regge}} \otimes A^{\text{elastic}}. \tag{V-6}$$

The distinguishing feature of the Argonne model is the existence of WSZ in A^{Regge}, not whether $\lambda \simeq 1.0$.

In the Argonne model then, structure in differential cross sections derives from WSZ in the pole input, modified by the cuts, which for a nonflip amplitude would be typically about 20% of the pole contribution at $t = 0$.

b. *The Michigan Model.* The distinguishing feature of this model is that A^{Regge} is held to have no WSZ. Structure in differential cross sections is attributed to a destructive pole-cut interference, producing a minimum of the amplitude modulus—or equivalently a complex zero of the amplitude. The designation *Michigan model* is appropriate, since the formulation of the model and most subsequent work on it has been carried out by Kane and Ross together with their co-workers at the University of Michigan (see Henyey *et al.*, 1969; Ross *et al.*, 1970).

In order to achieve sufficient destructive interference to produce structure, the cuts must be strong. The strength of the cuts in the Michigan model has two sources. First, the cuts for a Regge input without WSZ will naturally be stronger than those for an input with WSZ because the convolution (V-6) samples all t values of the Regge amplitude, and zeros of A^{Regge} clearly diminish the integral. This is only relevant in practice for odd-signature exchange (for example, ρ exchange). For even-signature exchange (for example, A_2 exchange) the question of WSZ does not arise in the t region of interest $(0 < -t < 1)$, and there is no distinction in input between the Argonne and Michigan models. The second source of the cut strength in the Michigan model is the utilization of the boost factor λ in Eq. (V-5).

The Michigan group argue that setting $\lambda > 1$ has a physical basis in that there are diffractive intermediate states not included in the simple discussion of the absorption model prescription (V-4). This is true, but it is equally far from clear how far their additional contributions will change λ from unity, and in what direction the change should go. Nonetheless, the Michigan group ascribe a nominal value $\lambda \simeq 2.0$, although this varies in given fitting procedures.

The pole-cut interference gives a structure in the modulus of the amplitude independent of whether the exchange is of odd or even signature. However, this structure differs characteristically according to the net amount of s-channel spin flip. The relative strength of cut to pole is higher in a nonflip amplitude than in a flip amplitude; therefore, structure appears at lower $-t$ values in nonflip amplitudes. A typical Michigan model for the nonflip amplitude with structure at $t = -0.2$ is illustrated in Fig. 16. Figure 17 represents the case of a single-flip amplitude showing structure at $t = -0.6$. Similar systematics are claimed for structure in u for baryon exchange amplitudes.

Clearly structure at $t \simeq -0.2$ in nonflip amplitudes and $t \simeq -0.6$ in flip amplitudes offers a convenient explanation for crossover phenomena and the dip in $\pi^- p \to \pi^0 n$, respectively. However, the t (or u) values at which structure appears is λ dependent, and it could be argued (Drago *et al.*, 1970) that the systematics claimed by the Michigan group for their model refer more properly to the systematics of the data and the flexibility of their model.

Summarizing the main qualitative difference between Michigan and Argonne models, in the Michigan model structure in differential cross sections has no relation to the signature of the exchange, while in the Argonne model structure in differential cross sections is always related to WSZ. The fact that both models have fitted much data demonstrates the ambiguity of the Regge-pole input in an absorptive-cut model.

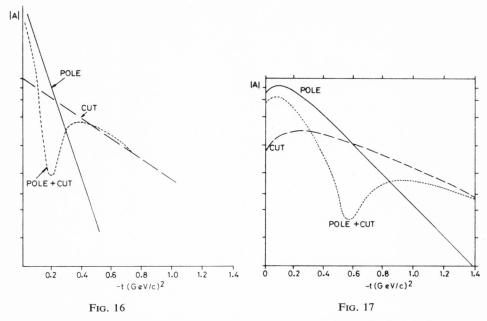

FIG. 16

FIG. 17

FIG. 16. Structure in the nonflip amplitude from pole-cut interference in the Michigan model.

FIG. 17. Structure in the single-flip amplitude from pole-cut interference in the Michigan model.

D. Specific Examples

In accord with our general approach, only a few examples of the quantitative applications of the Reggeized absorption model are given—we consider the examples to be of particular significance.

1. *Crossovers in* πN, KN, *and* NN *Scattering*

The basic physical idea of crossover phenomena is explained in Section V,A. Most of the quantitative work on this problem has been done on the πN system, and since the results are representative, examples will be confined to this system. However, the crossover in the KN system is a significantly bigger effect—perhaps more work should be done in this case.

As already mentioned in Section V,C the destructive absorption-model cuts can be used to help understand this problem. First, in the Argonne model, crossover is obtained by the destructive cut moving the WSZ in the ρ-exchange s-channel nonflip amplitude from $t = -0.6$ to a more forward t value—the crossover point of πN scattering is in the region of $0.05 \leqslant -t \leqslant 0.25$. This problem was first considered by Arnold and Blackmon (1968), White (1968,

1969) and Rivers and Saunders (1968). The essential result is that if the simple nonflip model of vacuum exchange is kept (as in Section V,A), corresponding to $A'^+/vB^+ \simeq 1$, then crossover occurs at the place where Im $A'^{(-)}$ has a zero. The above authors find that in the Argonne model the zero of Im $A'^{(-)}$ can only be induced to migrate from the WSZ point at $t = -0.6$ to $t = -0.4$. Consequently, with nonflip vacuum amplitudes a significant disagreement with data remains.

Continuous moment sum rules strongly indicate the above sign and magnitude for A'^+/vB^+ (see Section II,D,2), and A and R measurements are compatible with this assignment. This strongly suggests that the Argonne model does not have sufficiently strong destructive cuts to move the crossover far enough forward. In an exhaustive treatment of the problem White (1969) and Carreras and White (1970) show that a good value for the crossover can be obtained with $A'^+/vB^+ < 0$. This introduces a vacuum spin-flip term with suitable sign to interfere with the ρ-flip term and move the crossover further forward. Such a vacuum spin structure is in strong disagreement with CMSR results, and does not give good agreement with the present R measurements, although present errors on this difficult measurement do not completely rule out a negative ratio. We also note that FESR support a zero of Im $A'^{(-)}$ at $t \simeq -0.2$ (Section II,D,1), whereas the above mechanism accepts a zero of Im $A'^{(-)}$ at $t \simeq -0.4$ and modifies the dominant vacuum-exchange spin structure to produce the crossover at $t \simeq -0.2$.

Henyey et al., (1969), using a nonflip pomeron, find the Michigan model gives a crossover at $t = -0.3$. However, the boost factor for their nonflip ρ-cut $\lambda_{++} = 1.29$ is well below their nominal value of 2.0. By using $\lambda_{++} = 2.0$ they would obtain a better crossover, but it is not clear how this would affect their fits to $\pi^- p \to \pi^0 n$ differential cross sections and polarization.

2. Charged π Photoproduction

Reactions $\gamma p \to \pi^+ n$ and $\gamma n \to \pi^- p$ are extremely important in the Regge-pole-plus-cut model. The primary significance of these reactions is that, conspiracy aside, no combination of Regge poles can contribute in the forward direction. Of course, charged π photoproduction is maximum in the forward direction, so we know the magnitude of the s-channel nonflip cut contribution directly and the fact that it is negative with respect to the dominant pole (π) at small t.

For π exchange at small t the question of WSZ does not arise, so one expects both Argonne and Michigan models to agree in the peak region. Good fits with a reasonable π-trajectory slope, $0.5 < \alpha_\pi' < 1$ [see, for instance, Jackson and Quigg (1969) and the Michigan group, Kane et al. (1970], require very large cuts associated with π exchange—the boost factors required are roughly $\lambda = 3.5$ for the nonflip π cut. If one is determined not to amplify the π-absorption model, a fit can be obtained by using a very flat π trajectory —this was done by the Imperial College group (Collins et al., 1970).

There are strong theoretical and phenomenological hints that the π trajectory has a normal slope, for example, $\pi - B$ EXD in dual models, and CMSR results on photoproduction. Taken together with the crossover problem, perhaps this is stating that the relative strength of cut to pole in nonflip amplitudes should be stronger than the nominal values of either the Argonne or Michigan models.

The ratio of π^+- to π^--photoproduction cross sections is of great interest. At values of $|t|$ outside the sharp forward peak this ratio differs considerably from unity; indeed, for $0.3 < |t| < 1.2$, σ^-/σ^+ is approximately 0.25. In this t region, one knows through data from polarized photons that the dominant contribution to the cross section is from natural parity exchange. The data are beautifully summarized by Lübelsmeyer (1969). One therefore expects this very significant difference to be given by interference between the two leading natural parity contributions having $T = 1$ and opposite C. These, of course, are the ρ and A_2 amplitudes, with their associated cuts—denoted by $A_{``\rho"}$, and $A_{``A_2"}$, respectively. Thus

$$\Delta = [(d\sigma/dt)(\gamma p \to \pi^+ n) - (d\sigma/dt)(\gamma n \to \pi^- p)] \propto \text{Re}\{A_{``\rho"} A^*_{``A_2"}\}. \quad (\text{V-7})$$

Figure 18 shows the data together with a reasonable fit by the Michigan model (Kane et al., 1970). In this paper, the Michigan group argue that the charge difference Δ, smoothly varying with t, cannot be fitted in an Argonne model. The substance of their argument is as follows. Consider Eq. (V-7).

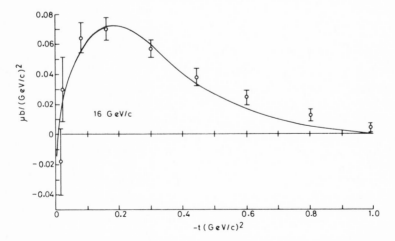

FIG. 18. The difference between π^+ and π^- photoproduction: $(d\sigma/dt)(\gamma p \to \pi^+ n) - (d\sigma/dt)(\gamma n \to \pi^- p)$. (From Kane et al., 1970.)

Neglecting cuts in zeroth order, the Argonne model will have a zero (or double zero) at $t = -0.6$, and thus give structure in Δ at $t = -0.6$. The addition of weak cuts will not change this significantly.

We consider the above argument to be fallacious. In an Argonne model the ρ and A_2 poles are exchange degenerate, and thus $\pi/2$ out of phase. Therefore, the pure pole contribution to Δ is zero everywhere, *and hence has no structure*. The addition of cuts then is not a small perturbation in such a model, but a vital contribution. One would expect the major contributions to Δ to come from ρ pole–A_2 cut, and ρ cut–A_2 pole interference. There is no reason for the latter term to have any structure and one might plausibly expect it to dominate at $t \simeq -0.6$.

Naturally the final answer as to whether an Argonne model can fit the charge difference must await detailed quantitative fits, which are expected soon (Gault *et al.*, 1971). What we have sought to show is that qualitative arguments pose no problem for the Argonne model explanation of cross-section differences in $\pi^{\pm}$ photoproduction.

3. *Exchange Degeneracy Breaking: A Challenge for the Argonne Model*

In Section III,C we have already noted that duality implies exchange degeneracy of both trajectories (weak exchange degeneracy, WEXD) and residues (strong exchange degeneracy, SEXD) and enumerated the experimental consequences of WEXD and SEXD for sets of reactions related by s-u crossing.

Just WEXD for K* and K** exchange gives equality for the differential cross sections $d\sigma^s/dt = d\sigma^u/dt$ in the reactions

(a) $K^- p \to \pi^- \Sigma^+$ (s), $\pi^+ p \to K^+ \Sigma^+$ (u),

(b) $\overline{K}^0 p \to \pi^+ \Lambda^0$ (s), $\pi^- p \to K^0 \Lambda^0$ (u).

I spin invariance relates $\overline{K}^0 p \to \pi^+ \Lambda^0$ to the more easily observed $\overline{K}^- n \to \pi^- \Lambda^0$. The additional constraint of SEXD demands that the s-channel amplitudes be purely real, and u-channel amplitudes have the phase $\exp[-i\pi\alpha(t)]$. Consequently, all polarizations are predicted to be zero.

The cross-section equality is violated by experiment for reactions (a) and (b):

$$d\sigma^s/dt > d\sigma^u/dt;$$

indeed for reaction (a)

$$d\sigma^s/dt \simeq 2d\sigma^u/dt$$

(see Figs. 19a and 19b). The vanishing of polarization demanded by SEXD also is not in accord with the data (see Figs. 20 and 21).

We have stressed in Section V,C that the Argonne model has a *dual Regge pole input*. An important question therefore arises: can the simple absorptive cut corrections to a SEXD pole input reconcile pole duality with the experimental results on cross sections and polarizations? For the cross sections the question was posed and answered by Michael (1969): cuts make things worse. Krzywicki and Tran Thanh Van (1969) found that cuts could help with polarization, although this required a rotating phase for A^{elastic}. We find it more interesting to discuss the difficulties, and so shall concentrate mainly on the cross-section problem. Michael (1969) showed that starting with a dual input giving $d\sigma^s/dt = d\sigma^u/dt$, simple absorptive-cut corrections produced $d\sigma^s/dt < d\sigma^u/dt$, increasing the discrepancy with data.

The basic argument is quite simple; we recall the schematic form of the absorptive-cut prescription

$$A(s, t) = A^{\text{Regge}}(s, t) + iA^{\text{Regge}}(s, t) \otimes A^{\text{elastic}}(s, t).$$

We remember $A^{\text{Regge}}(s, t)$ is purely real and $A^{\text{Regge}}(u, t)$ has the phase $\exp[-i\pi\alpha(t)]$. If the simplest diffractive form of A^{elastic} is taken, then iA^{elastic} is real and negative. Now the convolution $A^{\text{Regge}} \otimes A^{\text{elastic}}$ samples all negative t values of A^{Regge}. If A^{Regge} has a phase which rotates with t, it is not, in

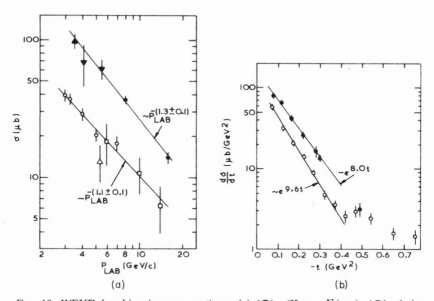

FIG. 19. WEXD breaking in cross sections: (a) (●) $\sigma(\text{K}^-p \to \Sigma^+\pi^-)$, (○) $\sigma(\pi^+p \to \Sigma^+\text{K}^+)$; (b) (●) 16-GeV/$c$ K$^-p \to \pi^-\Sigma^+$ (Carnegie Mellon Institute and Brookhaven National Laboratory); (○) 14-GeV/c $\pi^+p \to \text{K}^+\Sigma^+$ (Stony Brook and University of Wisconsin). (From Lai and Louie, 1970.)

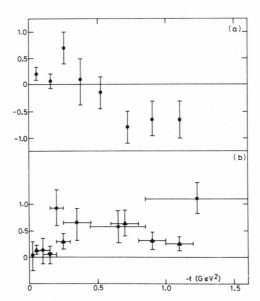

FIG. 20. SEXD breaking in polarization for $\pi^- p \to K^0 \Lambda^0$ and $K^- n \to \Lambda \pi^-$: (a) $P_{\pi^- p \to \Lambda K^0}$ at 2.9–3.3 GeV/c; (b) $P_{K^- n \to \Lambda \pi}$ with ($\bullet$) data at 3 GeV/c, ($\blacktriangle$) theory at 2.9–3.3 GeV/c. (From Krzywicki and Tran Thanh Van, 1969.)

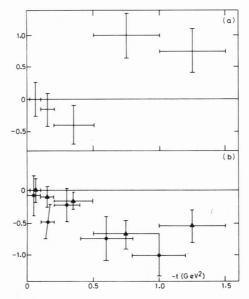

FIG. 21. SEXD breaking in polarization for $\pi^+ p \to K^+ \Sigma^+$ and $K^- p \to \pi^- \Sigma^+$: (a) $P_{\pi^- p \to \Sigma^+ K^+}$ at 3.23 GeV/c; (b) $P_{K^- p \to \Sigma^+ \pi^+}$ with ($\bullet$) data at 3 GeV/c, ($\blacktriangle$) theory at 3.23 GeV/c. (From Krzywicki and Tran Thanh Van, 1969.)

general expected that $A^{\text{Regge}} \otimes A^{\text{elastic}}$ be exactly π out of phase with A^{Regge}, although it will be approximately destructive.

The exception, of course, is the case where A^{Regge} is purely real; then the convolution will be exactly $180°$ out of phase with A^{Regge}, and thus give the maximum destructive interference. Therefore, one expects the dominant Pomeron-Regge cut corrections to give $A(s, t) < A(u, t)$, in contradiction to the data systematics. Michael further showed that Regge-Regge cuts could help in the reaction KN charge exchange (s) and $\overline{\text{K}}$N charge exchange (u), but would make things worse in reaction (a): $\text{K}^-p \to \pi^-\Sigma^+$, $\pi^+p \to \text{K}^+\Sigma^+$. In any case, Regge-Regge cuts could hardly explain the apparent energy independence of the cross-section ratios of reaction (a). Line reversal breaking falls off rapidly with energy in KN charge exchange (Fig. 22), and Regge-Regge cuts have been invoked here by O'Donovan (1970).

The difficult problem of reaction (a), which we know persists to 16 GeV/c, has been excellently reviewed by Krzywicki (1970), who gives three possible solutions:

1. Strong exchange degeneracy is badly broken for the pole input.

2. A^{elastic} for the cut in $\pi^+p \to \text{K}^+\Sigma^+$ should be significantly larger than A^{elastic} for $\text{K}^-p \to \pi^-\Sigma^+$. This would contradict pomeron factorization, which would give A^{elastic} identical in first order in both channels.

3. The absorptive cut model should have opposite sign. This, of course, would be disastrous for crossover explanations and π-exchange reactions.

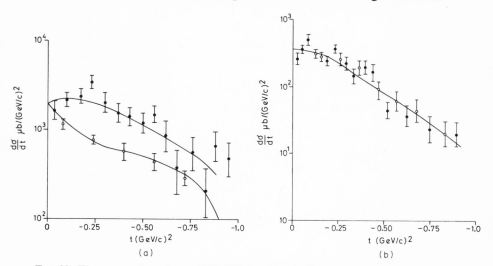

FIG. 22. The energy dependence of WEXD breaking in $\overline{\text{K}}$N and KN charge exchange: (a) at ~2.3 GeV/c with (●) $\text{K}^+n \to \text{K}^0p$ (2.3 GeV/c) and (○) $\text{K}^-p \to \overline{\text{K}}^0n$ (2.24 GeV/c); (b) at 5.5 GeV/c with (●) $\text{K}^+n \to \text{K}^0p$, (○) $\text{K}^-p \to \overline{\text{K}}^0n$. (From Lai and Louie, 1970.)

Of these three proposals we reluctantly prefer solution 1, in opposition to Krzywicki, who opted for solution 3.

Reaction (a) has been fitted by Meyers *et al.* (1970), who chose a particular form of SEXD breaking. They noted that one representation of SEXD for $SU(3)$-related reactions had an f/d ratio given by $\alpha = (1 + f/d)^{-1} = 0.75$. A slight perturbation of α has little effect on most reactions, except for reactions (a) and (b). In (a) and (b) the perturbation of α (+signature) from $0.75 \rightarrow 0.82$ changes the ratio γ_{K*}/γ_{K**} from 1.0 to 0.125.

This severe SEXD breaking is only used in the B amplitude, consequently asymptotically only in the s-channel nonflip amplitude. This SEXD breaking is so strong in this amplitude that $A(u, t)$ becomes purely real, while $A(s, t)$ now has the rotating phase. Not surprisingly this change helps the cuts give agreement with data.

It is far from clear that the work of Meyers and his co-workers is a solution to the problem, but we mention it since it shows that small SEXD breaking in an $SU(3)$ sense can give very significant SEXD breaking in the problematic channels.

Finally, it should be emphasized that SEXD breaking is only a problem for the Argonne model with its dual input. There is no obvious reason why a Michigan model for hypercharge-exchange reactions should have any difficulties.

4. *Backward πN Scattering*

Here we only try to explain the basic (and conflicting) approaches of the Argonne and Michigan models to backward πN scattering. For a detailed comparison of all pole models versus pole-plus-absorptive-cut models of backward πN scattering, the compendious work of Berger and Fox (1971) should be consulted.

Figure 23 shows the backward differential cross sections for $\pi^+ p \rightarrow p\pi^+$ and $\pi^- p \rightarrow p\pi^-$. At first sight these data appear dramatically to favor the

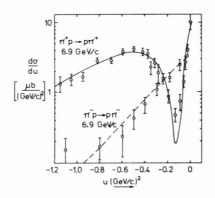

FIG. 23. Backward differential cross section for $\pi^- p$ and $\pi^+ p$ scattering at 6.9 GeV/c. (From Baker *et al.*, 1968.)

Argonne model. The distinctive features are the presence of a sharp dip at $u \simeq -0.14$ in $\pi^+ p \to p\pi^+$ and the absence of any structure in $\pi^- p \to p\pi^-$. By comparing the magnitudes of the $\pi^+ p \to p\pi^+$ differential cross section where the nucleon [N_α] and $\Delta(1236)$ [Δ_δ] can contribute with the $\pi^- p \to p\pi^-$ differential cross section where only Δ_δ can contribute, we see that N_α is the dominant contribution to $\pi^+ p \to p\pi^+$. As mentioned in Section I,C,2, the dominant nucleon trajectory is expected to pass through $\alpha = -\frac{1}{2}$ near $u = -0.14$, which, in the Argonne model, would give rise to a WSZ at $u \simeq -0.14$, Therefore, a natural explanation of the dip is obtained.

The Δ_δ trajectory is not expected to have a wrong-signature point for $0 < -u < 1.8$, and in the Argonne model no WSZ and consequent structure for $0 < -u < 1.8$ would be expected, in accord with the data. The Δ_δ may have a WSZ at $-u = 1.8$, but with this large $-u$ value the comparatively isotropic cut should dominate and thus obscure the zero of the pole amplitude.

However, the Michigan model is able to fit the backward πp data (see Kelly *et al.*, 1970). They have N_α exchange in $\pi^+ p \to p\pi^+$ as dominantly

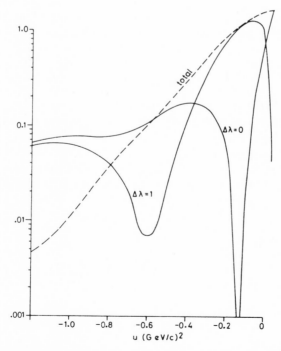

FIG. 24. Flip ($\Delta\lambda = 1$) and nonflip ($\Delta\lambda = 0$) cross sections, and their sum in a Michigan model for $\pi^- p \to p\pi^-$. (From Kelly *et al.*, 1970.)

nonflip and the destructive pole-cut interference at $u \simeq -0.14$ gives the structure in the differential cross section. The obvious problem is then to obtain a structureless differential cross section of $\pi^- p \to p\pi^-$. This is achieved in the Michigan model by making the flip ($\Delta\lambda = 1$) and nonflip ($\Delta\lambda = 0$) amplitudes from Δ_δ exchange of comparable strength. The structure in the flip and nonflip amplitude then complement one another to give a smooth cross section. This is illustrated in Fig. 24. It is worthwhile to remark that $\pi^- p \to p\pi^-$ differential cross section is remarkably small in view of the known coupling of Δ_δ at its pole position. Extrapolation to the pole in a pure Regge-pole model or an Argonne model gives values for the $\Delta(1236)$ width on the order of 2 MeV compared with the physical value of 120 MeV. This has not proved a problem in the Michigan model, since it can have a much larger pole contribution balanced by an equivalently large destructive cut contribution.

The Argonne and Michigan explanations of $\pi^- p \to p\pi^-$, then, are very different. The Argonne model explains this structureless differential cross section in terms of structureless amplitudes, while the Michigan model obtains a structureless differential cross section from two strongly structured spin amplitudes. One might expect polarization measurements to distinguish the two approaches; however, it turns out that the simple transverse polarization measurement does not distinguish between them (Berger and Fox, 1971). However, more elaborate, and unperformed, polarization measurements can distinguish between them, which leads naturally to Section V,E.

E. DISTINGUISHING BETWEEN MICHIGAN AND ARGONNE MODELS

As implied in the previous section, if one can look separately at the different spin amplitudes, it should be possible to make a clear distinction between the Michigan and Argonne models. Knowledge of just $d\sigma/dt$ and vector polariza-tion P (see Section I,B,2) has not been enough to distinguish between the very different (helicity) amplitude structure in the two models. Berger and Fox (1970) pointed out that measurement of the A and R spin-rotation para-meters in certain hypercharge exchange processes could make a clear distinc-tion between the Michigan and Argonne models. For $0^- + \frac{1}{2}^+ \to 0^- + \frac{1}{2}^+$ scattering, one can define

$$\hat{A} = (|F_{++}|^2 - |F_{+-}|^2)/(|F_{++}|^2 + |F_{+-}|),$$
$$\hat{R} = 2\,\mathrm{Re}(F_{++}F_{+-}^*)/(|F_{+-}|^2 + |F_{+-}|^2),$$

where $\hat{A}$ and $\hat{R}$ at high energies and small $-t$ values ($< |1|$) are essentially the same as the traditional Wolfenstein parameters A and R.

From Figs. 16 and 17 it is clear that in $0^- + \frac{1}{2}^+ \to 0^- + \frac{1}{2}^+$ reactions the Michigan model should predict a dramatic variation of $\hat{A}$ with t independent-ly of the exchanged meson trajectories. For $-t = 0$ only F_{++} survives; hence,

$\hat{A} = 1$. However, at $t \simeq -0.2$, $|F_{++}|$ becomes very small, thus requiring $\hat{A} \simeq -1$ here. Similarly, at $t \simeq -0.6$, the minimum of $|F_{+-}|$ implies $\hat{A} \simeq +1$. In an Argonne model it might be possible to obtain such behavior with odd-signature exchange, but not with even-signature exchange. The energies $\hat{A}$ and $\hat{R}$ are measured by using a target polarized in the scattering plane and measuring the polarization of the recoil baryon in the scattering plane. This is achieved most easily if the recoiling baryon is a Λ or Σ^+, and so analyzes its own polarization by its decay angular distribution. The reactions $\pi^- p \to K^0 \Lambda$, $\pi^+ p \to K^+ \Sigma^+$ and their meson-line-reversed partners are ideal since not only is the final baryon its own polarization analyzer but both odd (K*) and even (K**) signature exchange occur. Figure 25 shows the detailed predictions of A and R parameters in Michigan and Argonne models. The experiment, when finally performed, will be completely decisive.

It is not possible to produce a completely unambiguous test such as that described above using presently available data. However, one can make some interesting observations. For odd-signature exchange, for example ρ exchange in the πN system, the amplitude modulus may look very similar in the two models. However, the behavior of the real and imaginary parts of the amplitude near $\alpha = 0$ is very different, as illustrated in Fig. 26. Note particularly the behavior of the real part of the ρ-exchange amplitude near $\alpha = 0$ and $t \simeq -0.6$. In the Argonne model the WSZ in addition to the signature factor produces a double zero in the real part; weak destructive cuts change the double zero into two closely adjacent zeros. In the Michigan model the input only has a single zero in the real part (from the signature factor), and this is displaced by the cut. The displacement is small in the flip amplitude (F_{+-}). The question is: can one probe the structure of the real part of the ρ-exchange flip amplitude? Polarization in elastic πN-scattering gives this possibility. The most naive view of polarization in πN elastic scattering is that it comes from the interference of the dominantly imaginary, structureless nonflip P-exchange amplitude with the real part of the ρ-exchange contribution to the flip amplitude. This implies that one only need ask whether the polarization has a single or a double zero near $t \simeq -0.6$. However, knowledge of Re $F_{+-}(\rho)$ then depends on knowledge of the dominant vacuum exchanges. We (Phillips and Ringland, 1971) have used a thorough parametrization of the vacuum exchange P, P', P'', and demanded that this parametrization satisfies CMSR. It must be emphasized that we have not applied the Michigan model outside its context. The model has been used only to obtain the cut-corrected ρ-exchange amplitude, the vacuum exchange being inserted separately.

The results of Argonne and Michigan models are shown in Fig. 27. The results of the calculation clearly favor the Argonne model. We must say that our conclusions depend crucially on having a good representation of the vacuum exchanges, and in so far as our vacuum exchange parametrization

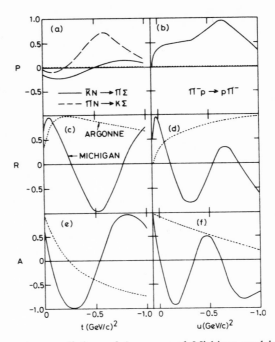

FIG. 25. Spin structure predictions of Argonne and Michigan models. (From Berger and Fox, 1970.)

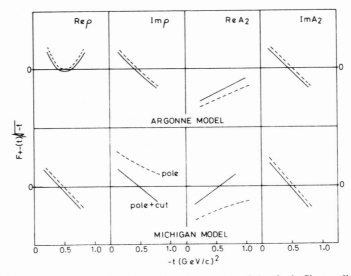

FIG. 26. The zero structure in real and imaginary parts of the single-flip amplitude near a wrong-signature point (ρ) and right signature point (A_2) in the Argonne and Michigan models.

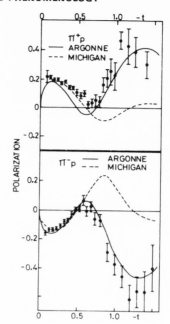

FIG. 27. Polarization in πp scattering at 6.0 GeV/c compared with the predictions of the Argonne and Michigan models, with (——) Argonne and (– – –) Michigan models, respectively: (a) $\pi^+ p$; (b) $\pi^- p$. (From Phillips and Ringland, 1971.)

is in doubt, so are our conclusions about the relative merits of the Argonne and Michigan models. However, we believe that it would be difficult to construct vacuum amplitudes for πN scattering that would allow the Michigan ρ-exchange amplitude to give agreement with polarization and simultaneously satisfy the CMSR for the vacuum-exchange amplitudes.

F. CONCLUSIONS

High energy phenomenology is in a state of flux and the following, with the possible exception of conclusion 1, must be tentative:

1. The pure Regge pole exchange model with factorization constraints is inadequate to explain much of the data.
2. The addition of absorptive-cut corrections helps to remove discrepancies, although the cut strength needs to be adjusted in specific instances.
3. There is a distressing ambiguity in the Regge-pole inputs to the absorptive-cut model. If the Michigan model turns out to be right, duality ideas will require drastic revision.
4. The measurement of A and R parameters in reactions like $\pi N \to K\Lambda$ should decisively reject one or both models. Polarization in πN elastic scattering appears to be a difficulty for the Michigan model.

REFERENCES

Adair, R. K. (1959). *Phys. Rev.* **113**, 338.

Ademollo, M., Rubinstein, H. R., Veneziano, G., and Virasoro, M. A. (1967). *Phys. Rev. Lett.* **19**, 1402.

Aderholz, M., *et al.* (1968). *Phys. Lett.* **27B**, 174.

Adler, S. (1965). *Phys. Rev.* **137**, B1022.

Alessandrini, V., Amati, D., and Squires, E. J. (1968). *Phys. Lett.* **27B**, 463.

Alessandrini, V., Amati, D., Le Bellac, M., and Olive, D. (1971). Dual multiparticle theory *Phys. Rep.* **1C**, 271.

Allaby, J. V., Bushnin, Yu. B., Denisov, S. P., Diddens, A. N., Dobinson, R. W., Donskov, S. V., Giacomelli, G., Gorin Yu, P., Klovning, A., Petrukhin, A. I., Prokoshkin, Yu. D., Shuvalov, R. S., Stahlbrandt, C. A., and Stoyanova, D. A. (1969). *Phys. Lett.* **30B**, 500.

Altarelli, G., and Rubinstein, H. R. (1969a). *Phys. Rev.* **178**, 2165.

Altarelli, G., and Rubinstein, H. R. (1969b). *Phys. Rev.* **183**, 1469.

Amati, D., Fubini, S., and Stanghellini, A. (1962). *Nuovo Cimento* **26**, 896.

Amblard, B., Cozzika, G., Ducros, Y., Hansroul, M., de Lesquen, A., Merlo, J., Movchet, J., and Van Rossum, L. (1969). Report submitted to the Lund conference.

Arbab, F., and Chiu, C. B. (1966). *Phys. Rev.* **147**, 1045.

Arnold, R. C. (1965). *Phys. Rev. Lett.* **14**, 657.

Arnold, R. C. (1967). *Phys. Rev.* **153**, 1523.

Arnold, R. C., and Blackmon, M. L. (1968). *Phys. Rev.* **176**, 2082.

Baacke, J., and Yvert, M. (1967). *Nuovo Cimento* **51A** 761.

Baker, W. F., Berkelman, K., Carlson, P. J., Fisher, G. P., Fleury, P., Harthill, D., Kalbach, R., Lundby, A., Mukhin, S., Nierhaus, R., Pretzl, K. P., and Woulds, J. (1968). *Phys. Lett.* **28B**, 291.

Ball, J. S., Frazer, W. R., and Jacob, M. (1968). *Phys. Rev. Lett.* **20**, 518

Barger, V., and Cline, D. (1966). *Phys. Rev. Lett.* **16**, 913.

Barger, V., and Cline, D. (1967). *Phys. Rev.* **155**, 1792.

Barger, V., and Cline, D. (1968). " Phenomenological Theories of High Energy Scattering." Benjamin, New York, 1968.

Barger, V., and Durand, L. D. (1967). *Phys. Rev. Lett.* **19**, 1295.

Barger, V., and Olsson, M. G. (1966). *Phys. Rev.* **151**, 1123.

Barger, V., and Phillips, R. J. N. (1968). *Phys. Lett.* **26B**, 730.

Barger, V., and Phillips, R. J. N. (1969a). *Phys. Rev.* **187**, 2210.

Barger, V., and Phillips, R. J. N. (1969b). *Phys. Lett.* **29B**, 676.

Barger, V., and Phillips, R. J. N. (1970). *Phys. Rev. Lett.* **24**, 291.

Barger, V., Michael, C., and Phillips, R. J. N. (1969). *Phys. Rev.* **185**, 1852.

Berger, E. L. (1968). *Phys. Rev.* **166**, 1525.

Berger, E. L. (1969). *Proc. Conf. $\pi\pi$ and πK Interactions, Argonne Illinois.*

Berger, E. L., and Fox G. C. (1969). *Phys. Rev.* **188**, 2120.

Berger, E. L., and Fox, G. C. (1970). *Phys. Rev. Lett.* **25**, 1783.

Berger, E. L., and Fox, G. C. (1971). *Nucl. Phys.* **B26**, 1.

Bertocchi, L. (1967). *Proc. Conf. Elementary Particles, Hiedelberg,* p. 197.

Bertocchi, L., and Ferrari, E. (1967). High energy strong interactions of particles *in* " High Energy Physics " (E. H. S. Burhop, ed.), Vol. II. Academic Press, New York.

Carreras, B., and White, J. N. J. (1970). *Nucl. Phys.* **B24**, 61.

Carroll, A. S., Corbett, J. F., Damerell, C. J. S., Middlemas, N., Newton, D., Clegg, A. B., and Williams, W. S. C. (1966). *Phys. Rev. Lett.* **16**, 288.

Chan, H. M. (1968). *Proc. Int. Conf. High Energy Physics, 14th, CERN, Geneva, 1968*, p. 391.
Chan, H. M. (1970). *Proc. Roy. Soc. (London)* **A318**, 379.
Chew, G. F., and Pignotti, A. (1968). *Phys. Rev. Lett.* **20**, 1079.
Chew, G. F., Goldberger, M. L., Low, F. E., and Nambu, Y. (1957). *Phys. Rev.* **106**, 1337.
Chiu, C. B., and Der Sarkissian, M. (1968). *Nuovo Cimento*, **55A**, 396.
Chiu, C. B., and Kotanski, A. (1968). *Nucl. Phys.* **B7**, 615.
Chiu, C. B., and Stirling, A. V. (1968). *Nuovo Cimento* **56A**, 805.
Cline, D., Matos, J., and Reeder, D. D. (1969). *Phys. Rev. Lett.* **23**, 1318.
Collins, P. A., Hartley, B. J., Jenkins, J. D., Moore, R. W., and Moriarty, K. J. M. (1970). *Phys. Rev.* **D1**, 2619.
Collins, P. D. B., and Squires, E. J. (1968). "Springer Tracts in Modern Physics," Vol. 45, p. 1. Springer, Berlin and New York.
Collins, P. D. B., Johnson, R. C. and Squires, E. J. (1968a). *Phys. Lett.* **27B**, 23.
Collins, P. D. B., Johnson, R. C., and Ross, G. G. (1968b). *Phys. Rev.* **176**, 1952.
Contogouris, A. P., Tran Thanh Van, J., and Lubatti, H. (1967). *Phys. Rev. Lett.* **19**, 1352.
Coulter. P. E., Ma, E., and Shaw, G. L. (1969). *Phys. Rev. Lett.* **23**, 106.
Dance, D. R., and Shaw, G. (1968). *Phys. Lett.* **28B**, 182.
Dass, G. V., Jacob, M., and Papageorgiou, S. (1970). *Nuovo Cimento* **67A**, 429.
Deck, R. T. (1964). *Phys. Rev. Lett.* **13**, 169.
Della Selva, A., Masperi, L., and Odorico, R. (1968). *Nuovo Cimento* **55A**, 602.
Di Vecchia, P., Drago, F., and Paciello, M. (1968a). *Nuovo Cimento* **55A**, 809.
Di Vecchia, P., Drago, F., Ferro-Fontan, C., Odorico, R., and Paciello, M. (1968b). *Phys. Lett.* **27B**, 296.
Dolen, R., Horn, D., and Schmid, C. (1967). *Phys. Rev. Lett.* **19**, 402.
Dolen, R., Horn, D., and Schmid, C. (1968). *Phys. Rev.* **166**, 1768.
Donnachie, A., and Kirsopp, R. G. (1969). *Nucl. Phys.* **B10**, 433.
Drago, F. (1970). *Phys. Rev. Lett.* **24**, 622.
Drago, F., Love, A., Phillips, R. J. N., and Ringland, G. A. (1970). *Phys. Lett.* **31B**, 647.
Drechsler, W. (1969). *Trieste Preprint* IC/69/39.
Drell, S. D. (1962). *Proc. Int. Conf. High Energy Physics, CERN, Geneva, 1962*, p. 897.
Durand, L. D., and Chiu, Y. T. (1965). *Phys. Rev.* **139**, 646.
Ferrari, E., and Violini, G. (1969). *Phys. Lett.* **28B**, 684.
Ferro-Fontan, C., Odorico, R., and Masperi, L. (1968). *Nuovo Cimento* **58A**, 435.
Firestone, A., Goldhaber, G., Hirata, A., Lissaner, D., and Trilling, G. H. (1970). *Phys. Rev. Lett.* **25**, 958.
Frazer, W. R. (1967). *Proc. Int. School of Physics, "Enrico Fermi," Varenna, Italy* **41**, 82.
Freund, P. G. O. (1968). *Phys. Rev. Lett.* **20**, 235.
Freund, P. G. O. (1969). *Nuovo Cimento Lett.* **1**, 928.
Fubini, S., and Veneziano, G. (1969). *Nuovo Cimento* **64A**, 811.
Galbraith, W., Jenkins, E. W., Kycia, T. F., Leontic, B. A., Phillips, R. H., Read, A. L., and Rubinstein, R. (1965). *Phys. Rev.* **138**, B913.
Gault, F., Kane, G. L., and Martin, A. D. (1971). *Nucl. Phys. B*, to be published.
Gilman, F., Harari, H., and Zarmi, Y. (1968). *Phys. Rev. Lett.* **21**, 323.
Gottfried, K., and Jackson, J. D. (1964). *Nuovo Cimento* **34**, 735.
Gribov, V. N. (1963). *Soviet Phys. JETP* **16** 1080.
Harari, H. (1968). *Phys. Rev. Lett.* **20**, 1395.
Harari, H. (1969a). *Phys. Rev. Lett.* **22**, 562.
Harari, H. (1969b). *Brookhaven Nat. Lab. Lectures*, Brookhaven Rep. BNL–50212.
Harari, H. (1970). *Proc. Roy. Soc. (London)* **A318**, 355.
Harari, H., and Zarmi, Y. (1969). *Phys. Rev.* **187**, 2230.

Henyey, F., Kane, G. L., Pumplin, J., and Ross, M. H. (1969). *Phys. Rev.* **182**, 1579.
Hohler, G., Baacke, J., and Eisenbeiss, G. (1966). *Phys. Lett.* **22**, 203.
Horn, D. (1969). *Acta Phys. Austr. Suppl.* **6**.
Igi, K. (1962). *Phys. Rev. Lett.* **9**, 76.
Igi, K., and Matsuda, S. (1967). *Phys. Rev. Lett.* **18**, 625.
Jackson, J. D. (1970). *Rev. Mod. Phys.* **42**, 12.
Jackson, J. D., and Quigg, C. (1969). *Phys. Lett.* **29B**, 236.
Jacob, M. (1969a). Schladming Lectures. *CERN Rep.* TH-1010; *Acta Physic Austr. Suppl.* **6**.
Jacob, M. (1969b). *Proc. Lund Intern. Conf. High Energy Physics*, p. 125.
Kane, G. L., Henyey, F., Richards, D. R., Ross, M. H., and Williamson, G. (1970). *Phys. Rev. Lett.* **25**, 1519.
Kelly, R. L., Kane, G. L., and Henyey, F. (1970). *Phys. Rev. Lett.* **24**, 1511.
Kikkawa, K., Sakita, B., and Virasoro, M. A. (1969). *Phys. Rev.* **184**, 1701.
Kirz, J. (1969). High energy collisions, *Intern. Conf. High Energy Physics, 3rd, Stony Brook, 1969*.
Kugler, M. (1970a). Schladming lectures. *Acta Phys. Austr. Suppl.* **7**.
Kugler, M. (1970b). *Phys. Lett.* **32B**, 107.
Krzywicki, A. (1970). *Proc. Meeting Moriond Electromagnetic Interactions, 5th, Orsay, 1970* **3**, 94.
Krzywicki, A., and Tran Thanh Van, J. (1969). *Phys. Lett.* **30B**, 185.
Lai, K. W., and Louie, J. (1970). *Nucl. Phys.* **B19**, 205.
Landshoff, P. V. (1969). Regge cuts: A review of the theory. *Cambridge Preprint* DAMTP69/30.
Leader, E. (1968). *Phys. Rev.* **166**, 1599.
Le Bellac, M. (1967). *Phys. Lett.* **25B**, 524.
Liu, Y. C., and Okubo, S. (1967). *Phys. Rev. Lett.* **19**, 190.
Logunov, A. A., Soloviev, L. D., and Tavkhelidze, A. N. (1967). *Phys. Lett.* **24B**, 181.
Lovelace, C. (1968). *Phys. Lett.* **28B**, 264.
Lovelace, C. (1969a). *Proc. Conf. $\pi\pi$ and πK Interactions, Argonne, Illinois.*, p. 562.
Lovelace, C. (1969b). *Nucl. Phys.* **B12**, 253.
Lovelace, C. (1970). *Proc. Roy. Soc. (London)* **A318**, 321.
Lovelace, C., and Wagner, F. (1970). *CERN Rep.* TH1251.
Lübelsmeyer, K. (1969). *Proc. Int. Symp. Electron and Photon Interactions at High Energies, 4th, Liverpool*, p. 45.
Mandelstam, S. (1963). *Nuovo Cimento* **30**, 1127, 1148.
Mandelstam, S. (1969). *Phys. Rev.* **184**, 1625.
Mandula, J., Weyers, J., and Zweig, G. (1969). *Phys. Rev. Lett.* **23**, 266.
Mandula, J., Weyers, J., and Zweig, G. (1970). *Ann. Rev. Nucl. Sci.* **20**, 289.
Martin, A. (1969). *Phys. Lett.* **29B**, 431.
Meshcheriakov, V. A., Rerikh, K. V., Tavkhelidze, A. N., and Shuravlev, V. I. (1967). *Phys. Lett.* **25B**, 341.
Meyers, C., Noirot, Y., Rimpault, M., and Salin, P. (1970). *Nucl. Phys.* **B23**, 99.
Michael, C. (1968). *Nucl. Phys.* **B8**, 431.
Michael, C. (1969). *Nucl. Phys.* **B13**, 644.
O'Donovan, P. J. (1970). Cut corrected exchange degeneracy and line reversed reactions, Tempe Preprint ASU-HEP-16, Arizona State Univ.
Olsson, M. G. (1968). *Nuovo Cimento* **57A**, 420.
Olsson, M. G., and Yodh, G. B. (1968). *Phys. Rev. Lett.* **21**, 1022.
Phillips, R. J. N. (1967). *Nucl. Phys.* **B2**, 394 (1967).
Phillips, R. J. N., and Rarita, W. (1965). *Phys. Rev.* **139**, B1336.

Phillips, R. J. N., and Ringland, G. A. (1969). *Nucl. Phys.* **B13**, 274.
Phillips, R. J. N., and Ringland, G. A. (1971). *Nucl. Phys.* **B32**, 131.
Rivers, R. J., and Saunders, L. M. (1968). *Nuovo Cimento* **58A**, 385.
Roberts, R. G. (1970). *Nucl. Phys.* **B21**, 528.
Roberts, R. G., and Wagner, F. (1969). *Phys. Lett.* **29B**, 368 (1969).
Robrish, P., Chamberlain, O., Field, R., Fuzesy, R., Gorn, W., Morehouse, G., Powell, T., Rock, S., Shannon, S., Shapiro, G., Weisberg, H., and Longo, M. (1970). *Phys. Lett.* **31B**, 617.
Roos, M., Bricman, C., Barbaro-Galtieri, A., Price, L. R., Rittenberg, A., Rosenfeld, A. H., Barash-Schmidt, N., Soding, P., Chien, C. Y., Wohl, C. G., and Lasinski, T. (1970). *Phys. Lett.* **33B**, 1.
Roskies, R. (1968). *Phys. Rev. Lett.* **21**, 1851.
Rosner, J. L. (1968). *Phys. Rev. Lett.* **21**, 950.
Rosner, J. L. (1969). *Phys. Rev. Lett.* **22**, 689.
Ross, M. H., Henyey, F. S., and Kane, G. L. (1970). *Nucl. Phys.* **B23**, 269.
Rothe, H. J. (1968). Ph.D. Thesis, Univ. California, Berkeley, UCRL-17068.
Rothe, H. J. (1967). *Phys. Rev.* **159**, 1471.
Rubinstein, H. R., Schwimmer, A., Veneziano, G. and Virasoro, M. A. (1968). *Phys. Rev. Lett.* **21**, 491.
Schmid, C. (1968a). *Phys. Rev. Lett.* **20**, 628.
Schmid, C. (1968b). *Phys. Rev. Lett.* **20**, 689.
Schmid, C. (1969). *Nuovo Cimento* **61A**, 288.
Schmid, C. (1970). *Proc. Roy. Soc. (London)* **A318**, 257.
Schwarz, J. H. (1967). *Phys. Rev.* **159**, 1269.
Singh, V. (1963). *Phys. Rev.* **129**, 1889.
Sopkovich, N. J. (1962). *Nuovo Cimento* **26**, 186.
Van Hove, L. (1966). *Proc. Intern. Conf High Energy Physics, 13th, Berkeley, California*, p. 253.
Volkov, D. V., and Gribov, V. N. (1963). *Soviet Phys. JETP* **17**, 720.
Veneziano, G. (1968). *Nuovo Cimento* **57A**, 190.
Walker, R. L. (1969). *Phys. Rev.* **182**, 1729.
Wagner, F. (1969). *Nuovo Cimento* **64A**, 189.
White, J. N. J. (1968). *Phys. Lett.* **27B**, 92.
White, J. N. J. (1969). *Nucl. Phys.* **B13**, 139.
Yen, W. L., Ammann, A. C., Carmony, D. D., Eisner, R. L., Garfinkel, A. F., Gutay, L. J., Lakshmi, R. V., Miller, D. H., and Tautfest, G. W. (1969). *Phys. Rev. Lett.* **22**, 963.

ANALYSIS OF PRESENT EVIDENCE ON THE VALIDITY OF QUANTUM ELECTRODYNAMICS

R. Gatto

Introduction

From the time of the preparation of the review on the tests of quantum electrodynamic in Volume II of *High Energy Physics* (Gatto, 1967, to be referred to as I in the following) important experimental and theoretical results have appeared, making necessary an updating of that work. It will appear, from the following summary, that no serious discrepancies exist at this time (May, 1970) between quantum electrodynamics and experiment.

We shall begin with the so-called low energy tests, which, particularly after the recent improvement of the Lamb-shift theory (Applequist and Brodsky, 1970), appear in satisfactory agreement with predictions.

We shall then review the high energy tests, including those with storage rings. For recent reviews, precedent however to the latest developments, see the articles by Farley (1969), Taylor *et al.* (1969), and Brodsky (1969), where additional references can be found. References to earlier work can be found in I (1967) and in general are not reported here. We shall also refer to I (1967) for a review of the theory and write the present updating in almost telegraphic style.

I. Value of the Fine Structure Constant

A precise determination of the fine structure constant is obtained through the expression (Taylor *et al.*, 1969)

$$\alpha^{-1} = \left(\frac{1}{4R_\infty} \frac{c\Omega_{ABS}}{\Omega_{NBS}} \frac{\mu_p'}{\mu_B} \frac{2e}{h} \frac{1}{\gamma_p'} \right)^{1/2},$$

where R_∞ is the Rydberg constant, Ω_{abs} is the absolute ohm, and Ω_{NBS} the ohm of National Bureau of Standards, μ_p' is the magnetic moment for protons in a spherical sample of water, μ_B is the Bohr magneton, and γ_p' is the gyromagnetic ratio of protons in water (spherical sample).

The errors on the "auxiliary constants" (see Table XI of Taylor *et al.*, 1969) are 0.10 ppm for R_∞, 0.39 ppm for the combination $c\Omega_{ABS}/\Omega_{NBS}$, 0.066 for μ_p'/μ_B, and 3.7 ppm in γ_p' [for the NBS average, Eq. (51) of Taylor *et al.* (1969)]. The ratio $2e/h$ is measured through the ac–Josephson effect (Parker *et al.*, 1967; Parker *et al.*, 1969; Petley and Morris, 1969; Denenstein *et al.*, in course of publication). Its determination by Parker *et al.* (1969) has an error of 2.4 ppm. Initial results towards a much more accurate determination of e/h have been reported from Finnegan *et al.* (1970). Their reported value is given within 0.46 ppm. Inserted into the above equation for α^{-1}, together with the values of Table XI of Taylor *et al.* (1969) for the other constants, and a weighted value of γ_p' [Eqs. (51) and (54) of Taylor *et al.* (1969)] it gives $\alpha^{-1} = 137.03610(22)$ (1.6 ppm) (Finnegan *et al.*, 1970).

For a most reliable determination of α^{-1} one must perform a full least-squares adjustment of the atomic constants. The latest least-squares fit [precedent to the determination of e/h by Finnegan *et al.* (1970)], derived from WQED (without quantum electrodynamic theory data), gives (Taylor *et al.*, 1969) .

$$\alpha^{-1} = 137.03608 \pm 0.00026 \qquad (1.9 \text{ ppm}).$$

II. The Fine Structure Interval in Hydrogen

Quantum electrodynamics provides for a very accurate expression for the $2P_{3/2} - 2P_{1/2}$ fine structure interval in hydrogen (see Taylor *et al.*, 1969). Using the WQED value for the fine structure constant ($\alpha^{-1} = 137.03608 \pm 0.00026$) the calculated interval is

$$10969.026 \pm 0.042 \quad \text{MHz.}$$

The latest experiment (Metcalf *et al.*, 1968; see Taylor *et al.*, 1969), making use of the Zeeman theory by Brodsky and Primack (1969), gives for the above interval

$$10969.127 \pm 0.095 \quad \text{MHz,}$$

in good agreement with the theoretical value.

TABLE I

Theoretical Contributions to the Lamb Interval $L = \Delta E(2S_{1/2} - 2P_{1/2})$ in Hydrogen[a]

Description	Order	Magnitude (MHz)
2nd order, self-energy	$\alpha(Z\alpha)^4 m\{\log Z\alpha, 1\}$	1079.32 ± 0.02
2nd order, vacuum polarization	$\alpha(Z\alpha)^4 m$	-27.13
2nd order, remainder	$\alpha(Z\alpha)^5 m$	7.14
	$\alpha(Z\alpha)^6 m\{\log^2 Z\alpha, \log Z\alpha, 1\}$	-0.38 ± 0.04
4th order, self-energy	$\alpha^2(Z\alpha)^4 m\begin{cases} F_1{}'(0) \\ F_2(0) \end{cases}$	0.45 ± 0.7 -0.10
	$\alpha^2(Z\alpha)^5 m$	± 0.02
4th order, vacuum polarization	$\alpha^2(Z\alpha)^4 m$	-0.24
Reduced-mass corrections	$\alpha(Z\alpha)^4 (m/M)m\{\log Z\alpha, 1\}$	-1.64
Recoil	$(Z\alpha)^5 (m/M)m\{\log Z\alpha, 1\}$	0.36 ± 0.01
Proton size	$(Z\alpha)^4 (mR_N)^2 m$	0.13

$$\alpha^{-1} = 137.03608(26), \quad L = \Delta E(2S_{1/2} - 2P_{1/2}) = 1057.91 \pm 0.16 \text{ (LE)}$$
$$\Delta E(2P_{3/2} - 2S_{1/2}) = 9911.12 \pm 0.22 \text{ (LE)}$$
$$\Delta E(2P_{3/2} - 2P_{1/2}) = 10969.03 \pm 0.12 \text{ (LE)}$$

[a] References may be found in Taylor *et al.* (1969) and Erickson and Yennie (1965).

TABLE II

Lamb Shift in H and D (in MHz)[a]

$L_{exp}(\pm 1\sigma)$[a]	Theory $(\pm \text{LE})$	Exp-Th $(\pm 1\sigma)$	Reference
H $(n = 2)$	1057.91 ± 0.16		
1057.77 ± 0.06		-0.14 ± 0.08	Triebwasser *et al.* (1953)
1057.90 ± 0.06		-0.01 ± 0.08	Robiscoe (revised, 1969)
(1057.65 ± 0.05)		-0.26 ± 0.07	Kaufman *et al.* (1969)
			$[(\Delta E - L)_{exp} = 9911.38 \pm 0.03]$
(1057.78 ± 0.07)		-0.13 ± 0.09	Shyn *et al.* (1969)
			$[(\Delta E - L)_{exp} = 9911.25 \pm 0.06]$
(1057.86 ± 0.06)		-0.05 ± 0.08	Cozens and Vorburger (1969)
			$[(\Delta E - L)_{exp} = 9911.17 \pm 0.04]$
D $(n = 2)$	1059.17 ± 0.22		
1059.00 ± 0.06		-0.17 ± 0.09	Triebwasser *et al.* (1953)
1059.28 ± 0.06		$+0.11 \pm 0.09$	Cozens (revised, 1969)

[a] The experimental values in parentheses are computed from the measured values of "$\Delta E - L$" $= \Delta E(2P_{3/2} - 2S_{1/2})$ and from the theoretical fine structure.

III. The Lamb Shift

Appelquist and Brodsky (1970) (see also Lautrup *et al.*, 1970; Barbieri *et al.*, 1970) have quite recently recalculated the fourth-order correction to the slope of the Dirac form factor of the free electron at the origin, finding a disagreement with previous work. The new result implies a revised theoretical value for the $2S_{1/2} - 2P_{1/2}$ separation in hydrogen and deuterium, which is 0.35 ± 0.07 larger than the previous theoretical value of 1057.56 ± 0.09 MHz (Taylor *et al.*, 1969). The new value for $L = \Delta E(2S_{1/2} - 2P_{1/2})$ in H is

$$L = 1057.91 \pm 0.16 \quad \text{MHz}.$$

The various contributions to the theoretical value of L are listed in Table I (Appelquist and Brodsky, 1970).

The experimental values (Cosens, 1968; Kaufman *et al.*, 1969; Lamb and Retherford, 1952; Robiscoe and Shyn, 1970; Shyn *et al.*, 1969; Triebwasser *et al.*, 1953; Cozens and Vorburger, 1969; Taylor *et al.*, 1969) of the Lamb shift in MHz in H and D are reported in Table II (see Appelquist and Brodsky, 1970).

The comparison between theory and experiment is reported in the third column of Table II. The one standard-deviation error limit in the comparison follows (Appelquist and Brodsky, 1970) from combining the one standard deviation experimental error by Taylor *et al.* (1969) with one-third of the theoretical limit of error (LE). The comparison is very good except perhaps for the value by Kaufman *et al.* (1969), which uses the "bottle method" instead of the atomic beam method.

For reference to theoretical work on the Lamb shift see I, Taylor *et al.* (1969), and Erickson and Yennie (1965). Experimental proposals (E. Zavattini, unpublished) and theoretical calculations exist (Di Giacomo, 1969) in view of possible measurement of the Lamb interval in muonic hydrogen.

IV. The Magnetic Moment of the Electron

Aldins *et al.* (1970) give for the electron anomalous magnetic moment

$$(a_e)_{\text{th}} = \tfrac{1}{2}\alpha/\pi - 0.32848(\alpha/\pi)^2 + 0.55(\alpha/\pi)^3.$$

The sixth-order term is not certain. It includes a calculated $(\Delta a_e)_{\gamma\gamma} = (0.36 \pm 0.04)(\alpha/\pi)^3$, combined with previously known contributions (Drell and Pagels, 1965; Parsons, 1968; Mignaco and Remiddi, 1969) or estimates (see I). The vacuum polarization insertions into the vertex at fourth order are not known.

Experiment (Wilkinson and Crane, 1963; Rich, 1967; Henry and Silver, 1969) gives

$$(a_e)_{exp} = (1159549 \pm 30) \times 10^{-9}$$
$$= \tfrac{1}{2}\alpha/\pi - 0.32848(\alpha/\pi)^2 - (7.0 \pm 2.4)(\alpha/\pi)^3$$

(with $\alpha^{-1} = 137.03608 \pm 0.00026$; see Section I).

Further experimental and theoretical work is required before drawing any conclusion as to the discrepancy (in sign and in magnitude) between $(a_e)_{th}$ and $(a_e)_{exp}$. Hadronic contributions to a_e are smaller than the remaining theoretical uncertainties.

V. The Magnetic Moment of the Muon

A recent calculation by Aldins *et al.* (1970) of three-photon exchange to the sixth-order anomalous g factor of the muon leads to a value

$$a_{th} = (116587 \pm 3) \times 10^{-8}.$$

The latest CERN measurement (Bailey *et al.*, 1968) gives

$$a_{exp} = (116616 \pm 31) \times 10^{-8}.$$

The agreement is within the experimental accuracy

$$a_{exp} - a_{th} = (250 \pm 290) \quad \text{ppm}.$$

The new calculation essentially adds to the previous theoretical value a photon-photon scattering contribution

$$\Delta a_{\gamma\gamma} = (18.4 \pm 1.1)(\alpha/\pi)^3 = (23.0 \pm 1.4) \times 10^{-8}.$$

For recent work [besides the papers quoted in I and in Picasso (1970)] relevant to the calculation of the previous theoretical value of a, see Elend (1966), Parsons (1968), Lautrup and de Rafael (1964, 1968), Mignaco and Remiddi (1969), and Drell (1966). A suspected cancellation between different graphs (Kinoshita, 1967) contributing to a does not take place. The theoretical value includes a contribution of

$$\Delta a_h = (6.5 \pm 0.5) \times 10^{-8}$$

from hadronic vacuum polarization, calculated from the Orsay data for e^+e^- annihilation into ρ, ω, and ϕ (Gourdin and de Rafael, 1969; see also Cabibbo and Gatto, 1961; Kinoshita and Oakes, 1967; Bell and de Rafael, 1969). In some models, weak interactions may contribute a term $\sim 1.10^{-8}$ (Shaffer, 1964; Brodsky and Sullivan, 1967; Burnett and Levine, 1967).

VI. Hyperfine Splitting in Muonium

Two new experimental results have appeared for the hyperfine splitting of the ground state of muonium. Crane *et al.* (1969) and Thompson *et al.* (1969) give

$$\nu_{\text{muonium}} = 4463.248 \pm 0.031 \quad \text{MHz}.$$

Ehrlich *et al.* (1969) give

$$\nu_{\text{muonium}} = 4463.317 \pm 0.021 \quad \text{MHz}.$$

The theory is clearly free of any uncertainty from strong interactions (for references see I).

Unfortunately the muon moment in Bohr magnetons is not known with sufficient accuracy and one makes use of the ratio μ_μ'/μ_p' in water, inserting a diamagnetic correction estimated at 10 ppm (Ruderman, 1966).

Using the WQED value, $\alpha^{-1} = 137.03608(26)$ one can predict (Taylor *et al.*, 1969)

$$\nu_{\text{muonium}} = 4463.272(61) \quad \text{MHz} \qquad (14 \text{ ppm}).$$

VII. Hyperfine Splitting in Hydrogen

The theoretical value of the hyperfine splitting in the hydrogen ground state depends, as well known (see I; Brodsky and Erickson, 1966), on a proton polarization correction δ. The correction is expected to be small in various models (Grotch an Yennie, 1969; Bjorken, 1966, 1967) although a correction ~ 5 ppm cannot be excluded (Drell and Sullivan, 1967). For additional theoretical work we refer to I and Brodsky and Erickson (1966).

The experimental accuracy in the determination of the interval is amazing. The technique uses the hydrogen maser (Crampton *et al.*, 1963). The most recent experiment (Vessot, 1966) gives

$$\nu_{\text{Hhfs}} = 1420.4057517864(17) \quad \text{MHz} \qquad (1.2/10^{12}).$$

The available accuracy can be taken as a symbol of theoretical impotence (as we have said the parameter δ introduces uncertainties at around 1 ppm).

With $\alpha^{-1} = 137.03608(26)$ and the auxiliary constants of their Table XI, Parker *et al.* (1969) compute from the above value

$$\frac{\nu_{\text{Hhfs}}(\text{exp}) - \nu_{\text{Hhfs}}(\text{theory})}{\nu_{\text{Hhfs}}(\text{exp})} = (2.5 \pm 4.0) \quad \text{ppm} \quad - \delta$$

or, differently said, $\delta = (2.5 \pm 4.0)$ ppm.

VIII. Tests with Colliding Beams

Electron-electron scattering at $2E = 1.1$ GeV was measured at Stanford (Barber *et al.*, 1965, 1968). The angular distribution was verified, but not the absolute cross section. For a modification of the form

$$\Lambda^2/\{q^2(\Lambda^2 - q^2)\}$$

of the photon propagator the result was

$$\Lambda > 4 \quad \text{GeV}$$

for 95% confidence limit.

Large angle Bhabha scattering

$$e^+ + e^- \to e^+ + e^-$$

has been measured at Orsay at energies around $2E = 1.02$ BeV (Augustin *et al.*, 1969). Luminosity was measured from double *bremsstrahlung*

$$e^+ + e^- \to e^+ + e^- + 2\gamma$$

with the two photons along the beam line (Bayer and Galitsky, 1964). The lowest order Bhabha scattering involves both spacelike and timelike photons. The experiment does not measure the charges of the final electrons and mainly tests the theory for spacelike momenta. The measured absolute cross section is compared with the theoretical prediction after inclusion of radiative corrections and a modified photon propagator

$$\Lambda^2/\{q^2(\Lambda^2 + \varepsilon q^2)\}.$$

The result is

$$-0.250 < \varepsilon/\Lambda^2 < 0.068$$

with 95% confidence limit or, in terms of the mass of the propagator pole, $\Lambda > 3.8$ GeV for $\varepsilon = + 1$, or $\Lambda > 2.0$ GeV for $\varepsilon = - 1$ (systematic errors are included).

Annihilation into $\mu^+\mu^-$

$$e^+ + e^- \to \mu^+ + \mu^-$$

only tests timelike momenta. With the above propagator modification, from measurements at $2E = 580, 644,$ and 704 MeV, Augustin *et al.* (1969a) have obtained

$$\Lambda^2 > 1.7 \quad (\text{GeV})^2$$

for 95% confidence limit.

Annihilation into two photons

$$e^+ + e^- \to 2\gamma$$

has been observed at Novosibirsk (Balakin *et al.*, 1969) to agree with theory within experimental errors. With a modification of the electron propagator by a factor

$$K^2/(K^2 + q^2),$$

they obtain $K > 1.5$ GeV for 95% confidence limit.

TABLE III

RECENT TESTS OF QUANTUM ELECTRODYNAMICS FROM
PAIR PRODUCTION AND *Bremsstrahlung*

Experiment	Parametrization	95% Confidence limit	References
$\gamma + C \to C + e^+ + e^-$ $m_{e+e-} \leqslant 900$ MeV/c^2	$\sigma_{exp}/\sigma_{th} = 1 + m^4/\Lambda^4$	$\Lambda > 1.6$ GeV/c^2	Alvensleben *et al.*, DESY-MIT (1968)
$\gamma + C \to C + e^+ + e^-$ $m_{e+e-} \leqslant 444$ MeV/c^2	$\sigma_{exp}/\sigma_{th} = 1 + m^4/\Lambda^4$	$\Lambda > 0.8$ GeV/c^2	Tenenbaum *et al.*, Harvard (1969)
$\gamma + p \to p + e^+ + e^-$ $m_{e+e-} \leqslant 490$ MeV/c^2	$\sigma_{exp}/\sigma_{th} = 1 + m^4/\Lambda^4$	$\Lambda > 0.7$ GeV/c^2	Biggs *et al.*, Daresbury (1969) Cohen *et al.*, MIT (1968) *See also* Eisenhandler *et al.*, Cornell (1967)
$\gamma + C \to C + \mu^+ + \mu^-$ $m_{\mu+\mu-} \leqslant 1225$ MeV/c^2	$\sigma_{exp}/\sigma_{th} = 1 - m^4/\Lambda^4$	$\Lambda > 1.5$ GeV/c^2	Hayes *et al.*, Cornell (1969) de Pagter *et al.* Northeastern (1967) *See also* Quinn, Stanford (1968)
$e^- + C \to e^- + C + \gamma$ $m_{e\gamma} \leqslant 1030$ MeV/c^2	$\sigma_{exp}/\sigma_{th} = 1 + m^4/\Lambda^4$	$\Lambda > 1.5$ GeV/c^2	Siemann *et al.*, Cornell (1969)
$\mu + C \to \mu + C + \gamma$ $m_{\mu\gamma} \leqslant 650$ MeV/c^2	$\sigma_{exp}/\sigma_{th} = 1 + m^4/\Lambda^4$	$\Lambda > 0.7$ GeV/c^2	Liberman *et al.*, Harvard-Case-McGill (1969) Bernardini *et al.*, (1968)

IX. Tests from Pair Production and *Bremsstrahlung*

Results of recent experiments on pair production and *bremsstrahlung* are summarized in Table I, taken from Brodsky's report at Daresbury (1969).

In Table III, m is the invariant mass of the pair or of the final lepton and photon. The theory is modified in order to keep gauge invariance (Kroll, 1966; McClure and Drell, 1965; Brodsky, 1969). The results are for 95% confidence limit without including systematic errors.

X. Comparison of the Muon and Electron Vertices

A recent comparison (Camilleri *et al.*, 1969) of μ-p and e-p elastic cross section, for $0.15 < q^2 < 0.85$ (GeV)^2, gives for Λ, defined through $\Lambda^{-2} = \Lambda_\mu^{-2} - \Lambda_e^{-2}$ (see I)

$$\Lambda > 2.4 \quad \text{GeV}$$

with 95% confidence. The muon and electron vertices are modified by the factors to $(1 + q^2/\Lambda_\mu^2)^{-1}$ and $(1 + q^2/\Lambda_e^2)^{-1}$, respectively. Unfortunately the comparison suffers from a systematic discrepancy in normalization between muon and electron data.

Tests of μ-e universality are also obtained from comparison of branching ratios of vector meson decays in muon and electron pairs (Asbury *et al.*, 1967; Augustin *et al.*, 1969; Auslander *et al.*, 1969; Moy *et al.*, 1969; Russell *et al.*, 1969). Additional experiments bearing on high energy tests of muon electrodynamics are described by Toner (1969), Russell *et al.* (1969), and Earles *et al.* (1969).

REFERENCES*

A dins, J., Brodsky, S. J., Dufn r, A. J., and Kinoshita, T. (1970). SLAC Pub. 701.
Alvensleben, H., Becker, U., Bertram, W. K., Binkley, M., Cohen, K., Jordan, C. L., Knasel, T. M., Marshall, R., Quinn, D. J., Rhode, R., Sanders, G. H., and Ting, S. C. (1968). *Phys. Rev. Lett.* **21**, 1501.
Appelquist, T., and Brodsky, S. J. (1970). *Phys. Rev. Lett.* **24**, 562.
Asbury, J. G., Becker, U., Bertram, W. K., Joos, P., Rohde, M., and Smith, A. J. S. (1967). *Phys. Rev. Lett.* **19**, 869.
Augustin, J. E., Buon, J., Delcourt, B., Haissinski, J., Jeanjean, J., Lalanne, D., Nguyen Ngoc, H., Perez-y-Jorba, J., Richard, F., Rumpf, F., and Treille, D. (1969a). *Proc. Symp. Electron and Photon Interactions, Daresbury,* p. 290. See also report by J. Perez-y-Jorba, p. 213.
Augustin, J. E., Bizot, J. C., Buon, J., Haissinski, J., Lalanne, D., Marin, P., Nguyen Ngoc, H., Perez-y-Jorba, J., Rumpf, F., Silva, E., and Tavernier, S. (1969b). *Phys. Lett.* **28B**, 508.
Augustin, J. E., Benaksas, D., Buon, J., Gracco, V., Haissinski, J., Lalanne, D., Laplanche, F., Lefrançois, J., Lehmann, P., Marin, P., Rumpf, F., Silva, E. (1969c). *Phys. Lett.* **28B**, 513.

* References to earlier work can be found in I.

Augustin, J. E., Bizot, J. C., Buon, J., Delcourt, B., Haissinski, J., Jeanjean, J., Lalanne, D., Marin, P. C., Nguyen Ngoc, H., Perez-y-Jorba, J., Richard, F., Rumpf, F., and Treille, D. (1969d). *Phys Lett.* **28B**, 517.

Auslander, V. L., Budker, G. I., Pakhtusova, E. V., Pestov, Yu. N., Sidorov, V. A., Skrinskii, A. N., and Khabakhapashev, A. G. (1969). *Yad. Fiz.* **9**, 114.

Bailey, J., Bartl, W., von Bochmann, G., Brown, R. C. A., Farley, F. J. M., Jötlein, H., Picasso, E., and Williams, R. W. (1968). *Phys. Lett.* **28B**, 287.

Balakin, V. E., Budker, G. I., Khabakhapashev, A. G., Korshunov, Yu. V., Mishnev, S. I., Pakhtusova, E. V., Pestov, Yu. N., Sidorov, V. A., Skrinskii, A. N., and Tumajkin, G. I. (reported by V. A. Sidorov) (1969). *Proc. Symp. Electron Photon Interactions, Daresbury*, p. 227.

Barber, W. C., Gittelman, B., O'Neill, G. K., and Richter, B. (1966). *Phys. Rev. Lett.* **16**, 1127.

Barber, W. C., Gittelman, B., O'Neill, G. K., and Richter, B. (1968). *Proc. Int. Conf. High Energy Physics, 4th, Vienna.*

Barbieri, R., Mignaco, J. A., and Remiddi, E. (1970). *Nuovo Cimento Lett.* **3**, 588.

Bayer, V. N., and Galitsky, V. M. (1964). *Phys. Lett.* **13**, 355.

Becker, U., Bertram, W. K., Binkley, M., Jordan, C. L., Knasel, T. M., Marshall, R., Quinn, D. J., Rohde, M., Smith, A. J. S., and Ting, S. C. C. (1968). *Phys. Rev. Lett.* **21**, 1504.

Bell, J. B., and de Rafael, E. (1969). *Nucl. Phys.* **B11**, 611.

Bernardini, C., Felicetti, F., Meneghetti, L., Vitale, V., Penso, G., Querzoli, R., Silvestrini, V., Vignola, G., Vitale, S. (1968). *Frascati Rep.* LNF-68/46.

Biggs, P. I., Braben, D. W., Clifft, R. W., Gabathuler, E., Kitching, P., and Rand, R. E. (1969). *Proc. Symp. Electron Photon Interactions, Daresbury*, p. 292. See also report by S. J. Brodsky, p. 3.

Bjorken, J. D. (1966). *Phys. Rev.* **148**, 1467.

Bjorken, J. D. (1967). *Phys. Rev.* **160**, 1582.

Brodsky, S. J. (1969). *Proc. Symp. Electron and Photon Interactions, Daresbury*, p. 3.

Brodsky, S. J., and Erickson, G. W. (1966). *Phys. Rev.* **146**, 26.

Brodsky, S. J., and Primack, J. R. (1969). *Ann. Phys. N.Y.* **52**, 315.

Brodsky, S. J., and Sullivan, J. D. (1967). *Phys. Rev.* **156**, 1644.

Burnett, T., and Levine, M. J. (1967). *Phys. Lett.* **24B**, 467.

Cabibbo, N., and Gatto, R. (1961). *Phys. Rev.* **124**, 1577.

Camilleri, L., Christenson, J. H., Kramer, M., Lederman, L. M., Nagashima, Y., and Yamanouchi, T. (1969). *Phys. Rev. Lett.* **23**, 153.

Cohen, K. J., Homma, S., Luckey, D., and Osborne, L. S. (1968). *Phys. Rev.* **173**, 1339.

Cozens, B. L. (1968). *Phys. Rev.* **173**, 49.

Cozens, B. L., and Vorburger, T. V. (1969). *Phys. Rev. Lett.* **23**, 1273.

Crampton, S. B., Kleppner, D., and Ramsey, N. F. *Phys. Rev. Lett.* **11**, 338.

Crane, P., Amato, J. J., Hughes, V. W., Lazams, D. M., zu Pulitz, G., and Thompson, P. A. (1969) *Proc. Symp. Electron Photon Interactions, Daresbury*, p. 309

Denenstein, A., Finnegan, T. F., Langenberg, D. N., Parker, W. H., and Taylor, B. N. (1970). *Phys. Rev.* **B1**, 4500.

de Pagter, J. K., Friedman, J. I., Glass, G., Chase, R. C., Gettner, M., von Goeler, E., Weinstein, R., and Boyarski, A. M. (1967). *Phys. Rev. Lett.* **17**, 767.

Di Giacomo, A. (1969). *Nucl. Phys.* **B11**, 411; Erratum CERN TH-1006.

Drell, S. D. (1966). *Proc. Int. Conf. High Energy Physics, 13th, Berkeley*, p. 93.

Drell, S. D., and Pagels, H. R. (1965). *Phys. Rev.* **140**, B1397.

Drell, S. D., and Sullivan, J. D. (1967). *Phys. Rev.* **154**, 1477.

Earles, D., von Briesen, Jr., H., Chase, R., Faissler, W., Gettner, M., Glass, G., von Goeler, E., Lutz. G., Parsons, R. G., Rothwell, P., and Weinstein, R. (1969). *Proc. Symp. Electron Photon Interactions, Daresbury*, p. 287.

Ehrlich, R. D., Hofer, H., Magnon, A., Stowell, D., Swanson, R. A., and Telegdi, V. L. (1969). *Chicago Univ. Preprint* EFINS-69-71.

Eisenhandler, E., Feigenbaum, J., Mistry, N., Mostek, P., Rust, D., Silverman, A., Sinclair, C., and Talman, R. (1967). *Phys. Rev. Lett.* **18**, 425.

Elend, H. H. (1966). *Phys. Lett.* **20**, 682; **21**, 720.

Ellsworth, R. W., Melissinos, A. G., Tinlot, J. H., von Briesen, Jr., H., Yamanouchi, T., Lederman, L. M., Tannenbaum, M. J., Cool, R. L., and Maschke, A. (1968). *Phys. Rev.* **165**, 1449.

Erickson, G. W., and Yennie, D. R. (1965). *Ann. Phys. N. Y.* **35**, 271, 447.

Farley, F. J. M. (1969). *Nuovo Cimento Suppl.* (*Special Issue*) **1**, 59.

Finnegan, T. F., Denenstein, H., and Langenberg, D. N. (1970). *Phys. Rev. Lett.* **24**, 738.

Gatto, R. (1967). Analysis of present evidence on the validity of quantum electrodynamics, *in* " High Energy Physics " (E. H. S. Burhop, ed.), Vol. II. Academic Press, New York.

Gourdin, M., and deRafael, E. (1969). *Nucl. Phys.* **B10**, 667.

Grotch, H., and Yennie, D. R. (1969). *Rev. Mod. Phys.* **41**, 350.

Hayes, S., Imlay, R., Joseph, P. M., Keizer, A. S., Knowles, J., Stein, P. C. (1969). *Phys. Rev. Lett.* **22**, 1134.

Henry, G. R., and Silver, J. E. (1969). *Phys. Rev.* **180**, 1262.

Kaufman, S. L., Lamb, W. E., Lea, K. R., and Leventhal, M. (1969). *Phys. Rev. Lett.* **22**, 507, 806.

Kinoshita, T. (1967) *Nuovo Cimento* **51B**, 140.

Kinoshita, T., and Oakes, R. J. (1967). *Phys. Lett.* **25B**, 143.

Kroll, N. M. (1966). *Nuovo Cimento* **45A**, 65.

Lamb, W. E., Jr., and Retherford, R. E. (1952). *Phys. Rev.* **86**, 1014.

Lautrup, B. E., and de Rafael, E. (1964). CERN-TH/1042.

Lautrup, B. E., and de Rafael, E. (1968). *Phys. Rev.* **174**, 1835.

Lautrup, B. E., Peterman, A. and de Rafael, E., (1970). *Phys. Lett.* **31B**, 577.

Lee, T. D., and Wick, G. C. (1969). *Nucl. Phys.* **B9**, 209.

Liberman, A. D., Hoffman, C. M., Engels, E., Jr., Imrie, D. C., Innocenti, P. G., Wilson, R., and Zajde, C. (1969a). *Phys. Rev. Lett.* **22**, 663.

Liberman, A. D., Hoffman, C. M., Engels, E., Jr., Imrie, D. C., Innocenti, P. G., Wilson, R., and Zajde, C. (1969b). *Proc. Symp. Electron Photon Interactions, Daresbury*, p. 298.

Metcalf, H., Brandenberger, J. R., and Baird, J. C. (1968). *Phys. Rev. Lett.* **21**, 165.

McClure, J. A., and Drell, S. D. (1965). *Nuovo Cimento* **37**, 1638.

Mignaco, J. A., and Remiddi, E. (1969). *Nuovo Cimento* **60A**, 519.

Moy, K. M., Faissler, W. L., Gettner, M. H., Lutz, G., von Briesen, H., and Weinstein, R. (1969). *Proc. Symp. Electron Photon Interactions, Daresbury*, p. 286.

Parker, W. H., Taylor, B. N., and Langenberg, D. N. (1967). *Phys. Rev. Lett.* **18**, 287.

Parker, W. H., Langenberg, D. N., Denenstein, A., and Taylor, B. N. (1969). *Phys. Rev.* **177**, 639.

Parsons, R. G. (1968). *Phys. Rev.* **168**, 1562.

Petley, B. W., and Morris, K. (1969). *Phys. Lett.* **29A**, 289.

Picasso, E. (1970). "High Energy Physics Nuclear Structure" (S. Devons, ed.), p. 615. Plenum, New York.

Quinn, D. J. Ph.D. thesis, Stanford Univ., unpublished.

Rich, A. (1967). *Phys. Rev. Lett.* **20**, 967; erratum **20**, 1221, 1968.

Robiscoe, R. T., and Shyn, T. W. (1970). *Phys. Rev. Lett.* **24**, 559.

Ruderman, M. A. (1966). *Phys. Rev. Lett.* **17**, 794.

Russell, J. J., Sah, R. C., Tenenbaum, M. J., Cleland, W. E., Ryan, D. G., and Stairs, D. G. (1969). *Proc. Symp. Electron Photon Interactions, Daresbury*, p. 282.

Siemann, R. H., Ash, W. W., Berkelman, K., Hartill, D. L., Lichtenstein, G. A., and Littaner, R. M. (1969). *Phys. Rev. Lett.* **22**, 421.

Shaffer, R. A. (1964). *Phys. Rev.* **135**, **B187**.

Shyn, T. W., Williams, W. L., Robiscoe, R. T., and Rebane, T. (1969). *Phys. Rev. Lett.* **22**, 1273.

Taylor, B. N. (1971). Unpublished data.

Taylor, B. N., Parker, W. H., and Langenberg, D. N. (1969). *Rev. Mod. Phys.* **41**, 375.

Tenenbaum, J., Eisner, A., Feldman, G., Lockeretz, W., Pipkin, F. M., and Randolph, J. K. (1969). *Proc. Symp. Electron Photon Interactions, Daresbury*, p. 309.

Thompson, P. A., Amato, J. J., Crane, P., Hughes, V. W., Mobley, R. M., zu Pulitz, G., and Rothberg, J. E. (1969). *Phys. Rev. Lett.* **22**, 163.

Toner, W. T. (1969). *Proc. Symp. Electron Photon Interactions, Daresbury*, p. 195.

Triebwasser, S., Dayhoff, E. S., and Lamb, W. E. (1953), *Phys. Rev.* **89**, 98, 106.

Vessot, R. (1966). *IEEE Trans. Intrum. Meas.*, IM 15, 165.

Wilkinson, D. T., and Crane, H. R. (1963). *Phys. Rev.* **130**, 852.

Note Added in Proof

The conclusion, stated in the introduction above, of no serious discrepancies between quantum electrodynamics and experiment has been reinforced by new theoretical and experimental work.

A supplementary list of more recent references is given below. Without going into a detailed review, we list the main conclusions.

VALUE OF THE FINE STRUCTURE CONSTANT

In Section I the value of Taylor *et al.* (1969) was reported:

WQED: $\alpha^{-1} = 137.03608 \pm 0.00026$ (1.9 ppm).

It is remarkable that a 2-ppm precision in α^{-1} now can be obtained from the comparison of the experimental $2P_{3/2} - 2S_{1/2}$ separation in hydrogen (Cozens and Vorburger, 1969) with theory (fine structure minus Lamb shift, see below). This QED value for α is

$$\alpha^{-1} = 137.03570 \pm 0.00027 \qquad (2.0 \text{ ppm}).$$

THE LAMB SHIFT

1. The self-energy correction to the electron of order α now is calculated by avoiding the expansion in powers and logs of $Z\alpha$ (Erickson, 1972; Desiderio and Johnson, 1971).

2. Fourth-order radiative corrections (essentially requiring knowledge of the radius of the electron Dirac form factor to the order α^2) are now well

known, including the Appelquist and Brodsky (1970) calculation. There are new calculations by Barbieri *et al.* (1971) (using dispersion methods) and Peterman (1971a, b).

The new theoretical value of 1057.911 ± 0.011 agrees well with the experimental value of 1057.90 ± 0.06 (Robiscoe and Shyn, 1970) (the agreement is not as good when compared with older experiments, see Section I).

THE MAGNETIC MOMENT OF THE ELECTRON

Recent theoretical work by Levine and Wright (1971), Brodsky and Kinoshita (1971), Calmet (1971), and others leads to a prediction for the electron anomalous magnetic moment of

$$(a_e)_{th} = \tfrac{1}{2}(\alpha/\pi) - 0.328479(\alpha/\pi)^2 + 1.46 \pm 0.25(\alpha/\pi)^3$$

(compare this with the value given in Section IV above).

The latest experiment (Wesley and Rich, 1970, 1972) gives

$$(a_e)_{exp} = \tfrac{1}{2}(\alpha/\pi) - 0.328479(\alpha/\pi)^2 + 1.68 \pm 0.33(\alpha/\pi)^3$$

(with $\alpha^{-1} = 137.03608 \pm 0.00026$), which differs from the previous value (see Section IV above). It is remarkable how the complex theoretical calculations (including vertex corrections with three exchanged photons) agree perfectly with experiment.

THE MAGNETIC MOMENT OF THE MUON

Recent theoretical work has been done by Brodsky and Kinoshita (1971) and Lautrup *et al.* (1971). The comparison is best made for the difference $a_\mu - a_e$, which is predicted to be 590.41×10^{-8} (fourth order) $+ (25.4 \pm 1.4) \times 10^{-8}$ (sixth order).

In addition, hadronic vacuum polarization contributions can be computed from e^+e^- data to be of about 8×10^{-8}. Forthcoming experiments should reach an accuracy of $\sim 1.5 \times 10^{-8}$ and thus test the hadronic vacuum polarization terms. In the meantime, evidence (said to be preliminary by the authors) for hadronic vacuum polarization has been obtained directly with e^+e^- colliding beams at Orsay (Augustin *et al.*, 1971). Augustin and his co-workers measured $e^+e^- \to \mu^+\mu^-$ near the φ-meson mass (1020 MeV) and found the oscillation in the total cross section, which is predicted around the φ mass from interference with the vacuum polarization term (see Section I). Hadronic vacuum polarization could be demonstrated by this experiment with a confidence level higher than 9.5%.

Accurate measurement of the muon anomalous moment, however, will be sensitive in a different way to such hadronic contributions (far from the mass shell). It is also to be stressed that the anomalous moment measurement eventually may test weak interaction terms (depending, however, on what the strong interactions actually are), whereas colliding beams, at most, will only test such contributions at energies higher than those available at present.

Hyperfine Splitting in Muonium

Recent experimental values are as follows:

$$\nu_{muonium} = 4463.302(9) \quad MHz$$

(De Voe et al., 1970) and

$$\nu_{muonium} = 4463.311(12) \quad MHz$$

(Crane et al., 1971). The relevant parameter μ_n/μ_p has been measured recently by Hague et al. (1970), in three different environments, as 3.183347(9) to the high precision of 2.8 ppm, showing, rather interestingly, the absence of a chemical correction (see I). Within error there is agreement with theory.

Similar remarks can be made for the hyperfine structure of positronium [see Theriot et al. (1970) for recent measurements]. A calculation of the contributions $\sim \alpha^4$ Ry is still lacking [for the contributions $\alpha^4 \log \alpha$ Ry, see Fulton et al. (1970a, b)].

Tests with Colliding Beams

In terms of the customary cutoff modifications higher limits on the various Λ have been determined experimentally, depending on the specific form adopted for the momentum dependence of the modifying factor (Alles Borelli et al., 1971; Bacci et al., 1971; Bartoli et al., 1970; Borgia et al., 1971).

References

Aldins, J., Kinoshita, T., Brodsky, S. J., and Dufner, A. (1970). *Phys. Rev.* **D1**, 2378.
Alles Borelli, V., Bernardini, M., Bollini, D., Massam, T., Monari L., Palmonari, F., and Zichichi, A. (1971). *Proc. Int. Conf. Elementary Particles, Amsterdam.*
Appelquist, T., and Brodsky, S. J. (1970). *Phys. Rev.* **A2**, 2293.
Augustin, J. E., Courau, A., Dudelzak, B., Fulda, F., Grosdidier, G., Haissinski, J. Masnou, J. L., Riskalla, R., Rumpf, F., and Silva, E. (1971). *Symp. Electron Photon Interactions, Ithaca, New York.*
Bacci, C., Baldini, R., Capon, G., Mencuccini, C., Murtas, G. P., Penso, G., Reale, A., Salvini, G., Spinetti, M., and Stella, B. (1971). Frascati Preprint LNF 71/16.
Bailey, J., Bartl, W., von Bochmann, G., Brown, R. C. A., Farley, F. J. M., Glesch, M., Jöstlein, H., van der Meer, S., Picasso, E., and Williams, R. W. (1971). CERN Preprint.
Barbieri, R., Mignaco, J., and Remiddi, E. (1971). *SNS* Preprint 71/3.

Bartoli, B., Coluzzi, B., Felicetti, F., Silvestrini, V., Goggi, G., Scannicchio, D., Marini, G., Massa, F., and Vanoli, F. (1970). *Nuovo Cimento* **70**, 603.

Borgia, B., Ceradini, F., Conversi, M., Paoluzi, L., Scandale, W., Barbiellini, G., Grilli, M., Spillantini, P., Visentin, R., and Mulachié, A. (1971). *Phys. Lett.* **35B**, 340.

Brodsky, S. J., and Drell, S. (1970). *Ann. Rev. Nucl. Sci.* **20**, 147.

Brodsky, S. J., and Kinoshita, T. (1971). *Phys. Rev.* **D3**, 356.

Calmet, J. (1971). Uibah University Preprint.

Crane, T., Casperson, D., Crane, P., Egan, P., Hughes, V. W., Stambaugh, R., Thompson, P. A., and zu Putlitz, G. (1971). *Phys. Rev. Lett.* **27**, 474.

De Rujula, A., and Zia, R. K. P. (1970). *Nucl. Phys.* **B19**, 224.

Desiderio. A. M., and Johnson, W. R. (1971). *Phys. Rev.* **A3**, 1267.

De Voe, R., McIntyre, P. M., Magnon, A., Stowell, D. Y., Swanson, R. A., and Telegdi, V. L. (1970). *Phys. Rev. Lett.* **25**, 1779.

Doncel, M. G., and de Rafael, E. (1971). IHES Preprint.

Earles, D. R., Faissler, W. L., Gettner, M., Lutz, G., Moy, K. M., Tang, Y. W., von Briesen, H., von Goeler, E., and Weinstein, R. (1970). *Phys. Rev. Lett.* **25**, 1312.

Erickson, G. W. (1972). To be published.

Fabjan, C., Pipkin, F., and Silverman, M. (1971). *Phys. Rev. Lett.* **26**, 347.

Fulton, T., Owen, D. A., and Repko, W. (1970a). *Phys. Rev. Lett.* **24**, 1035.

Fulton, T., Owen, D. A., and Repko, W. (1970b). *Phys. Rev. Lett.* **25**, 782 (E).

Fulton, T., Owen, D. A., and Repko, W. (1971). *Phys. Rev. Lett.* **26**, 61.

Gourdin, M., and de Rafael, E. (1969). *Nucl. Phys.* **10B**, 667.

Hague, J. F., Rothberg, J. E., Schenk, A., Williams, D. L., Williams, R. W., Young, K. K., and Crowe, K. M. (1970). *Phys. Rev. Lett.* **25**, 628.

Hayes, S., Imlay, R., Joseph, P. M., Keizer, A. S., Knowles, J., and Stein, P. C. (1970). *Phys. Rev. Lett.* **24**, 1369.

Hutchinson, D. P., Larsen, F. L., Schoen, N. C., Sober, D. I., and Kanofsky, A. S. (1970). *Phys. Rev. Lett.* **24**, 1254.

Jarlskog, C. (1971). CERN Preprint TH-1363.

Laland, D. (1970). LAL Preprint 1235 Orsay.

Lautrup, B. E. (1970). *Phys. Lett.* **32B**, 627.

Lautrup, B. E., Peterman, A., and de Rafael, E. (1971). *Nuovo Cimento* **1A**, 238.

Lee, T. D. (1970). Lectures at the Ettore Majorana School, Erice.

Leventhal, M., and Murnick, D. E. (1970). *Phys. Rev. Lett.* **25**, 1237.

Levine, M., and Wright, J. (1971). *Phys. Rev. Lett.* **26**, 1351.

Mader, O., Leventhal, M., and Lamb, W. E., Jr. (1971). *Phys. Rev.* **A3**, 1832.

Pascual, P., and, de Rafael, E. (1970). *Nuovo Cimento Lett.* **4**, 1144.

Peterman, A. (1971a). Preprint CERN TH-1354.

Peterman, A. (1971b). *Phys. Lett.* **35B**, 325.

Russell, J. J., Sah, R. C., Tenenbaum, M. J., Cleland, W. E., Ryan, D. G., and Stairs, D. (1971). *Phys. Rev. Lett.* **26**, 46.

Soto, M. F., Jr. (1970). *Phys. Rev.* **2A**, 734.

Theriot, E. D., Jr., Beers, R. H., Hughes, V. W., and Ziock, K. O. H. (1970). *Phys. Rev.* **2A**, 707.

Wesley, J. C., and Rich, A. (1970). *Phys. Rev. Lett.* **24**, 1320.

Wesley, J. C., and Rich, A. (1972). To be published.

AUTHOR INDEX

Numbers in italics refer to the pages on which the complete references are listed.

SUBJECT INDEX

PURE AND APPLIED PHYSICS

A Series of Monographs and Textbooks

Consulting Editors

H. S. W. Massey
University College, London, England

Keith A. Brueckner
University of California, San Diego
La Jolla, California

1. F. H. Field and J. L. Franklin, Electron Impact Phenomena and the Properties of Gaseous Ions. (Revised edition, 1970.)
2. H. Kopfermann, Nuclear Moments, English Version Prepared from the Second German Edition by E. E. Schneider.
3. Walter E. Thirring, Principles of Quantum Electrodynamics. Translated from the German by J. Bernstein. With Corrections and Additions by Walter E. Thirring.
4. U. Fano and G. Racah, Irreducible Tensorial Sets.
5. E. P. Wigner, Group Theory and Its Application to the Quantum Mechanics of Atomic Spectra. Expanded and Improved Edition. Translated from the German by J. J. Griffin.
6. J. Irving and N. Mullineux, Mathematics in Physics and Engineering.
7. Karl F. Herzfeld and Theodore A. Litovitz, Absorption and Dispersion of Ultrasonic Waves.
8. Leon Brillouin, Wave Propagation and Group Velocity.
9. Fay Ajzenberg-Selove (ed.), Nuclear Spectroscopy. Parts A and B.
10. D. R. Bates (ed.), Quantum Theory. In three volumes.
11. D. J. Thouless, The Quantum Mechanics of Many-Body Systems. (Second edition, 1972.)
12. W. S. C. Williams, An Introduction to Elementary Particles. (Second edition, 1971.)
13. D. R. Bates (ed.), Atomic and Molecular Processes.
14. Amos de-Shalit and Igal Talmi, Nuclear Shell Theory.
15. Walter H. Barkas. Nuclear Research Emulsions. Part I.
Nuclear Research Emulsions. Part II. *In preparation*
16. Joseph Callaway, Energy Band Theory.
17. John M. Blatt, Theory of Superconductivity.
18. F. A. Kaempffer, Concepts in Quantum Mechanics.
19. R. E. Burgess (ed.), Fluctuation Phenomena in Solids.
20. J. M. Daniels, Oriented Nuclei: Polarized Targets and Beams.
21. R. H. Huddlestone and S. L. Leonard (eds.), Plasma Diagnostic Techniques.
22. Amnon Katz, Classical Mechanics, Quantum Mechanics, Field Theory.
23. Warren P. Mason, Crystal Physics in Interaction Processes.
24. F. A. Berezin, The Method of Second Quantization.
25. E. H. S. Burhop (ed.), High Energy Physics. In five volumes. (*Volume 5 — In preparation*)

26. L. S. Rodberg and R. M. Thaler, Introduction to the Quantum Theory of Scattering.

27. R. P. Shutt (ed.), Bubble and Spark Chambers. In two volumes.

28. Geoffrey V. Marr, Photoionization Processes in Gases.

29. J. P. Davidson, Collective Models of the Nucleus.

30. Sydney Geltman, Topics in Atomic Collision Theory.

31. Eugene Feenberg, Theory of Quantum Fluids.

32. Robert T. Beyer and Stephen V. Letcher, Physical Ultrasonics.

33. S. Sugano, Y. Tanabe, and H. Kamimura, Multiplets of Transition-Metal Ions in Crystals.

34. Walter T. Grandy, Jr., Introduction to Electrodynamics and Radiation.

35. J. Killingbeck and G. H. A. Cole, Mathematical Techniques and Physical Applications.

36. Herbert Überall, Electron Scattering from Complex Nuclei. Parts A and B.

37. Ronald C. Davidson, Methods in Nonlinear Plasma Theory.

In preparation

O. N. Stavroudis, Analytic Foundations of Optics.

Arthur A. Sagle and Ralph E. Walde, Introduction to Lie Groups and Lie Algebras.

DATE DUE

GAYLORD			PRINTED IN U.S.A